Statistics *for* Environmental Engineers

Statistics *for* Environmental Engineers

Paul Mac Berthouex
Linfield C. Brown

LEWIS PUBLISHERS
Boca Raton   New York

Library of Congress Cataloging-in-Publication Data

Berthouex, P. Mac (Paul Mac), 1940–
    Statistics for environmental engineers / P. Mac Berthouex, Linfield C. Brown.
      p.   cm.
    Includes bibliographical references and index.
    ISBN 1-56670-031-0
    1. Environmental engineering–Statistical methods. I. Brown, Linfield C. II. Title.
TD153.B47 1994
001.4′22′024628—dc20
                                           94-10699
                                              CIP

No claim to original U.S. Government works
International Standard Book Number 1-56670-031-0
Library of Congress Card Number 94-10699
Printed in the United States of America   4 5 6 7 8 9 0
Printed on acid-free paper

PREFACE

When one is confronted with a new problem that involves the collection and analysis of data, two crucial questions exist: "How will using statistics help solve this problem?" and "Which techniques should be used?" This book is intended to help environmental engineers answer these questions in order better to understand and design systems for environmental protection.

This book is not about the environmental systems, except incidentally. It is about how to extract information from data and how informative data are generated in the first place. A selection of practical statistical methods are applied to the kinds of problems that we encountered in our work. We have not tried to discuss every statistical method that is useful for studying environmental data. To do so would mean including virtually all statistical methods, an obvious impossibility. Likewise, it is impossible to mention every environmental problem that can or should be investigated by statistical methods. Each reader, therefore, will find gaps in our coverage. When this happens, we hope that other authors will have filled the gap. Indeed some topics have been omitted precisely because we know they are discussed in other well-known books.

There are so many excellent books on statistics that one reasonably might ask, "Why write another book that targets environmental engineers?" A statistician may look at this book and correctly say, "Nothing new here." We have seen book reviews that were highly critical because "this book is much like book X with the examples changed from biology to chemistry." Does "changing the examples" have some benefit? We feel it does (although we hope the book does something more than just change the examples).

It is important to encourage engineers to see statistics as a professional tool used in familiar examples that are similar to those faced in one's own work. For most of the examples in this book, the environmental engineer will have a good idea how the test specimens were collected and how the measurements were made. The data thus have a special relevance and reality that should make it easier to understand special features of the data and the potential problems associated with the data analysis.

This book is organized into short chapters. The goal is for each chapter to stand alone so that one need not study the book from front to back or in any other particular order. Total independence of one chapter from another is not always possible, but the reader is encouraged to "dip in" where the subject of the case study or the statistical method stimulates interest. For example, an engineer whose current interest is fitting a kinetic model to some data can get some useful ideas from Chapter 25 without first reading the preceding 24 chapters. To most readers, Chapter 25 is not conceptually more difficult than Chapter 12. Chapter 40 can be understood without knowing anything about t-tests, confidence intervals, regression, or analysis of variance.

A number of people helped with this book. Our good friend, the late William G. Hunter, suggested the format for the book. He and George Box were our teachers, and the book reflects their influence on our approach to engineering and statistics. Lars Pallesen, engineer and statistician, worked on an early version of the book and is in spirit a co-author. A. (Sam) James provided encouragement and advice during some delightful and productive weeks in northern England. J. Stuart Hunter reviewed the manuscript at an early stage and helped to "clear up some muddy waters." We thank them all.

P. Mac Berthouex
Madison, Wisconsin

Linfield C. Brown
Medford, Massachusetts

THE AUTHORS

Paul Mac Berthouex is Professor of Civil and Environmental Engineering at the University of Wisconsin-Madison and Director of the Wisconsin Consortium for Applied Water Pollution Control Research. He hold B.S. and M.S. degrees in Civil Engineering from the University of Iowa, and a Ph.D. from the University of Wisconsin. He is a Registered Professional Engineer and a member of the American Society of Civil Engineers, Water Environment Federation, International Association on Water Quality, American Water Works Association, American Chemical Society, International Environmetrics Society, Sigma Xi, Chi Epsilon, and Tau Beta Pi.

He has held teaching positions in the Department of Preventive Medicine and Environmental Health, University of Iowa, the Department of Civil Engineering, University of Connecticut, and has been at the University of Wisconsin since 1971. Work experience includes two years (1969–71) as Chief Research Engineer for GKW Consult, Federal Republic of Germany; a major project was the Lagos Water Supply Rehabilitation and Expansion Project, Nigeria. He later worked as Project Manager on the Han River Master Plan Project in Korea (1981), Project Engineer on the Apia Water Supply, Western Samoa (1984–85), and Project Manager of the Industrial Pollution Control Sector Project in Indonesia (1990). Other international experience includes Project Director, Surabaya Institute of Technology Project, Indonesia (1976–81) and UNESCO Team for Engineering Education in Developing Countries, Air Pollution Control Group (1993). He has been visiting professor or visting scholar at the National Environmental Engineering Research Institute, India (1978), University of Newcastle upon Tyne, England (1985), Technical University of Denmark (1985), Gadjah Mada University, Indonesia (1992), and National Chung Hsing University, Taiwan (1992). He has been a consultant for the Tennessee Valley Authority, Canada Centre for Inland Waters (Environment Canada), The Soap and Detergent Association, and the National Council for Air and Stream Improvement (1990–91).

Professor Berthouex's research in the area of water pollution control includes many projects on process design and treatment plant operation, and a special interest in mathematical and statistical methods for process modeling. He has more than 100 published articles and conference presentations. His research contributions have been recognized by the 1971 Harrison Prescott Eddy Medal from the Water Pollution Control Federation and the Rudolph Hering Medal given by the American Society of Civil Engineers in 1975 and in 1991. The Central States Water Pollution Control Association gave him the Radebaugh Award in 1989 and 1991. He served on the National Research Council, National Academy of Science, and the Committee on National Statistics in 1975–76. He is co-author of *Strategy of Pollution Control* (Wiley, 1977; Russian translation 1980, Chinese translation 1989).

Linfield C. Brown is Professor and former Chair of the Department of Civil and Environmental Engineering at Tufts University. Professor Brown earned his BSCE and MS from Tufts, and joined the CEE faculty in 1970, after completing his Ph.D. in Sanitary Engineering at the University of Wisconsin-Madison. For more than two decades he has taught a variety of graduate and undergraduate courses in environmental engineering, including environmental process design and modeling, water chemistry, engineering statistics, hydrology, and industrial waste treatment. Dr. Brown's research work has covered a broad range of topics in oxygen transfer and reaeration, sampling strategies, flow equalization, and most recently, uncertainty analysis in water quality modeling and multi-response parameter estimation.

He has served as consultant to both industry and government. As a research engineer for the National Council for Air and Stream Improvement (NCASI), he developed their national research program on mathematical water quality modeling. While on sabbatical leave at the USEPA Center for Exposure Assessment Modeling (CEAM), he developed a computational framework for incorporating uncertainty analysis into the water quality model QUAL2E. He is the author of over 50 technical papers and reports covering the fields of environmental engineering and statistics, and has worked internationally in Spain, Poland, England, and Hungary on water quality modeling and control.

Professor Brown was Chairman of the Civil and Environmental Engineering Department at Tufts for 11 years. During that time he was founder and academic director of an innovative multi-disciplinary Masters program in Hazardous Materials Management, and initiated a similar program in Environmental Science and Management for mid-career professionals, targeted specifically for women and minorities. He received the prestigious Lillian Liebner Award for excellence in teaching and advising from Tufts in 1976. He is a member of the Water Environment Federation, The International Association on Water Quality, the American Society of Civil Engineers, The American Water Works Association, and the Association of Environmental Engineering Professors.

CONTENTS

1   Environmental Problems and Statistics 1
2   A Brief Review of Statistics 7
3   Plotting Data 25
4   Smoothing Data 35
5   Seeing the Shape of a Distribution 41
6   External Reference Distributions 49
7   Using Transformations 57
8   Estimating Percentiles 65
9   The Limit of Detection 71
10   Simple Methods for Analyzing Data that Are Below the Limit of
        Detection 81
11   Estimating the Mean of Censored Samples 91
12   Assessing Conformance with a Standard 97
13   Assessing the Average of Differences 103
14   Assessing the Difference of Two Averages 111
15   Assessing the Difference of Proportions 115
16   Multiple Paired Comparisons of k Averages 123
17   Analysis of Variance to Compare k Averages 129
18   Estimating Variance Components in Experimental Measurements 137
19   Multiple Factor Analysis of Variance 145
20   Factorial Experimental Designs 151
21   Fractional Factorial Experimental Designs 161
22   Screening of Important Variables 171
23   Correlation Coefficients 181
24   Assessing Serial Correlation 185
25   Estimating Parameters Using the Method of Least Squares 191
26   The Precision of Estimated Parameters 201
27   Calibration 213
28   Empirical Model Building by Linear Regression 221
29   The Coefficient of Determination, R^2 229
30   Regression Analysis with Categorical Variables 235
31   The Effect of Autocorrelation on Regression 243
32   The Iterative Approach to Modeling 251
33   Seeking Optimum Conditions by Response Surface Methodology 257
34   Designing Experiments to Estimate Parameters in Nonlinear Models 265
35   Why Linearization Can Bias Parameter Estimates 273
36   Fitting Models to Multiresponse Data 279
37   A Problem in Model Discrimination 287
38   Adjustment of Survey Data 295
39   How Measurement Errors Are Transmitted into Calculated Values 301
40   Using Simulations to Study Statistical Problems 309
41   Intervention Analysis 315
       Appendix — Statistical Tables 325
       Index 329

Chapter 1

Environmental Problems and Statistics

As the world gets more crowded and technology continues to develop, environmental problems multiply. They cannot be ignored. There are many aspects of these problems — economic, political, psychological, medical, scientific, and technological. Addressing such problems often involves certain quantitative aspects, in particular the acquisition and analysis of data. Treating these quantitative problems effectively involves the use of statistics. Statistics can be viewed as the prescription for making the quantitative learning process effective.

Many different substantive problems arise and many different statistical techniques exist, ranging from making simple plots of data to iterative model building and parameter estimation. When one is confronted with a new problem, a two-part question of crucial importance is, "How will using statistics help solve this problem and which techniques should be used?"

A variety of statistical techniques are available. Some problems can be solved by subjecting the available data to a particular analytical method. More often the analysis must be stepwise. As anyone knows who has ever coaxed a stubborn animal, small steps, often separated by intervals of frustration, are sometimes the only way to progress at all. This is frequently true when doing statistics on environmental problems. If our problem is both important and complex, our progress may be in small steps. Even though the data may contain large amounts of information, we may discover it in small bits and at intervals. The goal of statistics is to make that discovery process efficient. Analyzing data is part science, part craft, and part art. Skills and talent help and experience counts. Tools are necessary. This book illustrates some of the statistical tools that we have found useful. What is useful will vary from problem to problem. As Sir Ronald Fisher said, ". . . a statistician ought to strive above all to acquire versatility and resourcefulness, based on a repertoire of tried procedures, always aware that the next case he wants to deal with may not fit any particular recipe." We hope this book provides some useful tools and encourages environmental engineers to develop the necessary craft and art.

STATISTICS AND ENVIRONMENTAL LAW

Recently published environmental laws and regulations are about toxic chemicals, water quality criteria, air quality criteria, and so on, but they are also about *statistics* because they are laced with statistical terminology and concepts. For example, the *limit of detection* is a statistical concept used by chemists. In environmental biology, *acute and chronic toxicity criteria* are developed from complex data collection and statistical estimation procedures, *safe and adverse conditions* are differentiated through statistical comparison of control and exposed populations, and cancer potency factors are estimated by extrapolating models that have been fitted to dose-response data.

As an example, the 1989 Wisconsin laws on toxic chemicals in the aquatic environment specifically mention the following statistical terms: *geometric mean, ranks, cumulative probability, sums of squares, least squares regression, data transformations, normalization of geometric means, coefficient of determination, standard F test at a 0.05 level, representative background concentration, representative data, arithmetic average, upper 99th percentile, probability distribution, lognormal distribution, serial correlation, mean, variance, standard deviation, standard normal distribution,* and Z value. The U.S. Environmental Protection Agency (EPA) guidance documents on statistical analysis of bioassay test data mention *arc-sine transformation, probit analysis, non-normal distribution, Shapiro-Wilks test, Bartlett's test, homogeneous variance, heterogeneous variance, replicates, t-test with Bonferroni adjustment, Dunnett's test, Steel's rank test,* and *Wilcoxon rank sum test.* Terms mentioned in U.S. EPA guidance documents on groundwater monitoring at Resource Conservation and Recovery Act (RCRA) sites include *ANOVA, tolerance units, prediction intervals, control charts, confidence intervals, Cohen's adjustment, nonparametric ANOVA, test of proportions, alpha error, power curves,* and *serial correlation.* Air pollution standards and regulations also rely heavily on statistical concepts and methods.

One burden of these environmental laws is a huge investment on collecting environmental data. The cost of making measurements in support of existing environmental programs was estimated as $40 billion in 1987 and is forecasted to become $55 billion by the year 2000. Scheduled reduction in pollutant emissions will increase these costs, and it is projected that 5% of the U.S. household budget will be devoted to environmental protection by the year 2000 (Scott, 1990). No nation can afford to invest such huge amounts of money in programs and designs that are generated from badly designed sampling plans or by laboratories that have insufficient quality control. The cost of poor data is not only the price of collecting the sample and making the laboratory analyses, it also results in wasted investment in remedies for non-problems and in damage to the environment when real problems are not detected. One way to eliminate these inefficiencies in the environmental measurement system is to learn more about statistics.

TRUTH AND STATISTICS

Intelligent decisions about the quality of our environment and how it should be used and protected can be made only when information in suitable form is put before the decision makers. They, of course, want facts. They want *truth.* They may grow impatient when we explain that at best we can only make inferences about the truth. By making carefully planned measurements and using them properly, our level of knowledge is gradually elevated. Unfortunately, regardless of how carefully experiments are planned and conducted, the data produced will be imperfect and incomplete. The imperfections are due to unavoidable random variation in the measurements. The data are incomplete because we seldom know, let alone measure, all the influential variables. These difficulties and others prevent us from ever observing the truth exactly.

The relation between truth and inference in science is similar to that between guilty and not guilty in criminal law. A verdict of not guilty does not mean that innocence has been proven; it means only that guilt has not been proven. Likewise, the truth of a hypothesis cannot be firmly established. We can only test to see whether the data dispute its likelihood of being true. If the hypothesis seems plausible, in light of the available data, we must make decisions based on the likelihood of the hypothesis being true. Also, we must assess the consequences of judging a true, but unproven, hypothesis to be false. If the consequences are serious, action may be taken even when the scientific facts have not been established. Such decisions to act without scientific agreement fall into the realm of mega-tradeoffs, otherwise known as politics.

Statistics are numerical values that are calculated from imperfect observations. A suitable statistic estimates a quantity that we need to know about, but cannot observe directly. Using statistics should help us move toward the truth, but it cannot guarantee that we will reach it, nor will it tell us whether we have done so. It can help us make scientifically honest statements about the likelihood of certain hypotheses being true.

THE LEARNING PROCESS

An experiment is like a window through which we view nature (Box, 1974). Our view is never perfect. The observations that we make are distorted. The imperfections that are included in observations are "noise." A statistically efficient design reveals the magnitude and characteristics of the noise. It increases the size and improves the clarity of the experimental window. Using a poor design is like seeing blurred shadows behind the window curtains or, even worse, like looking out the wrong window.

Learning is an iterative process, the key elements of which are shown in Figure 1.1 The cycle begins with expression of a working hypothesis, which is typically based on *a priori* knowledge about the system. The hypothesis is usually stated in the form of a mathematical model that will be tuned to the present application, while at the same time be placed in jeopardy by experimental verification. Whatever form the hypothesis takes, it must be probed and given every chance to fail as data become available. Hypotheses that are not "put to the test" are like good intentions that are never implemented. They remain hypothetical.

Learning progresses most rapidly when the experimental design is statistically sound. If it is poor, so little will be learned that intelligent reformulation of the hypothesis and the data collection process may be impossible. A statistically efficient design may literally let us learn more with 8 well-planned experimental trials than with 80 that are badly placed. Good designs usually involve studying several variables simultaneously in a group of experimental runs (instead of changing one variable at a time). Iterating between data collection and data analysis provides the opportunity for improving precision by shifting emphasis to different variables, making repeated observations, and adjusting experimental conditions.

We strongly prefer working with experimental conditions that are statistically designed, and it is comparatively easy to arrange designed experiments in the laboratory.

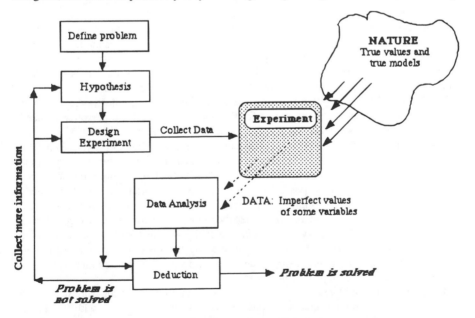

Figure 1.1 Nature is viewed through the experimental window and learning progresses by iterating between experimental design, data collection, and data analysis. In each cycle, one may formulate a new hypothesis, add or drop variables, change experimental settings, and try new methods of data analysis.

Unfortunately, in studies of natural systems and treatment facilities, it may be impossible to manipulate the independent variables to create conditions of special interest. A range of conditions can be observed only by spacing observations or field studies over a long period of time, perhaps several years. We may need to use historical data to assess changes that have occurred over time, and often the available data were not collected with a view toward assessing these changes. A related problem is not being able to replicate experimental conditions. These are huge stumbling blocks, and it is important for us to recognize how they block our path toward discovery of the truth. Hopes for successfully extracting information from such happenstance data are not often fulfilled.

SPECIAL PROBLEMS

Introductory statistics courses commonly deal with linear models and assume that available data are normally distributed and independent. There are some problems in environmental engineering where these fundamental assumptions are satisfied. Often the data are not normally distributed, they are serially or spatially correlated, or nonlinear models are needed (Hunter, 1980; Hunter, 1977; Berthouex et al., 1981; Hunter, 1982). Some specific problems encountered in data acquisition and analysis are as follows.

Aberrant values — Values that stand out from the general trend are fairly common. They may occur because of gross errors in sampling or measurement, or they may be mistakes in data recording. If we think only in these terms, it becomes too tempting to discount or throw out such values. However, rejecting any value out of hand may lead to serious errors. Some early observers of stratospheric ozone concentrations failed to detect the hole in the ozone layer because their computer had been programmed to screen incoming data for "outliers." The values that defined the hole in the ozone layer, therefore, were disregarded. This is a reminder that rogue values may be real. Indeed, they may be the only values that contain important information.

Censored data — Great effort and expense are invested in measurements of toxic and hazardous substances, which should be absent or else be present in only trace amounts. The analyst handles many specimens for which the concentration is reported as "not detected" or "below the the analytical method detection limit." This method of reporting *censors* the data at the limit of detection and condemns all lower values to be qualitative. This manipulation of the data may seem reasonable to the analytical chemist, but it creates severe problems for the data analyst and the person who needs to use the data to make decisions.

Large amounts of data — Every treatment plant, river basin authority, and environmental control agency has accumulated a mass of data in filing cabinets or computer databases. Most of this is *happenstance data*. It was collected for one purpose; later it is considered for another purpose. Happenstance data are often ill suited for model building. They may be ill suited for detecting trends over time or for testing any hypothesis about system behavior because (1) the record is not consistent and comparable from period to period, (2) all variables that affect the system have not been observed, and (3) the range of variables has been restricted by the system's operation. In short, happenstance data often contain surprisingly little information. No amount of analysis can extract information that does not exist.

Large measurement errors — Many biological and chemical measurements have large random measurement errors, despite the usual care that is taken with instrument calibration, reagent preparation, and personnel training.

Lurking variables — Sometimes important variables are not measured for a variety of reasons. Such variables are called "lurking variables." The problems they can cause are discussed by Box (1966) and Joiner (1981). A related problem occurs when a truly influential variable is carefully kept within a narrow range with the result that the variable appears to be insignificant if it is used in a regression model.

Nonconstant variance — The error associated with measurements is often nearly proportional to the magnitude of their measured values rather than approximately constant over the range of the measured values. Many measurement procedures and instruments introduce this property. Dilution and concentration steps in an analytical method will create this feature in data.

Nonnormal distributions — We are strongly conditioned to think of data being symmetrically distributed about their average value in the bell shape of the normal distribution. Environmental data seldom have this distribution. The typical pattern is a distribution that has a long tail toward high values.

Serial correlation — Many environmental data occur as a sequence of measurements taken over time or space. The order of the data is critical. In such data, it is common that the adjacent values are not statistically independent of each other because the natural continuity over time (or space) tends to make neighboring values more alike than randomly selected values. This property, called serial correlation, violates the assumptions on which many statistical procedures are based, and even low levels of serial correlation may distort estimation and hypothesis testing procedures.

THE AIM OF THIS BOOK

Learning statistics is not difficult, but engineers often dislike their introductory statistics course. Why? One reason may be that the introductory course is largely a sterile examination of textbook data, usually from a situation of which they have no intimate knowledge or deep interest. We hope this book, by presenting statistics in a familiar context, will make the subject more interesting and palatable.

The book is organized into short chapters, each dealing with one essential idea that is usually developed in the context of a case study. We hope that using statistics in relevant and realistic examples will make it easier to understand peculiarities of the data and the potential problems associated with its analysis. The goal is for each chapter to stand alone, so that the book does not need to be studied from front to back or in any other particular order. This is not always possible, but the reader is encouraged to "dip in" where the subject of the case study or the statistical method stimulates interest.

Most chapters have the following format:

- Introduction to the general kind of engineering problem and the statistical method to be discussed.
- Case Study introduces a specific environmental example, including actual data.
- Method gives a brief explanation of the statistical method that is used to prepare the solution to the case study problem. Statistical theory has been kept to a minimum. Sometimes it is condensed to an extent that reference to another book is mandatory for a full understanding. Even when the statistical theory is abbreviated, the objective is to explain the broad concept sufficiently for the reader to recognize situations when the method is likely to be useful, even though all details required for their correct application are not understood.
- Analysis shows how the data suggest and influence the method of analysis and gives our solution. Many solutions are developed in detail, but we do not always show all calculations. Most problems were solved using commercially available computer programs (e.g., Minitab, SYSTAT, and Staview 512+).
- Comments provide guidance to other chapters and statistical methods that could be useful in analyzing a problem of the kind presented in the chapter. We also attempt to expose the sensitivity of the statistical method to assumptions and to recommend alternate techniques that might be used when the assumptions are violated.

* References to selected articles and books are given at the end of each chapter. Some cover the statistical methodology in greater detail, while others provide additional case studies.

SUMMARY

In order to gain from what statistics offers, we must proceed with an attitude of letting the data reveal the critical properties and of selecting statistical methods that are appropriate to deal with these properties. Environmental data often have troublesome characteristics. If this were not so, this book would be unnecessary. All useful methods would be published in introductory statistics books. This book has the objective of bringing together, primarily by means of examples, appropriate methods with real data and real problems. Not all useful statistical methods are included, and not all widely encountered problems are discussed. Still, we hope the range of material covered will contribute to improving the state of the practice of statistics in environmental engineering and will provide guidance to relevant publications in statistics and engineering.

REFERENCES

Berthouex, P. M., W. G. Hunter, and L. Pallesen (1981). "Wastewater Treatment: A Review of Statistical Applications," in *ENVIRONMETRICS 81—Selected Papers*, pp. 77–99, Philadelphia, Society for Industrial and Applied Mathematics.

Box, G. E. P. (1966). "The Use and Abuse of Regression," *Technometrics*, 8, 625–629.

Box, G. E. P. (1974). "Statistics and the Environment," *J. Wash. Acad. Sci.*, 64, 52–59.

Box, G. E. P., W. G. Hunter, and J. S. Hunter (1978). *Statistics for Experimenters: An Introduction to Design, Data Analysis, and Model Building*, New York, Wiley Interscience.

Gilbert, R. O. (1987). *Statistical Methods for Environmental Pollution Monitoring*, New York, Van Nostrand Reinhold.

Hunter, J. S. (1977). "Incorporating Uncertainty into Environmental Regulations," in *Environmental Monitoring*, Washington, D.C., National Academy of Sciences.

Hunter, J. S. (1980). "The National Measurement System," *Science*, 210, 869–874.

Hunter, W. G. (1982). "Environmental Statistics," in *Encyclopedia of Statistical Sciences*, Vol. 2, Eds. Kotz and Johnson, New York, John Wiley & Sons.

Hunter, W. G. and J. J. Crowly (1979). "Hazardous Substances, the Environment and Public Health: A Statistical Overview," *Environ. Health Perspect.*, Oct. 1979, 241–254.

Joiner, B. L. (1981). "Lurking Variables: Some Examples," *Am. Statistician*, 35, 227–233.

Millard, S. P. (1987). "Environmental Monitoring, Statistics, and the Law: Room for Improvement," *Am. Statistician*, 41, 249–259.

Scott, F. L. (1990). "IAETL-3: Exploring the Challenges of Environmental Testing," *Am. Environ. Lab.*, 2, 12, 29–30.

Chapter 2

A Brief Review of Statistics

Key words: accuracy, average, bias, central limit effect, confidence interval, degrees of freedom, dot diagram, error, mean, histogram, hypothesis test, independence, noise, normal distribution, parameter, population, precision, probability density function, sample, random variable, significance, standard deviation, statistic, t distribution, t statistic, variance

---

It is assumed that most readers have some understanding of the basic statistical concepts and computations. Even so, a brief review of some definitions and basic concepts may be helpful.

POPULATION AND SAMPLE

The person who collects a specimen of river water speaks of that specimen as a sample. The chemist, when given this specimen, says that he has a sample to analyze. When people ask, "How many samples shall I collect?", they usually mean "On how many specimens collected from the population shall we make measurements?" They correctly use "sample" in the context of their discipline. The statistician uses it in another context with a different meaning. The *sample* is a group of n observations actually available. A *population* is a very large set of n observations (or data values) from which the sample can be imagined to have come.

RANDOM VARIABLE

The term *random variable* is widely used in statistics, but, interestingly, many statistics books do not give a formal definition for it. A practical definition by Watts (1991) is "the value of the next observation in an experiment." He also said, in a plea for terminology that is more descriptive and evocative, that "A random variable is the soul of an observation" and the converse, "An observation is the birth of a random variable."

EXPERIMENTAL ERRORS

A guiding principle of statistics is that any quantitative result should be reported with an accompanying estimate of its error. Replicated observations of some physical, chemical, or biological characteristic that has the true value η will not be identical, even though the analyst has tried to make the experimental conditions as identical as possible. This relation between the value η and the observed (measured) value y_i is $y_i = \eta + e_i$, where e_i is an error or disturbance.

Error, experimental error, or *noise* refer to the fluctuation or discrepancy in replicate observations from one experiment to another. In the statistical context, *error* does not imply fault, mistake, or blunder. It refers to variation that is often unavoidable resulting from such factors as measurement fluctuations due to instrument condition, sampling imperfections, variations in ambient conditions, skill of personnel, and many other factors. Such variation always exists, and even though in certain cases it may have been minimized, it should not be ignored entirely.

7

Example 2.1. A laboratory's measurement process for nitrate concentration was assessed by randomly inserting 27 specimens having a known concentration of 8.0 mg/L into the normal flow of work over a period of two weeks. A large number of measurements were being done routinely and any of several chemists might be assigned any sample specimen. The chemists were "blind" to the fact that performance was being assessed. The "blind specimens" were identical to all other specimens passing through the laboratory. The work was arranged so observed values would be random and independent. The results, as milligrams per liter, in the order of observation were 6.9, 7.8, 8.9, 5.2, 7.7, 9.6, 8.7, 6.7, 4.8, 8.0, 10.1, 8.5, 6.5, 9.2, 7.4, 6.3, 5.6, 7.3, 8.3, 7.2, 7.5, 6.1, 9.4, 5.4, 7.6, 8.1, and 7.9.

The *population* is all specimens having a known concentration of 8.0 mg/L. The *sample* is the 27 observations (measurements). The *sample size* is n = 27. The *random variable* is the measured concentration in each specimen having a known concentration of 8.0 mg/L. *Experimental error* has caused the observed values to vary about the true value of 8.0 mg/L.

PLOTTING THE DATA

A useful first step is to plot the data. Figure 2.1 shows the data from Example 2.1 plotted in time order of observation, with a dot diagram plotted on the right-hand side. Dots that fall on top of previously plotted points are stacked to indicate frequency. A dot diagram starts to get crowded when there are more than about 20 observations.

For a large number of points (a large sample size), it is convenient to group the dots into intervals and represent a group with a bar, as shown in Figure 2.2. This plot shows the empirical (realized) distribution of the data. Plots of this kind are usually called *histograms*, but the more suggestive name of *data density plot* has been suggested (Watts, 1991).

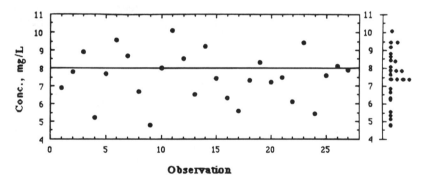

Figure 2.1 Time plot and dot diagram (right-hand side) of the example data.

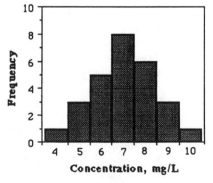

Figure 2.2 Frequency diagram (histogram).

The ordinate of the histogram can be the actual count of occurrences in an interval or it can be the relative frequency. Relative frequency, f_i, is the actual count, n_i, divided by the total number of values, n, used to construct the histogram: $f_i = n_i/n$. This is equivalent to choosing the vertical scale so that the area under the histogram is unity. It also provides an estimate of the probability that an observation will fall within a particular interval.

Another useful plot of the raw data is the cumulative frequency distribution. Here, the data are rank ordered, usually from the smallest (rank = 1) to the largest (rank = n), and plotted vs their rank. Figure 2.3 shows this plot of the nitrate data from Example 2.1. This plot serves as the basis of the *probability plots*, which are discussed in Chapter 5.

PROBABILITY DISTRIBUTIONS

As the sample size, n, becomes very large, the frequency distribution becomes smoother and approaches the shape of the underlying *population frequency distribution*. This distribution function may represent *discrete random variables* or *continuous random variables*. A discrete random variable is one that has only point values (often integer values). A continuous random variable is one that can assume any value over a range. A continuous random variable may appear to be discrete as a manifestation of the sensitivity of the measuring device or because an analyst has rounded off the values that actually were measured.

The mathematical function used to represent the population frequency distribution of a continuous random variable is called the *probability density function*. The ordinate p(y) of the distribution is not a probability itself. It is the probability density. It becomes a probability when it is multiplied by an appropriate interval on the horizontal axis, i.e., $P = p(y) \Delta$, where Δ is the size of the interval. Probability is always given by the area under the probability density function. The laws of probability require that the area under the curve equal one (1.00). This concept is illustrated by Figure 2.4, which shows the probability density function known as the *normal distribution*.

THE AVERAGE, VARIANCE, AND STANDARD DEVIATION

We distinguish between a quantity that represents a population and a quantity that represents a sample. A *statistic* is a realized quantity calculated from data that are taken to represent a population. A *parameter* is an idealized quantity associated with the population. Parameters cannot be measured directly unless the entire population can be observed. Therefore, *parameters* are estimated by *statistics*. Parameters are usually

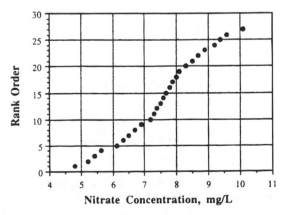

Figure 2.3 Cumulative distribution plot of the nitrate data from Example 2.1.

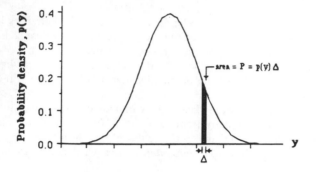

Figure 2.4 The normal probability density function.

designated by Greek letters (e.g., α, β, γ) and statistics by Roman letters (e.g., a, b, x). Parameters are constants (often unknown in value), and statistics are random variables computed from data.

Given a population of a very large set of N observations from which the sample is to come, the *population mean* is η,

$$\eta = \frac{\Sigma\, y:}{N}$$

where y_i is an observation. The summation, indicated by Σ, is over the population of N observations. We can also say that the mean of the population is the *expected value* of y, which is written as $E(y) = \eta$, when N is very large.

The *sample of n observations* actually available from the population is used to calculate the *sample average,*

$$\bar{y} = \frac{1}{n} \Sigma\, y_i$$

which *estimates* the mean η.

The *variance* of the population is denoted by σ^2. The measure of how far any particular observation is from the mean η is $y_i - \eta$. The variance is the mean value of the square of such deviations taken over the whole population,

$$\sigma^2 = \frac{\Sigma\, (y_i - \eta)^2}{N}$$

The *standard deviation* of the population is a measure of spread that has the same units as the original measurements and as the mean. The standard deviation is the square root of the variance,

$$\sigma = \sqrt{\frac{\Sigma\, (y_i - \eta)^2}{N}}$$

The true values of the population parameters σ and σ^2 are often unknown to the experimenter. They can be estimated by the *sample variance,*

$$s^2 = \frac{\Sigma(y_i - \bar{y})^2}{n - 1} = \frac{\Sigma y_i^2 - n\,\bar{y}^2}{n - 1}$$

where n is the size of the sample and $\bar{y}$ is the sample average. The *sample standard deviation* is the square root of the sample variance,

$$s = \sqrt{\frac{\Sigma (y_i - \bar{y})^2}{n - 1}}$$

Here the denominator is n – 1 rather than n. The n – 1 is the *degrees of freedom* of the sample.

One degree of freedom (the –1) is consumed because the average must be calculated to estimate s. The deviations of n observations from their sample average must sum exactly to zero. This implies that any n – 1 of the deviations or *residuals* completely determine the one remaining residual. The n residuals, and hence their sum of squares and sample variance, are said, therefore, to have n – 1 degrees of freedom. Degrees of freedom will be denoted by the Greek letter ν. For the sample variance, $\nu = n - 1$.

Most of the time "sample" will be dropped from sample standard deviation, sample variance, and sample average. It should be clear from the context that it is the calculated statistics that are being discussed. The Roman letters, for example s^2, s, and $\bar{y}$, will indicate quantities that are statistics. Greek letters indicate parameters.

Example 2.2. For the 27 nitrate observations in Example 2.1, the sample average is

$$\bar{y} = (6.9 + 7.8 + \ldots + 8.1 + 7.9)/27 = 7.51 \text{ mg/L}$$

The sample variance is

$$s^2 = (6.9\text{-}7.51)^2 + \ldots + (7.9\text{-}7.51)^2/(27\text{-}1) = 1.9138 \ (\text{mg/L})^2$$

and the sample standard deviation is

$$s = \sqrt{1.9138} = 1.38 \text{ mg/L}$$

The sample variance and sample standard deviation have $\nu = n - 1 = 27 - 1 = 26$ degrees of freedom.

The data were reported with two significant figures. The average of several values should be calculated with at least one more figure than that of the data. The standard deviation should be computed to at least three significant figures (Taylor, 1987).

ACCURACY, BIAS, AND PRECISION

Accuracy is a function of both *bias* and *precision*. As illustrated by Example 2.3 and Figure 2.5, bias measures systematic errors and precision measures the degree of scatter in the data. Methods that give accurate measurements have good precision and near zero bias. Inaccurate methods can have poor precision, unacceptable bias, or both.

Example 2.3. Five samples containing the analyte in the known amount of 8.0 mg/L were given to four analysts. The results are shown in Figure 2.5. Two separate kinds of errors have occurred in A's work: (1) random errors cause the individual results to be "scattered" about the average of the five results and (2) a fixed component in the measurement error, a systematic error or bias, makes the observations too high. Analyst B has poor precision, but little observed bias. Analyst C has poor accuracy and poor precision. Only Analyst D has little bias and good precision.

Example 2.4. For the 27 nitrate observations of Example 2.1, we estimate the bias in the measurement method as the difference between the sample average and the known value.

$$\text{Bias} = \bar{y} - \eta = 7.51 - 8.0 = -0.49 \text{ mg/L}$$

The precision of the measurements is given by the sample standard deviation.

$$\text{Precision} = s = \pm 1.38 \text{ mg/L}$$

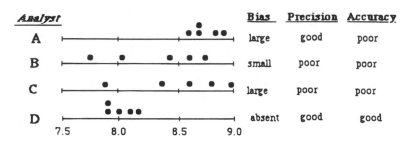

Figure 2.5 Accuracy is a function of bias and good precision.

Bias (systematic errors) can be removed, once identified, by careful checks on experimental technique and equipment. It cannot be averaged out by making more measurements. Often bias cannot be estimated because the true value of the underlying value is unknown.

Precision has to do with the scatter between repeated measurements. This scatter is caused by random errors in the measurements. Precise results have small random errors. The standard deviation, σ, is often used as an index of precision (or imprecision). When σ is large, the measurements are imprecise. Random errors can never be eliminated, though by careful technique they can be minimized. Their effects can be reduced by making repeated measurements and averaging them. Making replicate measures also provides the means to quantify the measurement errors and evaluate their importance. Systematic error (bias) cannot be eliminated by replication or averaging. It can be removed, once identified, by careful checks on experimental technique and equipment.

Reproducibility and *repeatability* are often used as synonyms for precision. However, a distinction should be made between these words. Suppose an analyst made the five replicate measurements in rapid succession, say within an hour or so, using the same set of reagent solutions and glassware throughout. Temperature, humidity, and other laboratory conditions would be nearly constant. Such measurements would estimate repeatability, which might also be called *within-run precision*. If the same analyst did the five measurements on five different occasions, differences in glassware, lab conditions, reagents, etc. would be reflected in the results. This set of data would give an indication of *reproducibility*, which might also be called *between-run precision*. We expect that the between-run precision will have greater spread than the within-run precision. Therefore, repeatability and reproducibility are not the same and it would be a misrepresentation if they were not clearly distinguished and honestly defined. We do not want to underestimate the total variability in a measurement process. Error estimates based on sequentially repeated observations are likely to give a false sense of security about the precision of the data. The quantity of practical importance is reproducibility, which refers to differences in observations recorded from *replicate experiments* performed in random sequence.

NORMALITY, RANDOMNESS, AND INDEPENDENCE

The three important properties on which many statistical procedures rest are normality, randomness, and independence. Of these, normality is the one about which many people worry most. It is not always the most important.

Normality means that the error term in a measurement, e_i, is assumed to come from a normal probability distribution. There is a tendency for error distributions that result from many *additive component errors* to be "normal-like." This is the *central limit effect*. It rests on the assumption that there are several sources of error, that no single source dominates, and that the overall error is a linear combination of independently distributed errors. These conditions seem very restrictive, but they often (but not always) exist. Even when they do not exist, lack of normality is not necessarily a serious problem. Transformations are available to make non-normal errors normal like. Many commonly used statistical procedures, including those that rely directly on comparing averages (such as t-tests to compare two averages and analysis of variance tests to compare several averages), are *robust* to deviations from normality. *Robust* means that they will tend to yield correct conclusions even when applied to data that are not normally distributed.

Random means that the observations are drawn from a population in a way that gives every element of the population an equal chance of being drawn. Randomization of sampling is the best form of insurance that observations will be independent.

> **Example 2.5.** In the context of the nitrate laboratory test, it is interesting to check whether the measurement errors are random. One easy way to do this is to plot the errors, $e_i = y_i - \eta$, to see if any pattern exists. Figure 2.6 is such a plot, showing e_i as a function of specimen number. The plot does not suggest any reason to believe the errors are not random.

Imagine ways in which the errors of the nitrate measurements might be nonrandom. Suppose, for example, that the measurement process drifted such that early measurements tended to be high and later measurements low. A plot of the errors against time of analysis would show a trend (positive errors followed by negative errors), indicating that an element of nonrandomness had entered the measurement process. Or, suppose that two different chemists had worked on the specimen and that one analyst always measured values that tended too high and the other always too low. In this case, a plot

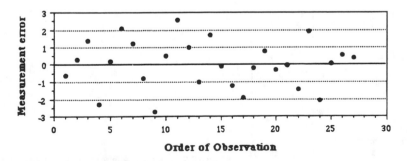

Figure 2.6 Plot of nitrate measurement errors indicates randomness.

of errors by the chemist would indicate this nonrandom error component. Note that this kind of nonrandom error would not be revealed by Figure 2.6. It is a good idea to check randomness with respect to each identifiable factor that could influence the measurement process.

Independence means that the simple multiplicative laws of probability work (that is, the probability of the joint occurrence of two events is given by the product of the probabilities of the individual occurrence). In the context of a series of observations, suppose that unknown causes produced experimental errors that tended to persist over time so that whenever the first observation y_1 was high, the second observation y_2 was also high. In such a case, y_1 and y_2 are not statistically independent. They are dependent in time or serially correlated. The same effect can result from cyclic patterns or slow drift in a system. Lack of independence can seriously distort the variance estimate and thereby make probability statements based on the normal or t distributions very much in error.

> **Example 2.6.** In the context of the nitrate laboratory test, it is interesting to check whether the measurement errors are independent. Figure 2.7 plots one measurement, y_i, against the previous observation, y_{i-1}. This plot shows no pattern (the correlation coefficient is –0.077) and indicates that the measurements are independent of each other.

Independence is often lacking in environmental data because (1) it is inconvenient or impossible to randomize the sampling or (2) it is undesirable to randomize the sampling because it is the cyclic or otherwise dynamic behavior of the system that needs to be studied. We, therefore, cannot automatically assume that observations are independent. When they are not, special methods are needed to account for the correlation in the data.

THE NORMAL DISTRIBUTION

Repeated observations that differ because of experimental error often vary with regard to some central value with a bell-shaped probability distribution that is symmetric and in which small deviations occur much more frequently than large ones. A continuous population frequency distribution that represents this condition is the *Gaussian* or *normal distribution*. Figure 2.8 shows a normal distribution for a random variable with $\eta = 8$ and $\sigma^2 = 1$. The normal distribution is characterized completely by its mean and variance and is often described by the notation $N(\eta, \sigma^2)$, which is read "a normal distribution with mean η and variance σ^2."

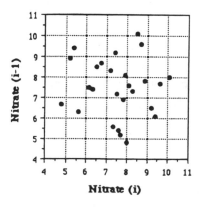

Figure 2.7 Plot of measurement y_i vs measurement y_{i-1} shows a lack of serial correlation.

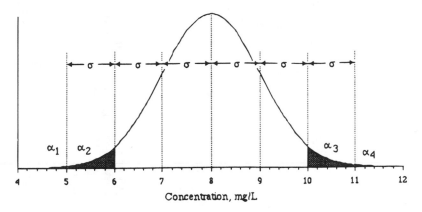

Figure 2.8 A normal distribution centered at mean $\eta = 8$. Because of symmetry, the areas $\alpha_1 = \alpha_4$ and $\alpha_1 + \alpha_2 = \alpha_3 + \alpha_4$.

The geometry of the normal curve is as follows:

1. The values of the random variable on the abscissa range from $-\infty$ to $+\infty$.
2. The vertical axis (probability density) is scaled such that area under the curve is unity (1.0).
3. The standard deviation σ measures the distance from the mean to the point of inflection.
4. The probability that a positive deviation from the mean will exceed one standard deviation is 0.1587, or roughly 1/6. This is the area to the right of 9 mg/L in Figure 2.8. The probability that a positive deviation will exceed 2σ is 0.0228 (roughly 1/40) which is area $\alpha_3 + \alpha_4$ in Figure 2.8. The chance of a positive deviation exceeding 3σ is 0.0013 (roughly 1/750), which is the area α_4.
5. Because of symmetry, the probabilities are the same for negative deviations and $\alpha_1 = \alpha_4$ and $\alpha_1 + \alpha_2 = \alpha_3 + \alpha_4$.
6. The chance that a deviation in either direction will exceed 2σ is $2(0.0228) = 0.0456$ (roughly 1/20). This is the sum of the two small areas under the extremes of the tails, $(\alpha_1 + \alpha_2) + (\alpha_3 + \alpha_4)$.

It is convenient to work with standardized normal deviates, $z = (y - \eta)/\sigma$, where z has the distribution $N(0,1)$ because the areas under the standardized normal curve are tabulated. This merely scales the data in terms of the standard deviation instead of the original units of measurement (e.g., concentration). A portion of this table is reproduced in Table 2.1. For example, the probability of a standardized normal deviate exceeding 1.57 is 0.0582, or 5.82%.

THE *t* DISTRIBUTION

Standardizing a normal random variable requires that both η and σ are known. In practice, however, we cannot calculate $z = (y - \eta)/\sigma$ because σ is unknown. Instead we substitute s for σ and calculate $t = (y - \eta)/\sigma$. For now consider that η is either known (for example, because it is the concentration of a primary standard against which the average of samples will be compared to check the performance of an analytical method) or assumed (for example, the difference between two treatments is assumed to be zero). Under certain conditions, which are given below, t has a known distribution, called the Student's t distribution, or simply the t distribution.

The shape, like the normal distribution, is bell-shaped and symmetric. The precise form of the t distribution depends on the degree of uncertainty in s^2, which is measured by the degrees of freedom v on which this estimate of σ^2 is based. When $v = \infty$, i.e.,

16

Table 2.1 Tail Area of the Unit Normal Distribution

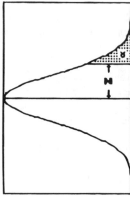

z	α = 0.00	0.01	0.02	0.03	0.04	0.05	0.06	0.07	0.08	0.09
0.0	0.5000	0.4960	0.4920	0.4880	0.4840	0.4801	0.4761	0.4721	0.4681	0.4641
...	...	...	...	...	...	...	...	...	...	...
1.5	0.0668	0.0655	0.0643	0.0630	0.0618	0.0606	0.0594	0.0582	0.0571	0.0559
1.6	0.0548	0.0537	0.0526	0.0516	0.0505	0.0495	0.0485	0.0475	0.0465	0.0455
1.7	0.0446	0.0436	0.0427	0.0418	0.0409	0.0401	0.0392	0.0384	0.0375	0.0367
1.8	0.0359	0.0351	0.0344	0.0366	0.0329	0.0322	0.0314	0.0307	0.0301	0.0294
1.9	0.0287	0.0281	0.0274	0.0268	0.0262	0.0256	0.0250	0.0244	0.0239	0.0233
2.0	0.0228	0.0222	0.0217	0.0212	0.0207	0.0202	0.0197	0.0192	0.0188	0.0183

when the sample size is infinite, there is no uncertainty in s^2 (because $s^2 = \sigma^2$) and the t distribution becomes the standard normal distribution. When v is small, however, the possibility of variation in s^2 increases. This is reflected by the spread of the t distribution increasing as the number of degrees of freedom of s^2 decreases. The tail area under the bell-shaped curve of the t distribution is the probability of t exceeding a given value. A portion of the t table is reproduced in Table 2.2.

The conditions under which the quantity $t = (y - \eta)/s$ has a t distribution with v degrees of freedom are (1) y is normally distributed about η with variance σ^2; (2) s is distributed independently of the mean, i.e., the variance of the sample must be independent of the magnitude of the mean; and (3) the quantity s^2, which has v degrees of freedom, is calculated from normally and independently distributed observations having variance σ^2.

THE SAMPLING DISTRIBUTION OF THE AVERAGE AND THE VARIANCE

All calculated statistics are random variables and, as such, are characterized by a probability distribution having an expected value (mean) and a variance. First, we consider the *sampling distribution* of the average, $\bar{y}$. Suppose that many random samples of size n were collected from a population and that the average was calculated for each sample. Many different average values would result, and these averages could be plotted in the form of a probability distribution. This would be the sampling distribution of the average (that is, the distribution of $\bar{y}$ values computed from different samples). If discrepancies in the observations y_i about the mean are random and independent, then the sampling distribution of $\bar{y}$ has mean η and variance σ^2/n. The quantity σ^2/n is the variance of the average. Its square root is the *standard error of the mean*,

$$\sigma_{\bar{y}} = \frac{\sigma}{\sqrt{n}}$$

The subscript $\bar{y}$ reminds us that this parameter describes the spread of the sample average

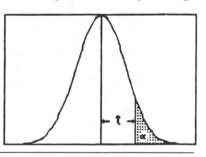

Table 2.2 **Values of t for Several Different Tail Probabilities and Degrees of Freedom**

v	**Tail Area Probability**				
	$\alpha = 0.1$	0.05	0.025	0.01	0.005
10	1.372	1.812	2.228	2.764	3.169
20	1.325	1.725	2.086	2.528	2.845
25	1.316	1.708	2.060	2.485	2.787
26	1.315	1.706	2.056	2.479	2.779
27	1.314	1.703	2.052	2.473	2.771
40	1.303	1.684	2.021	2.423	2.704
∞	1.282	1645	1.960	2.326	2.576

$\bar{y}$ about η, while σ describes the spread of the sample observations y about η. That is, $\sigma_{\bar{y}}$ indicates the spread we would expect to observe in calculated average values if we could repeatedly draw samples of size n at random from a parent population that has mean η and variance σ^2. We note that the sample average has smaller variability about η than does the sample data.

The sample standard deviation is

$$s = \sqrt{\frac{\Sigma(y_i - \bar{y})^2}{n - 1}}$$

and the estimate of the *standard error of the mean* is

$$s_{\bar{y}} = \frac{s}{\sqrt{n}}$$

This statistic is called the *standard error* to distinguish it from the sample standard deviation, s.

Example 2.7. The average for the n = 27/nitrate measurements is $\bar{y}$ = 7.51, and the sample standard deviation is s = 1.383. The estimated *standard error* of the mean is

$$s_{\bar{y}} = s/\sqrt{n} = 1.383/\sqrt{27} = 0.266 \text{ mg/L}$$

If the parent distribution is normal, the sampling distribution of $\bar{y}$ will be normal. If the parent distribution is non-normal, the distribution of $\bar{y}$ will be more nearly normal than the parent. As the number of observations n used in the average increases, the distribution of $\bar{y}$ becomes increasingly more normal. This fortunate property is the central limit effect. This means that we can use the normal distribution with mean η and variance σ^2/n as the reference distribution to make probability statements about $\bar{y}$ (e.g., the probability that $\bar{y}$ is less than or greater than a particular value or that it lies in the interval between two particular values).

If the population variance, σ^2, is not known, which is usually the case, we cannot use the normal distribution as the reference distribution for the sample average. Instead, we substitute $s_{\bar{y}}$ for $\sigma_{\bar{y}}$ and use the t distribution. Thus, when the parent distribution (i.e., the distribution from which the data are drawn) is normal and the population variance is estimated by s^2, the standardized mean t (also called the t statistic)

$$t = \frac{\bar{y} - \eta}{s/\sqrt{n}}$$

will have a t distribution with $\nu = n - 1$ degrees of freedom. If the parent population is not normal, but the sampling is random, t will tend toward the t distribution just as the distribution of $\bar{y}$ tends to being normal.

If the parent population is normally distributed with η and variance σ^2, and assuming once again that the observations are random and independent, the sample variance s^2 has especially attractive properties. For these conditions, the sample variance s^2 is distributed independently of y in a scaled χ^2 (chi-square) distribution. The scaled quantity is $\chi^2 = \nu s^2/\sigma^2$. This distribution is skewed to the right. The exact form of the χ^2 distribution depends on the number of degrees of freedom, ν, on which s^2 is based. The spread of the distribution increases as υ increases. The tail area under the χ^2 distribution is the probability of a value of $\chi^2 = \upsilon \, s^2/\sigma^2$ exceeding a given value.

Figure 2.9 illustrates these properties of the sampling distributions of $\overline{y}$, s^2, and t for a random sample of size n = 5.

Example 2.8. For the nitrate data, the sample mean concentration of $\overline{y}$ = 7.51 mg/L lies a considerable distance below the true value of 8.0 mg/L (Figure 2.10). If the true mean of the sample is 8.0 mg/L and the laboratory is measuring accurately, an estimated mean as low as 7.51 would occur by chance only about 4 times in 100. This is established as follows. For the example data t = (7.51 − 8.00)/0.266 = −1.853 with υ = 26 degrees

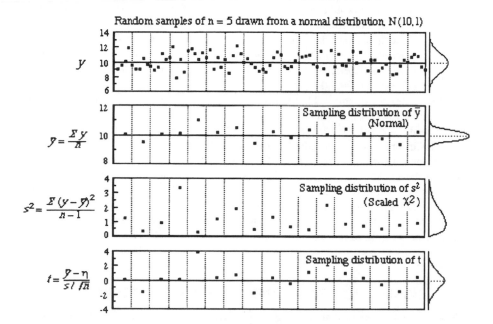

Figure 2.9 Random sampling from normal distributions to produce the sampling distributions of $\overline{y}$, s^2, and t. (From Box et al., 1978).

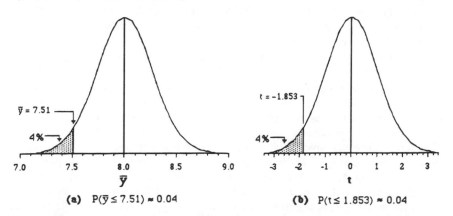

Figure 2.10 The $\overline{y}$ and t reference distributions for the sample average of the nitrate data of Example 2.1.

of freedom. To find the probability of such a value of t occurring, refer to the tabulated tail areas of the t distribution given in Appendix A. Because of symmetry, this table serves for negative as well as positive values. For $v = 26$, the tail areas are 0.05 for $t = -1.706$, 0.025 for $t = -2.056$, and 0.01 for $t = -2.479$. Plotting these and drawing a smooth curve as an aid to interpolation gives Prob(t $<$ -1.853) $\approx$ 0.04, or only about 4%. This low probability suggests that there may be a problem with the measurement method in this laboratory.

The assessment given in Example 2.8 can also be made by examining the t reference distribution of $\bar{y}$ as sketched in Figure 2.10. The distribution shows the relative frequency at which values of $\bar{y}$ would be expected to occur if the laboratory assessment was repeated many times on samples of side $\eta = 27$. The distribution of $\bar{y}$ is centered about $\eta = 8.0$ mg/L with standard deviation s = 0.266 mg/L. The value of $\bar{y}$ observed for this particular experiment is 7.51 mg/L. The shaded area to the left of $\bar{y} = 7.51$ in Figure 2.10a is the same as the area to the left of $t = -1.853$ in Figure 2.10b. Thus, P(t $\leq$ -1.583) = P($\bar{y}$ $\leq$ 7.51) $\approx$ 0.04.

In the context of Example 2.8, the investigator is considering a particular result, say the result that $\bar{y} = 7.51$ mg/L in a laboratory assessment based on 27 blind measurements on specimens known to have concentration $\eta = 8.0$ mg/L. A relevant reference distribution is needed in order to decide whether the result is easily explained by mere chance variation or whether it is exceptional. This reference distribution represents the set of outcomes that could occur by chance. The t distribution is a relevant reference distribution under certain conditions which have already been identified. An outcome that falls on the tail of the distribution can be considered exceptional. If it is found to be exceptional, it is declared statistically significant. Significant in this context does not refer to scientific importance, but only to its statistical plausibility in light of the data.

SIGNIFICANCE TESTS AND CONFIDENCE INTERVALS

In Example 2.8, we knew that the nitrate population mean was truly 8.0 mg/L and asked, "How likely are we to get a sample mean as small as 7.51 mg/L from the analysis of 27 specimens?" If this result is highly unlikely, we might decide that the sample did not represent the population, probably because the measurement process was biased to yield concentrations below the true value. Or, we might decide that the result, although unlikely, should be accepted as occurring due to chance rather than due to an assignable cause (like bias in the measurements).

Statistical inference involves making an assessment from experimental data about an unknown population parameter (e.g., a mean or variance). Consider that the true mean, instead of being known as in Example 2.8, is unknown and we wish to ask, "If a sample mean of 7.51 mg/L is estimated from measurements on 27 specimens, what is the likelihood that the true population mean is 8.00 mg/L?" Confidence intervals and significance tests are two commonly used methods for making such statistical inferences.

The significance test typically takes the form of a hypothesis test. The hypothesis to be tested is often designated H_o. In this case, H_o is that the true value of the population mean is $\eta = 8.0$ mg/L. This is sometimes more formally written as H_o: $\eta = 8.0$. This is the *null hypothesis*. The alternate hypothesis is H_a: $\eta < 8.0$ or $\eta > 8.0$, or $\eta \neq 8.0$. A significance level, α, is selected at which the null hypothesis will be rejected. The significance level, α, represents the risk of falsely rejecting the null hypothesis.

The relevant t statistic is

$$t = \frac{\text{statistic} - E(\text{statistic})}{\sqrt{V(\text{statistic})}}$$

where E(statistic) denotes the expected value of the statistic being estimated and V(statistic) denotes the variance of this statistic.

Example 2.9. Use the nitrate data in Example 2.1 to test the hypothesis that at the $\alpha = 0.05$ significance level, the true mean concentration of the 27 specimens is 8.0 mg/L. The appropriate hypotheses are H_o: $\eta = 8.0$ and H_a: $\eta < 8.0$. This is a one-sided test, since the alternate hypothesis involves η on the lower side of 8.0.

The hypothesis test is made using

$$t = \frac{\text{statistic} - E(\text{statistic})}{\sqrt{V(\text{statistic})}} = \frac{\bar{y} - \eta}{s_{\bar{y}}}$$

where the "statistic" is $\bar{y} = 7.51$, E(statistic) is $\eta = 8.0$, and $\sqrt{V(\text{statistic})} = s_{\bar{y}} = 0.266$. The quantity $t = (\bar{y} - \eta)/s_{\bar{y}}$ is $t = (7.51 - 8.0)/0.266 = -1.842$. The null hypothesis will be rejected if the computed t is less than the value of the lower tail t statistic having a probability of $\alpha = 0.05$. This "critical" value of t with $\nu = 26$ degrees of freedom is obtained from tables: $t_c = t_{\alpha = 26, \nu = .05} = -1.706$. The computed value of $t = -1.853$ is smaller than $t_c = -1.706$, and the decision is to reject the null hypothesis in favor of the alternate hypothesis.

The differences between Examples 2.8 and 2.9 may appear to be a matter of semantics, but they are important and exemplify the distinction between a sampling distribution and a statistical inference problem. In Example 2.8, the experimenter assumes the population parameter to be known and asks whether the sample data can be construed to represent the population. In Example 2.9, the experimenter assumes the sample data are representative and asks whether the assumed population value reasonably characterizes the population. Mathematically, they are equivalent. In practice, the experimental context will usually suggest one approach as the more comfortable interpretation.

Example 2.9 illustrated a *one-sided hypothesis test*. It evaluated the hypothesis that the sample mean was truly to one side of 8.0. This particular example was interested in the mean being less than the hypothesized value. A *two-sided hypothesis test* would consider the statistical plausibility of both the positive and negative deviations from the mean.

Example 2.10. Use the nitrate data to test the null hypothesis that H_o: $\eta = 8.0$ and H_a: $\eta \neq 8.0$. Here, the alternate hypothesis considers both deviations on the positive and negative sides of the population mean, which makes this a *two-sided* test. Both the lower and upper tail areas of the t reference distribution must be used. Because of symmetry, these tail areas are equal. For a test at the $\alpha = 0.05$ significance level, the sum of the upper and lower tail areas equals 0.05. The area of each tail is $\alpha/2 = 0.05/2 = 0.025$. The "critical" t value for $\alpha/2 = 0.025$ and $\nu = 26$ degrees of freedom is $t_c = \pm 2.056$. The computed t value is the same as in Example 2.9, that is, $t = -1.853$. Because the computed t value is not outside the range of the critical t values, there is not sufficient evidence to reject the null hypothesis at the stated level of significance.

Notice that the hypothesis tests in Examples 2.9 and 2.10 reached different conclusions even though they used the same data, the same significance level, and the same null hypothesis. The only difference was the alternate hypothesis. The two-sided alternative hypothesis implies interest in detecting both negative and positive deviations from the assumed mean by dividing the rejection probability α between the two tails. Thus, a decision to reject the null hypothesis takes into account differences between the sample mean and the assumed population mean that are both significantly smaller and significantly larger than zero. The consequence of this is that in order to be declared "statistically significant," the deviation $\bar{y} - \eta$ must be larger in a two-sided test than in a one-sided test.

Which test — the one-sided test or the two-sided test — was the correct one to make? The question cannot be answered in general, but often the decision-making context will indicate which test is appropriate. In a case where a positive deviation is a problem, but a negative deviation is not, a one-sided test would be indicated. Examples would be (1) judging compliance with an environmental protection limit where *high* values indicate a violation, and (2) an experiment intended to investigate whether adding chemical A to the process *increases* the efficiency. If the experimental question is whether adding chemical A changes the efficiency (either for better or worse), a two-sided test would be indicated.

Hypothesis testing can be overdone. It is often more informative to state an interval within which the value of a parameter would be expected to lie. A 1-α *confidence interval* for the population mean can be constructed using the appropriate value of t as:

$$\bar{y} - t_{\alpha/2}\, s_{\bar{y}} < \eta < \bar{y} + t_{\alpha/2}\, s_{\bar{y}}$$

where $t_{\alpha/2}$ and $s_{\bar{y}}$ have $\nu = n - 1$ degrees of freedom. This confidence interval is bounded by a lower and an upper limit. The meaning of the 1-α probability is, "If a series of random sets of n observations is sampled from a normal distribution with mean η and fixed σ, and a 1-α confidence interval $\bar{y} \pm t_{\alpha/2}\, s_{\bar{y}}$ is constructed from each set, a proportion, 1-α, of these intervals will include the value η and a proportion, α, will not" (Box et al., 1978). (Another interpretation, a Bayesian interpretation, is that there is a 1-α probability that the true value falls within this confidence interval.)

Example 2.11. The confidence limits for the true mean of the test specimens are constructed for $\alpha = 0.05/2 = 0.025$, which gives a 95% confidence interval. For $\nu = 26$, $t_{26,\,0.025} = 2.056$, $\bar{y} = 7.51$ and $s_{\bar{y}} = 0.266$, the upper and lower 95% confidence limits are

$$7.51 - 2.056(0.266) < \eta < 7.51 + 2.056(0.266)$$
$$6.96 < \eta < 8.05$$

This interval contains $\eta = 8.0$ so we conclude that the difference is not so large that random measurement error should be rejected as a plausible explanation.

This use of a confidence interval is equivalent to making a two-sided test of the null hypothesis, as done in Example 2.10. Figure 2.11 shows the two-sided 90% and 95% confidence intervals for η.

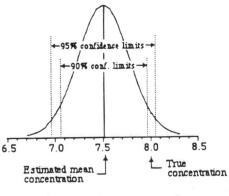

Figure 2.11 The 90 and 95% confidence intervals for the true mean of the nitrate data using the t reference distribution.

SUMMARY

This chapter has reviewed basic definitions, assumptions, and principles. The key points are listed below.

A sample is a sub-set of a population and consists of a group of n observations taken for analysis. Populations are characterized by parameters, which are usually known and unmeasurable because we cannot measure every item in the population. Parameters are estimated by statistics that are calculated from the sample. Statistics are random variables and are characterized by a probability distribution that has a mean and a variance.

All measurements are subject to experimental (measurement) error. Accuracy is a function of both bias and precision. The role of statistics in scientific investigations is to quantify and characterize the error and take it into account when the data are used to make decisions.

Given a *normal parent distribution* with mean η and variance σ^2 and for random and independent observations, the sample average $\bar{y}$ has a normal distribution with mean η and variance σ^2/n. The sample variance s^2 has expected value σ^2. The statistic $t = (\bar{y} - \eta)/(s/\sqrt{n})$ with $\nu = n - 1$ degrees of freedom has a t distribution.

Statistical procedures that rely directly on comparing means, such as t tests to compare two means and analysis of variance tests to compare several means, are robust to non-normality but may be adversely affected by a lack of independence.

Hypothesis tests are useful methods of statistical inference, but they are often unnecessarily complicated in making simple comparisons. Confidence intervals are a statistically equivalent alternative to hypothesis testing, and they are simple and straightforward. They give the interval (range) within which the population parameter value is expected to fall.

These basic concepts are discussed in any introductory statistics book (e.g., Devore, 1991; Miller et al., 1991). A careful discussion of the material in this chapter, with special attention to the importance regarding normality and independence, is found in Chapters 2, 3, and 4 of Box et al. (1978).

REFERENCES

Box, G. E. P., W. G. Hunter, and J. S. Hunter (1978). *Statistics for Experimenters: An Introduction to Design, Data Analysis, and Model Building*, New York, Wiley Interscience.

Devore, J. (1991). *Probability and Statistics for Engineers*, 3rd ed., New York, Brooks-Cole.

Miller, I., J. E. Freund, and R. A. Johnson (1991). *Probability and Statistics for Engineers*, 4th ed., Englewood, NJ, Prentice-Hall.

Watts, D. G. (1991). "Why Is Introductory Statistics Difficult To Learn? And What Can We Do to Make It Easier?" *Am. Statistician*, 45, 4, 290–291.

Taylor, J. K. (1987). *Quality Assurance of Chemical Measurements*, Chelsea, MI, Lewis Publishers.

Chapter 3

Plotting Data

Key words: box and whisker plot, digidot plot, error bars, residual plots, scatterplot, Youden plot

---

"The most effective statistical techniques for analyzing environmental data are graphical methods. They are useful in the initial stage for checking the quality of the data, highlighting interesting features of the data, and generally suggesting what statistical analyses should be done. Interesting enough, graphical methods are useful again after intermediate quantitative analyses have been completed and again in the final stage for providing complete and readily understood summaries of the main findings of investigations" (Hunter, 1988).

The first step in data analysis should be to plot the data. Graphing data should be an interactive experimental process (Chatfield, 1988, 1991; Tukey, 1977). Do not expect your first graph to reveal all interesting aspects of the data. Make a variety of graphs to view the data in different ways. Doing this may (1) reveal the answer so clearly that little more analysis is needed, (2) point out properties of the data that would invalidate a particular statistical analysis, (3) reveal that the sample contains unusual observations, (4) keep you from doing something foolish, (5) save time in subsequent analyses, and (6) suggest an answer that you had not expected.

The time spent making some different plots almost always rewards the effort. Many top notch statisticians like to plot data by hand, believing that the physical work of the hand stimulates the mind's eye. Whether you adopt this work method or use one of the many available computer programs, the goal is to free your imagination by trying a variety of graphic forms. Keep in mind that some computer programs offer a restricted set of plots and thus could limit rather than expand the imagination.

MAKE THE ORIGINAL DATA RECORD A PLOT

Since the best way to display data is in a plot, it makes little sense to make the primary data record a table of values. Instead, plot the data directly on a *digidot plot*, which is Hunter's (1988) innovative combination of a *time sequence* plot with a *stem-and-leaf* plot (Tukey, 1977) and is extremely useful for a modest-sized collection of data.

The graph is illustrated in Figure 3.1 for a time series of 36 hourly observations (time, in hours, is measured from left to right).

30	27	41	38	44	29	43	21	15	33	33	28
49	16	22	17	17	23	27	32	47	71	46	42
34	34	34	44	27	32	28	25	36	22	29	24

As each observation arrives, it is placed as a dot on the time sequence plot and simultaneously recorded with its final digit on a stem-and-leaf plot. For example, the first observation was 30. The last digit, a zero, is written in the "bin" between the tick marks for 30 and 35. As time goes on, this bin also accumulates the last digits of observations having the values of 33, 33, 32, 34, 34, 34, and 32. The analyst thus generates a complete visual record of the data: a display of the data distribution, a display of the data time history, and a complete numerical record for later detailed arithmetic analysis.

THE YOUDEN PLOT

Sometimes making the right plots eliminates the need to make statistical calculations. Figure 3.2 is a variation on the Youden plot (named after the scientist who introduced

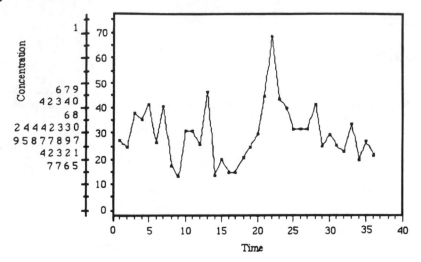

Figure 3.1 Digidot plot shows the sequence and distribution of the data.

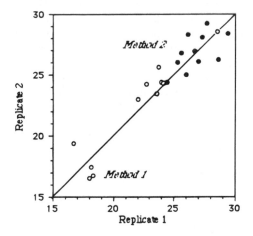

Figure 3.2 Youden plot of suspended solids data.

it) that compares two methods of measuring suspended solids in wastewater. Eleven different analysts participated, and each made two replicate measurements using each method. Each point represents a pair of replicate values produced by one analyst. The solid points are produced by measurement Method 2; open points show Method 1 results. If the work is under statistical control, the plotted points will fall within a small region oriented along the diagonal line. The work of four analysts is noticeably different from the others, as shown by the four points in the lower left corner of the graph. Even if these four points were disregarded, it seems sufficiently clear that the two methods are not equivalent; Method 2 tends to give higher values than Method 1. No statistical calculations are needed.

IN SEARCH OF TRENDS

Figure 3.3 is a plot of 558 pH observations on a small stream in the Great Smoky Mountains. The data cover the period from 1971 to mid-1982, as shown by the values on the top of the plot. Time measured in weeks is on the bottom abscissa.

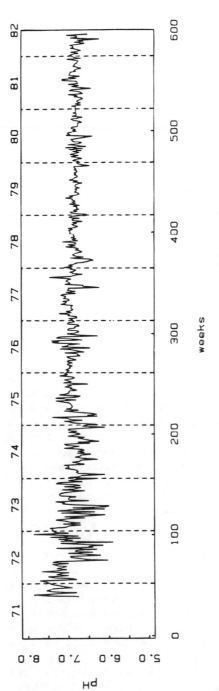

Figure 3.3 Time series plot of data measured on a small mountain stream.

The data were submitted (on computer tape) to an agency that intended to do a trend analysis to assess possible changes in water quality related to acid precipitation. The data were plotted before any regression analysis or time series modeling was begun. This plot was not expected to be useful in showing a trend, since any trend would be small (subsequent analysis indicated that there was no trend). The purpose of plotting the data was to reveal any peculiarities in the data.

The plot served this purpose by showing that the variance, which was large early in the series, decreased at about 150 weeks and seemed to decrease again at about 300 weeks. This stimulated the data analyst to ask two questions. Was there any natural phenomenon to explain this pattern of variability? Is there anything about the measurement process that could explain it? From this questioning, it was discovered that different measuring instruments had been used. The pH meter was replaced at about the beginning of 1974 with a more precise instrument, which was itself replaced by an improved model in about 1976.

The presence of nonconstant variance influenced the subsequent data analysis. For example, if ordinary linear regression were used to assess the existence of a trend, the large variance would have given the early data more "weight" or "strength" in determining the position and slope of the trend line. This weighting is not desirable because the latter data are the most precise.

Failure to plot the data initially might not have been fatal. The nonconstant variance might have been discovered later in the analysis, perhaps by plotting the residuals, but by then considerable work would have been invested. But, it might be overlooked because an analyst who does not start by plotting the data is not likely to make residual plots either. If the problem were overlooked, an improper analysis would be reported.

SCATTERPLOTS

It has been estimated that 75% of the graphs used in science are scatterplots (Tufte, 1983). Simple scatterplots are often made before any other data analysis is considered. The insights gained may lead to more elegant and informative graphs or suggest a promising model. Linear or nonlinear relations are easily seen and so are outliers or other aberrations in the data. For an example of how a scatterplot can spark the scientific imagination, read about the spectacular success of a scatterplot made by the American astronomer Henry Norris Russell (Spence and Garrison, 1993).

A more routine use of scatterplots is illustrated with data from a study of how phosphorus removal by a wastewater treatment plant was related to influent levels of phosphorus, flow, and other characteristics of wastewater. The scatterplots shown in Figure 3.4 were made as a guide to constructing the first tentative models, There are no scales shown on these plots because we are looking for patterns; the numerical levels are unimportant at this stage of work. The computer automatically scales each two-variable scatterplot to best fill the available area of the graph. Each paired combination of the variables is plotted to reveal possible correlations. For example, it is discovered that effluent total phosphorus (TP-out) is correlated rather strongly with effluent suspended solids (SS-out) and effluent biochemical oxygen demand (BOD-out), moderately correlated with flow, BOD-in, and not correlated with SS-in and TP-in. Effluent soluble phosphorus (SP-out) is correlated only with SP-in and TP-out. These observations provide a starting point for model building.

The values plotted in Figure 3.4 are actually logarithms of the original variables. Making this transformation was advantageous in showing extreme values, and it simplified interpretation by giving linear relations between variables. It is often helpful to use transformations in analyzing environmental data. The logarithmic and other transformations are discussed in Chapter 7.

JONES ISLAND (logs)

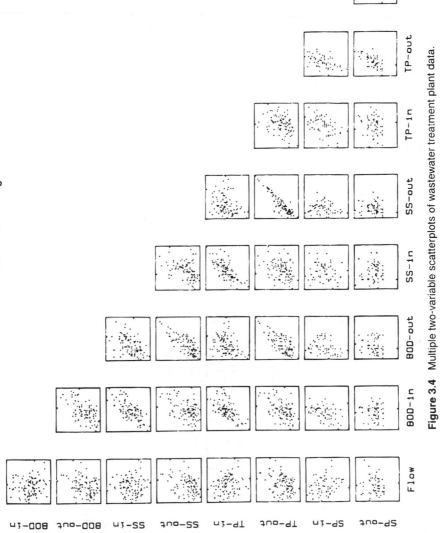

Figure 3.4 Multiple two-variable scatterplots of wastewater treatment plant data.

PLOTS OF RESIDUALS

Graphing residuals is an important method that has applications in all areas of data analysis and model building. Residuals are the difference between the observed values and the smooth curve constructed from a model of the data. If the model fits the data, the residuals represent the measurement error.

The visual impression in the top panel in Figure 3.5 is that the curve fits the data fairly well, but that the vertical deviations of points from the fitted curve are greater for small values of time than at the longer times. The graph of residuals in the bottom plot shows the opposite is true. The curve does not fit well at the shorter times, and in this region they are large and predominantly positive. Tukey (1977) calls this process of plotting residuals "flattening the data." He emphasizes its power to shift our attention from the fitted line to the discrepancies between prediction and observation. It is these discrepancies that contain the information that is needed to improve the model.

Make it a habit to examine the residuals of a fitted model, including deviations from a simple mean. Check for normality by making a dot diagram or histogram. Plot the residuals against the predicted values, against the predictor variables, and as a function of the time order in which the measurements were made. If the residuals appear to be random and of uniform variance in all these plots, they provide persuasive evidence that the model has no serious deficiencies. If they show a trend, it is evidence that the model is inadequate. If the residuals spread out, it suggests that a data transformation is probably needed.

STATISTICAL VARIATION

Measurements vary, and one important function of graphs is to show the variation. There are two very different ways of showing variation. One is to show the value of the data in the form of a histogram or a box graph. The other is by using error bars to show statistics such as standard deviations, standard errors, or confidence intervals.

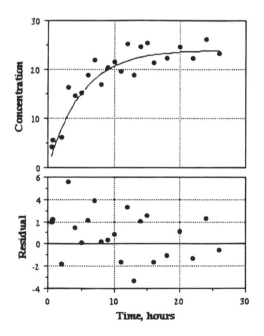

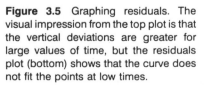

Figure 3.5 Graphing residuals. The visual impression from the top plot is that the vertical deviations are greater for large values of time, but the residuals plot (bottom) shows that the curve does not fit the points at low times.

A histogram shows the shape of the frequency distribution and the range of values; it also gives an impression of central tendency and shows symmetry or lack of it. A box graph is a designed to convey the few primary features of a set of data. One form of box graph, the so-called *box and whisker plot*, is used in Figure 3.6 to compare the effluent quality of 12 identical trickling filter pilot plants that were operated in parallel for 35 weeks (Gameson et al., 1961). It shows the median (50th percentile) as a center bar, the quartiles (25th and 75th percentiles) as a box, and adds whiskers that extend to the 10th and 90th percentiles values in the data set. Values above the 90th percentile and below the 10th percentile are plotted as points. The simplicity of the plot makes a convenient comparison of the performance of the 12 replicate filters.

SHOWING PRECISION OF REPLICATED MEASUREMENTS

Often repeated observations of the dependent variable are made at the settings of the independent variable. In this case, it is desirable that the plot show the average value of the replicate measured values and some indication of their precision or variation. This is done by plotting a symbol to locate the sample average and adding to it *error bars* to show statistical variation. Too often authors fail to tell the reader what the error bars represent. They could be sample standard deviations, standard errors of the means, or confidence intervals of the means. Whichever is used, the meaning of the error bars must be clearly defined or the author will introduce confusion where the intent is to clarify the meaning of the data.

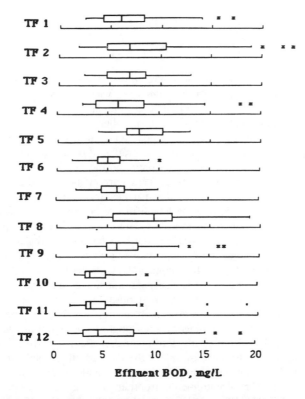

Figure 3.6 Box and whisker plots to compare the performance of 12 identical trickling filters operating in parallel. Each panel summarizes 35 measurements.

Showing averages and sample standard deviations is often a poor way to convey the variation in data. A serious weakness of bars marking the sample standard deviation is that they imply that the data are distributed symmetrically about the mean. To most readers, suggesting symmetry also will suggest that the replicates were normally distributed even when this may not be true. If the purpose of using error bars is to show the empirical distribution of the replicate data, consider using box graphs.

If the error bars are intended to show the precision with which the average of the replicate values has been estimated one can plot the standard error or a confidence interval for the average. This has weaknesses as well. The average of replicates does tend to be normally distributed so there is less chance of seriously misleading the reader by showing one-standard-error bars. Nevertheless, it is better to show confidence intervals. If all plotted averages were based on the same number of observations, one-standard-error bars would convey an approximate 68% confidence interval. This is not a particularly interesting interval. If the averages are calculated from different numbers of values, the confidence intervals would be different multiples of the standard error bars (according to the appropriate degrees of freedom of the t distribution). Cleveland (1985) suggests using *two-tiered error bars*. The inner error bars would show the 50% confidence interval, a middle range analogous to the box of a box graph. The outer interval of the two-tiered error bar would simply reflect the 95% confidence interval.

SUMMARY

Graphical methods are obviously useful for *initial data analysis* and *exploratory data analysis*, but they also serve us well in the final analysis. "A picture is worth a thousand words" is a cliche, but still powerfully true. The right graph may reveal all that is important. If it only tells part of the story, that is the part that is most likely to be remembered.

Tables of numbers camouflage the interesting features of data. The human mind, which is remarkably well adapted to so many and varied tasks, is simply not capable of extracting useful information from tabulated numbers. Why, then, are data so often reported as tables when putting these same numbers in appropriate graphical form completely changes the situation? The informed human mind can then operate efficiently with these graphs as inputs. In short, suitable graphs of data and the human mind are an effective combination; endless tables of data and the mind are not.

It is extremely important that plots be kept current, because the first purpose of keeping these plots is to help monitor and, if necessary, to troubleshoot difficulties as they arise. The plots do not have to be beautiful nor be computer drafted, in order to be useful. Make simple plots by hand as the data are available. If the plots are made at some future date to provide a record of what happened in the distant past, it will be too late to take appropriate action to improve performance. The second purpose is to have an accurate record of what has happened in the past, especially if the salient information is in such a form that it is easily communicated and readily understood. If they are kept up to date and used for the first purpose, they can also be used for the second. On the other hand, if they are not kept up to date, they may be useful for the second only. In the interests of efficiency, they ought to serve double duty.

Intelligent data analysis begins with plotting the data. Be imaginative. Use a collection of different graphs to see different aspects of the data. Plotting graphs in a notebook is not as useful as making plots large and visible. Plots should be displayed in a prominent place so that those concerned with the environmental system can review them readily.

We close with Tukey's (1977) declaration: "**The greatest value of a picture** is when it *forces* us to notice *what we never expected to see.*" (Emphasis and italics in the original.)

REFERENCES

Chatfield, C. (1988). *Problem Solving: A Statistician's Guide*, London, Chapman and Hall.

Chatfield, C. (1991). "Avoiding Statistical Pitfalls," *Stat. Sci.,* 6, 3, 240–268.

Cleveland, W. S. (1985). *The Elements of Graphing Data*, Monterey, CA, Wadsworth.

Cleveland, W. S., B. Kleiner, J. E. McRae, and J. L Warner (1976). *Science*, 191, 179–181.

Gameson, A. L. H., G. A. Truesdale, and M. J. Van Overdijk (1961). "Variation in Performance of Twelve Replicate Small-Scale Percolating Filters," *Water Waste Treat. J.*, 9, 342–350.

Hunter, J. S. (1988). "The Digidot Plot," *Am. Statistician*, 42, 54.

Spence, I. and R. F. Garrison (1993). "A Remarkable Scatterplot," *Am. Statistician*, 47, 12–19.

Tufte, E. R. (1983). *The Visual Display of Quantitative Information*, Cheshire, CT, Graphics Press.

Tukey, J. W. (1977). *Exploratory Data Analysis*, Reading, MA, Addison-Wesley.

Chapter 4

Smoothing Data

Key words: moving average, exponentially weighted moving average, weighting factors, smoothing, median smoother

---

Mark Twain once said, "You cannot see the truth when your imagination is out of focus." Smoothing helps more clearly focus the imagination of persons responsible for identifying such things as long-term trends, cyclical fluctuations around a trend, and sudden changes in a system's behavior. They need to (1) realize that something has gone wrong, (2) identify when the trouble started, (3) pinpoint the cause of the trouble, and (4) rectify the situation. Because treatment plants and other environmental systems are complicated, the data from them is highly variable, and it is difficult in practice to proceed through these four steps. Experience has shown that smoothing and plotting the data in various simple ways can facilitate the successful completion of these four steps.

Smoothing is drawing a smooth curve through data in order to eliminate the roughness (scatter) that blurs the fundamental underlying pattern. It helps sharpen our focus by unhooking our eye from the irregularities.

Smoothing can be thought of as a decomposition of the data. In curve fitting, this decomposition has the general relation *data = fit + residuals*. In smoothing, the analogous expression is *data = smooth + rough*. Since the *smooth* is intended to be smooth (as the "fit" is smooth in curve fitting), we usually show its points connected. Similarly, we show the *rough* (or residuals) as separated points, if we show them at all. We may choose to show only those rough (residual) points that stand out markedly from the smooth (Tukey, 1977).

We will discuss several methods of smoothing. Graphs of certain simple statistics are effective and economical in terms of time and money. The methods are well established and have a long history of successful use in industry and econometrics. They are applicable to all kinds of environmental data. They should not be overlooked merely because they are simple. They are useful to everyone, regardless of statistical expertise. Only elementary arithmetic is needed to make graphs that are useful in clarifying the data and drawing our attention to the most important features of the record. A computer may be helpful, but one is not needed. A handheld calculator will serve perfectly well.

In statistics and quality control literature, one finds mathematics and theory that can embellish these graphs. A formal statistical analysis, such as adding control limits, can become quite complex because often the assumptions on which such tests are usually based are violated rather badly by environmental data. These embellishments are omitted for now, not because they may be complex, but because we strongly believe that measurable improvements in data interpretation are possible without them.

SMOOTHING METHODS

One method of smoothing would be to fit a straight line or polynomial curve to the data. Aside from the computational bother, this is not a useful general procedure because the very fact that smoothing is needed means that we cannot see the underlying pattern clearly enough to know what particular polynomial would be useful. The methods we will consider are computationally simple and can be used without any specification of polynomial curves. They are to reexpress the data before plotting (for example, plot the logarithm of y instead of y itself) and to plot moving averages (MA) or exponentially weighted moving averages (EWMA).

An MA gives equal weight to a sequence of past values; the weight depends on how many values are to be remembered. The EWMA gives more weight to recent events and progressively forgets the past. (We can easily control how quickly the past is forgotten.) The EWMA will follow the current observations more closely than the MA. Often this is desirable, but this responsiveness is purchased by a loss in smoothing.

The running median methods require no computations in contrast to the moving average methods, which require the (admittedly simple) calculation of averages. The median of a set of numbers is not shifted by extreme values that might be part of the set, whereas an average will reflect all values. This means that the median smoothing methods have the additional advantage of not being affected by extreme values. The extremes, which may have great importance, show up in the residuals.

REEXPRESSING THE DATA

The top panel of Figure 4.1 is a plot of influent copper concentration at a wastewater treatment plant. The primary features of this record are a few high values. The bottom panel shows the same data plotted on a logarithmic scale. This method of plotting makes the process behavior appear more consistent, and it therefore focuses attention on the main features: the long run of stable efficient performance and a value at day 70 which is above the average (on the log scale) by about the same amount as the value on day 69 is below the average.

Is this high value so extraordinary as to deserve special attention? Or is it an outlier (i.e., a rogue value that can be disregarded)? This question cannot be answered without knowing the underlying distribution of the data. If this distribution is believed to be normal, then the value on day 70 almost certainly is not from the population that produced the rest of the data series. If, on the other hand, the data are produced by a lognormal distribution, this value is not so noteworthy.

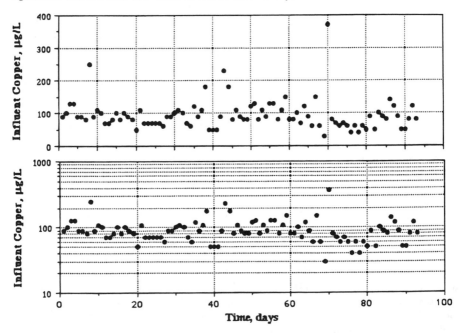

Figure 4.1 Copper data plotted on arithmetic (top) and logarithmic (bottom) scales give a different impression about the high values.

THE MOVING AVERAGE

Many standards for environmental quality have been written for an average of 30 consecutive days. The language is something like the following: "Average daily values for 30 consecutive days shall not exceed . . . ." This is commonly interpreted to mean a monthly average, probably because dischargers submit monthly reports to the regulatory agencies, but one should note the great difference between the moving 30-day average and the monthly average as an effluent standard. There are only 12 monthly averages in a year of the kind that start on the first day of a month, but there are a total of 365 moving 30-day averages that can be computed. One very bad day could make a monthly average exceed the limit. This same single value would be used to calculate 30 other moving averages, and several of these might exceed the limit. These two statistics — the strict monthly average and the 30-day moving average — imply different effects on the environment, even though the effluent and the environment are the same.

The length of time over which a moving average is calculated can be adjusted to represent the memory of the environmental system as it responds to pollutants. This is done in ambient air pollution monitoring, where a short averaging time (1 hr) is used for ozone, to take one example.

The moving average is the sum of the most recent k data divided by k, that is, the simple average of the most recent k data points. Thus, for example, a 7-day moving average (MA7) uses the latest seven data points, a 10-day average (MA10) uses ten points, and so on. Each data point has been given equal weight in computing the average.

A k-day moving average plot consists of graphing the average

$$\bar{y}_i(k) = \frac{1}{k} \sum_{j=i-k+1}^{i} y_j \qquad i = k, k+1, \ldots, n$$

As each new observation is made, the summation will drop one term and add another term, giving the simple updating formula

$$\bar{y}_i(k) = \bar{y}_{i-1}(k) + \frac{1}{k}(y_{i-k})$$

By smoothing random fluctuations, the moving average sharpens the focus on recent performance levels. Figure 4.2 shows the MA7 and MA30 for some polychlorinated biphenyl (PCB) data. Both moving averages help general trends in performance show up more clearly because random variations are averaged and smoothed. The MA30 highlights long-term changes in performance and clarifies low frequency cycles in performance. The MA7, which is more reflective of short-term variations, has special appeal in being a weekly average.

EXPONENTIALLY WEIGHTED MOVING AVERAGE

In the simple moving average, recent values and long past values are weighted equally. For example, in an MA30, the performance four weeks ago is reflected in this statistic to the same degree as yesterday's, even though the receiving environment may have "forgotten" the event of four weeks ago. The EWMA weights the most recent event heavily and each event going into the past proportionately less.

Since the EMWA forgets the past, it may give a more realistic representation of the actual threat of the pollutant on the environment. For example, the biochemical oxygen demand (BOD) discharged into a freely flowing stream is important the day it is discharged. A 2- or 3-day average might be important as well because a 2- or 3-day

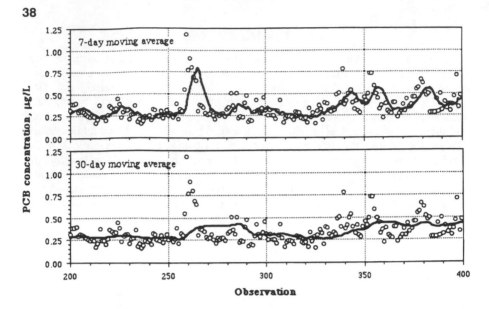

Figure 4.2 Seven-day and 30-day moving averages of PCB data.

depression of dissolved oxygen could be disastrous, while a 1-day shock might be tolerable to aquatic organisms. A 30-day average of BOD could be a less informative statistic about the threat to fish than a short-term average, but it may be needed to assess the long-term trend in plant performance.

For suspended solids that settle on a stream bed and form sludge banks, a long-term average might be related to the depth of the sludge bed and, therefore, be an informative statistic. If the solids do not settle, the daily values may be more descriptive of potential damage. For a pollutant that could be ingested by an organism and later excreted or metabolized, the EWMA might be a good statistic.

Conversely, some pollutants may not exhibit their effect for years. Carcinogens are an example where the long-term average could be important. Long-term in this context is years, so the 30-day average would not be a particularly useful statistic. The first ingested (or inhaled) irritants may have more importance than recently ingested material. Asbestos accumulations in the lungs may be an example of this. If so, perhaps past events should be weighted more heavily than recent events if a statistic is to relate a source of pollution to present effect. Care is needed to select the proper statistic for use as an effluent limit. If an inappropriate statistic is reported, it may mean that masses of data have reduced value for biologists, epidemiologists, and others who seek to relate pollutant discharges to effects on organisms.

The EWMA is calculated as

$$\overline{Z}_i = (1 - \phi) \sum_{j=0}^{\infty} \phi^j y_{i-j} \qquad i = 1, 2, \ldots$$

where ϕ is a suitably chosen constant between 0 and 1 that determines the length of the EWMA's memory. Of course, one does not include infinitely many terms in the summation, because for reasonable choices of ϕ the weight $(1 - \phi)\phi^j$ rapidly approaches 0 as j increases. In practice, one may only need a few terms. For example, if $\phi = 0.3$,

$$\overline{Z}_0 = (1 - 0.3)y_0 + (1 - 0.3)(0.3)y_{-1} + (1 - 0.3)(0.3)^2 y_{-2} + \ldots$$
$$= 0.7 \, y_0 + 0.21 \, y_{-1} + 0.063 \, y_{-2} + 0.019 \, y_{-3} + \ldots$$

The small coefficient of y_{-3} shows that values more than three days into the past are essentially forgotten since.

The EWMA can be easily updated using

$$\overline{Z}_i = \phi \, \overline{Z}_{i-1} + (1 - \phi) \, y_i$$

where $\overline{Z}_{i-1}$ is the EWMA at the last sampling time and $\overline{Z}_i$ is the updated value that is computed when the new observation of y_i becomes available.

Figure 4.3 shows how the weight given to past times depends on the selected value of ϕ. The parameter ϕ indicates how much smoothing is done. As ϕ increases from 0 to 1, the smoothing increases, and long-term cycles and trends stand out more clearly. When ϕ is small, the "memory" of the EWMA is short, and the weights a few days past rapidly shrink toward zero. We have found that $\phi = 0.5$ to 0.3 often gives a useful balance between smoothing and responsiveness. Values in this range will roughly approximate a simple 7-day moving average, as shown in Figure 4.4. Note that the EWMA ($\phi = 0.3$) lags the current observations less than the MA7.

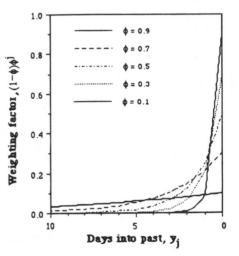

Figure 4.3 Weights for exponentially weighted moving average (EWMA).

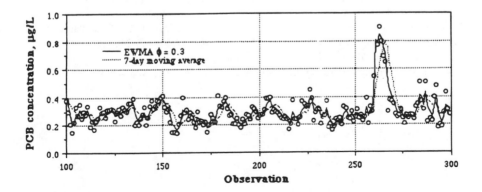

Figure 4.4 Comparison of 7-day moving average and an exponentially weighted moving average with $\phi = 0.3$.

SUMMARY

Suitable graphs of data and the human mind are an effective combination. "Suitable graphs" often will show the smooth along with the rough. This prevents the eye from being distracted by unimportant details. The smoothing methods illustrated here are ideal for *initial data analysis* (Chatfield, 1988, 1991) and *exploratory data analysis* (Tukey, 1977). Their application is straightforward, fast, and easy. We have not tried to identify the best value of ϕ in the EWMA. It is possible to do this by fitting time series models (Box and Jenkins, 1976; Cryer, 1986). This becomes important if the smoothing function is to be used to predict future values, but it is not necessary if we just want to clarify the general underlying pattern of variation.

The simple moving averages (7-day, 30-day, etc.) effectively smooth out random and other high frequency variation. The longer the averaging period, the smoother the moving average becomes, and the more slowly it reacts to changes in the underlying pattern of change. That is, in order to gain smoothness, response to short-term change is sacrificed.

Exponentially weighted moving averages can smooth effectively while also being responsive. This is because they give more relative weight (influence) to recent events and dilute or forget the past. The rate of forgetting is determined by the value of the smoothing factor, ϕ.

An alternative to the moving average smoothers are median smoothers (Tukey, 1977). A *median of 3* smooth is constructed by plotting the middle value of three consecutive observations. It can be constructed without computations, and it is entirely resistant to occasional extreme values. The computational simplicity is an insignificant advantage, however, since the moving averages, especially the exponentially weighted average, are so easy to compute.

Missing values in the data series might seem to be a barrier to smoothing, but for practical purposes they usually can be filled in using some simple ad hoc method. For purposes of smoothing to clarify the general trend, several methods of filling in missing values can be used. The simplest is linear interpolation between adjacent points. Other alternatives are to fill in the most recent moving average value or to replicate the most recent observation. The general trend will be nearly the same regardless of the choice of method, and the user should not be unduly worried about this so long as missing values occur only occasionally.

REFERENCES

Box, G. E. P. and G. M. Jenkins (1976). *Time Series Analysis: Forecasting and Control*, rev. ed., San Francisco, Holden Day.

Chatfield, C. (1988). *Problem Solving: A Statistician's Guide,* London, Chapman and Hall.

Chatfield, C. (1991). "Avoiding Statistical Pitfalls," *Stat. Sci.*, 6, 3, 240–268.

Cryer, J. D. (1986). *Time Series Analysis*, Boston, Duxbury Press.

Tukey, J. W. (1977). *Exploratory Data Analysis*, Reading, MA, Addison-Wesley.

Chapter 5

Seeing the Shape of a Distribution

Key words: dot diagram, histogram, probability distribution, cumulative probability distribution, frequency diagram

---

Let us be clear about which distribution we are studying in this chapter. The data in a sample have some distribution, perhaps normal and perhaps skewed. The statistics (mean, variance, etc.) computed from these data also have some distribution. For example, it was mentioned in Chapter 2 that the distribution of the mean tends to be normal regardless of the distribution of the sample data used in its computation. This chapter is about the shape of the distribution of the data in the sample and not the distribution of statistics computed from the sample. It does not deal with the problem of fitting different distribution functions to the sample in order to identify the best distributional model. It does show how to work with the normal and lognormal distributions.

Many times the first analysis done on a set of data is to compute mean and standard deviation, no doubt because these two statistics fully characterize a normal distribution. Unfortunately, we cannot assume that environmental data will be normally distributed. Experience shows that stream quality data, wastewater treatment plant influent and effluent data, and air quality data typically are not normal. They are more likely to have a positive skewness, that is, to have a long tail of high values. Therefore, a more complete description of the distribution is often needed. Suppose, for example, that one needs to estimate level of performance of a process that is achieved 90% of the time (i.e., to estimate the 90 percentile). The 90th percentile level computed assuming a normal distribution will be much lower than that computed if a lognormal distribution is assumed. Clearly, in this case, one cannot *assume* the distribution, it must be discovered from the data. On the other hand, if the problem is to establish a 95% confidence interval on the mean of the sample, it is not important that the sample is lognormally distributed because the distribution of the mean tends to be normal regardless of the sample's distribution.

To large extent, simple plots will reveal the sample's distribution. Some of these have already been discussed in Chapters 2 and 3. *Dot diagrams* are particularly useful. These simple plots often have been overlooked. Environmental engineering references are likely to advise, by example if not by explicit advice, the construction of a *probability plot* (also known as the *cumulative frequency plot*). Probability plots can be useful. Their construction and interpretation and the ways in which such plots can be misused will be discussed.

CASE STUDY — INDUSTRIAL WASTE SURVEY DATA ANALYSIS

Imagine that our job is is to clarify the structure of a set of 5-day biochemical oxygen demand (BOD$_5$) data, given in Table 5.1, that were obtained from an industrial wastewater survey (U.S. EPA, 1973). There are 99 observations, each measured on a 4-hr composite sample, giving 6 observations daily for 16 days plus 3 observations on the 17th day. The survey was undertaken to estimate the average BOD and to estimate the concentration that is exceeded some small fraction of the time (say 10%). This information is needed to design a treatment process. The pattern of variation also needs to be seen because it will influence the feasibility of using an equalization process to reduce the variation in BOD loading. There may be other properties of the data that are important to the user, so the data presentation should be complete, clear, and not open to misinterpretation.

Table 5.1 **BOD Data from an Industrial Survey (EPA, 1973)**

Date	Time of Day					
	4 am	8 am	12 pm	4 pm	8 pm	12 am
2/10	717	946	623	490	666	828
2/11	1135	241	396	1070	440	534
2/12	1035	265	419	413	961	308
2/13	1174	1105	659	801	720	454
2/14	316	758	769	574	1135	1142
2/15	505	221	957	654	510	1067
2/16	329	371	1081	621	235	993
2/17	1019	1023	1167	1056	560	708
2/18	340	949	940	233	1158	407
2/19	853	754	207	852	318	358
2/20	356	847	711	1185	825	618
2/21	454	1080	440	872	294	763
2/22	776	502	1146	1054	888	266
2/23	619	691	416	1111	973	807
2/24	722	368	686	915	361	346
2/25	1110	374	494	268	1078	481
2/26	472	671	556	—	—	—

ANALYSIS — DOT DIAGRAMS

Figure 5.1 is a time sequence plot of the data. The concentration fluctuates rapidly with more or less equal variation above and below the average, which is 687 mg/L. The range is from 207 to 1185 mg/L. There may be a some cyclic pattern, but it is not very regular. There is little else to be see from this plot.

A *dot diagram* shown in Figure 5.2 gives a better picture of the variability. The data seem to be distributed uniformly between 200 and 1200 mg/L. Any value within this

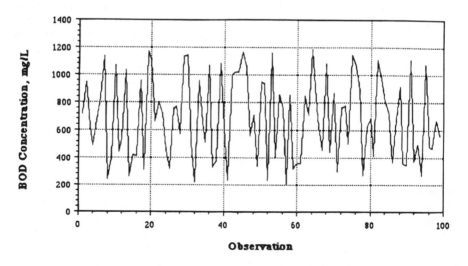

Figure 5.1 Time series plot of the BOD data.

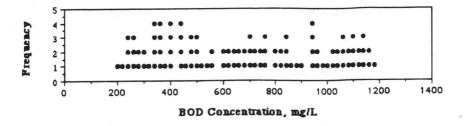

Figure 5.2 Dot diagram of the 99 BOD observations.

range seems equally likely. The dot diagrams in Figure 5.3 show that there is no time of day that has consistently high or consistently low values. The observed values cover the full range regardless of the time of day.

Given the uniform pattern of variation, the extreme values take on a different meaning than if the data were clustered around the average, as they would be in a normal distribution. If the distribution were normal, the extreme values would be relatively rare in comparison to other values. Here, they are no more rare than values near the average. The designer may feel that the rapid and apparently random fluctuation with no tendency to cluster toward one average or central value is the most important feature of the data.

The elegantly simple dot diagram and time series plot have beautifully described the data. No numerical summary could transmit the same information as efficiently and clearly. The conventional assumption of a "normal-like" distribution having the computed average and standard deviation would be very misleading.

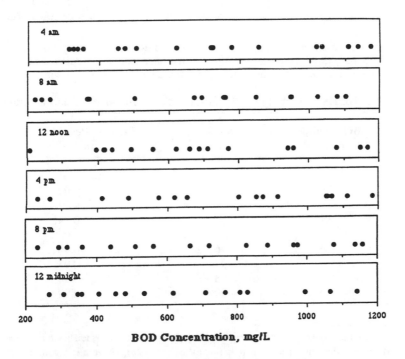

Figure 5.3 Dot diagrams of the data for each sampling time.

ANALYSIS — PROBABILITY PLOTS

A probability plot is not needed to interpret the data in Table 5.1 because the dot diagram exposes the important characteristics of the data, but for comparison these data will be used to illustrate how such a plot is constructed, how its shape is related to the shape of the frequency distribution, and how it could be misused to estimate population characteristics.

The probability plot, or cumulative frequency distribution, shown in Figure 5.4 is constructed by ranking the observed values from small to large; assigning each value a rank, which will be denoted by i; and calculating the plotting position of the probability scale as $p = i/(n + 1)$, where n is the total number of observations. A portion of the ranked data and their calculated plotting positions are shown in Table 5.2. The relation $p = i/(n + 1)$ has traditionally been used by engineers. Statisticians seem to prefer $p = (i - 0.5)/n$, especially when n is small.[1] The major differences in plotting position values

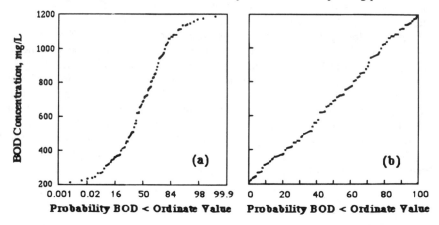

Figure 5.4 Probability plots of the uniformly distributed BOD data. Plot (a) has the abscissa scaled so normally distributed data would plot as a straight line (these data are not normally distributed). Plot (b) is scaled so uniformly distributed data would plot as a straight line.

Table 5.2 **Probability Plotting Positions for the n = 99 Values Listed in Table 5.1**

BOD Value mv/L	Rank i	Plotting Position $p = i/(n + 1)$
207	1	1/100 = 0.01
221	2	0.02
223	3	0.03
235	4	0.04
. . .	. . .	. . .
1158	96	0.96
1167	97	0.97
1174	98	0.98
1185	99	0.99

---

[1] There are still other possibilities for the probability plotting positions (see Hirsch and Stedinger, 1987). Most have the general form of $p = (i - a)/(n + 1 - 2a)$, where a is a constant between 0.0 and 0.5. Some values are a = 0 (Weibull), a = 0.5 (Hazen), and a = 0.375 (Blom).

computed from these formulas occur in the tails of the distribution (high and low ranks). These differences diminish in importance as the sample size increases.

Figure 5.4a is a *normal probability plot* of the data, so named because the probability scale (the abscissa) is arranged in a special way to give a straight-line plot when the data are normally distributed. Any frequency distribution that is not normal will plot as a curve on the normal probability scale used in Figure 5.4a. The ordinate is an arithmetic scale showing the BOD concentration. The abscissa is a cumulative probability scale on which the calculated *p* values are plotted to show the probability that the BOD is less than the value shown on the ordinate.

Clearly, the BOD data are distributed symmetrically, but not in the form of a normal distribution. The S-shaped curve is characteristic of distributions having "heavy tails," i.e., having more observations on the tails than predicted by the normal distribution. (A data set that is light tailed (peaked) or skewed will also have an S shape, but with different curvature.)

While it may be nice if the plot produces a straight line, there is often no particular need to make this happen. If the purpose of making the cumulative probability plot is to estimate the median and the 90th percentile value, Figure 5.4a is satisfactory. If a probability plot on which the sample BOD data plot as a straight line were wanted for this data, a simple arithmetic plot of *p* vs BOD will do, as shown by Figure 5.4b. The linearity in this case indicates that the data are uniformly distributed over the range of observed values, which agrees with the impression drawn from the dot plots.

Probability paper is available with a logarithmic scale as the ordinate; on this paper, the plot will produce a straight line if the data are lognormally distributed. Drawing a straight line on such a plot suggests that you believe the data are at least approximately lognormally distributed. If a straight line appears to describe the data, draw such a line on the graph "by eye." If a straight line does not appear to describe the points and you feel that a line needs to be drawn to emphasize the pattern, draw a smooth curve.

The probability plot could be constructed in terms of normal scores, also known as rankits. These can be looked up in standard statistical tables. An advantage of this plot is that linear regression can be done on the rankit scores. An example of this procedure is discussed in Chapter 10 on analyzing censored data.

THE USE AND MISUSE OF PROBABILITY PLOTS

Engineering texts often show how to estimate the mean and sample standard deviation from probability plots, saying that the mean is located at p = 50% on a normal probability graph and the standard deviation is the distance from the p = 50% to p = 84.1% (or, because of symmetry, from p = 15.9% to p = 50%). These graphical estimates *are valid only* when the data are normally distributed. One way that probability plots are misused is to make the graphical estimates when the distribution is not normal. For example, if the data are lognormally distributed, p = 50% is the median and not the arithmetic mean, and the distance from p = 50% to p = 84.1% is not the sample standard deviation.

Since few environmental data sets are normally distributed, this graphical method has little utility for estimating the mean and standard deviation. The plot is useful, however, to estimate the median (p = 50%) and to read directly any percentile of special interest.

RANDOMNESS AND INDEPENDENCE

Data can be normally distributed without being random or independent. Furthermore, randomness and independence cannot be perceived or proven using a probability plot.

This plot does not provide any information regarding serial dependence or randomness, both of which may be more critical than normality in the statistical analysis.

The 52 weekly BOD loading values plotted in Figure 5.5 will plot as a straight line on a normal probability plot. It could be said, therefore, that the sample of 52 observations is normally distributed. This characterization is uninteresting and misleading since the data are not randomly distributed about the mean and there is a strong trend with time (i.e., serial dependence). The time series plot, Figure 5.5, shows these important features. In contrast, the probability plot and dot plot, while excellent for certain purposes, obscure these features. To be sure all important features of the data are revealed, a variety of plots must be used, as recommended in Chapter 3.

COMMENTS

We are almost always interested in knowing the shape of a sample's distribution. Often it is important to know whether a set of data are distributed symmetrically about a central value or whether there is a tail of data toward a high or a low value. It may be important to know what fraction of time a critical value is exceeded.

The useful graphical tools for seeing the shape of a distribution are dot plots and probability plots. To reduce the odds of having a probability plot misinterpreted, use it only in conjunction with other plots. Make dot diagrams and, if the data are sequential in time, a time series plot. Sometimes these graphs provide all the important information, and the probability plot is unnecessary.

Probability plots are convenient for estimating percentile values, especially the median (50th percentile) and extreme values. It is not necessary for the probability plot to be a straight line to do this. If it is straight, fine; then draw a straight line. But, if it is not straight, draw a smooth curve and go ahead with the estimation.

Do not use probability plots to estimate the mean and standard deviation except in the very special case when the data give a linear plot on normal probability paper. This special case is common in textbooks, but rare with real environmental data. If the data plot as a straight line on log-probability paper, the 50th percentile value is not the mean (it is the geometric mean), and furthermore, there is no distance that can be measured on the plot to estimate the standard deviation.

Probability plots may be useful in discovering the distribution of the data in a sample. Sometimes the analysis is not clear cut. Because of random sampling variation, the curve can have a substantial amount of "wiggle" when the data actually are normally distributed. When the number of observations approaches 50, the shape of the probability

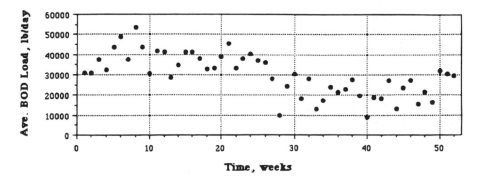

Figure 5.5 This sample of 52 observations will give a linear normal probability plot, but such a plot would hide the important time trend and the serial correlation.

distribution becomes much more clear than when the sample is small (say, 20 observations). Hahn and Shapiro (1967) point out the following:

1. The variance of points in the tails (extreme low or high plotted values) will be larger than that of the points at the center of the distribution. Thus, the relative linearity of the plot near the tails of the distribution will often seem poorer than at the center, even if the correct model for the probability density distribution has been chosen.
2. The plotted points are ordered and hence are not *independent*. Thus, we should not expect them to be *randomly* scattered about a line. For example, the points immediately following a point above the line are also likely to be above the line. Hence, even if the chosen model is correct, the plot may consist of a series of successive points (known as runs) above and below the line.
3. A model can never be *proven* to be adequate on the basis of sample data. Thus, the probability of a small sample taken from a near-normal distribution will frequently not differ appreciably from that of a sample from a normal distribution.

If the data have positive skew, it is often convenient to use graph paper that has a log scale on one axis and a normal probability scale on the other axis. If the logarithms of the data are normally distributed, this kind of graph paper will produce a straight-line probability plot. The log scale may provide a convenient scaling for the graph even if it does not produce a straight-line plot, for example, when the data are bacterial counts that range from 10 to 100,000.

REFERENCES

Hahn, G. A. and S. S. Shapiro (1967). *Statistical Methods for Engineers*, New York, John Wiley & Sons.

Hirsch, R. M. and J. D. Stedinger (1987). "Plotting Positions for Historical Floods and Their Precision," *Water Resour. Res.*, 23, 4, 715–727.

Mage, D. T. (1982). "An Objective Graphical Method for Testing Normal Distributional Assumptions Using Probability Plots," *Am. Statistician*, 36, 116–120.

U.S. EPA (1973). *Monitoring Industrial Wastewater*, Washington, D.C.

Chapter 6

External Reference Distributions

Key words: histogram, reference distribution, moving average, serial correlation, t-test

---

When data are analyzed to decide whether conditions are as they should be or whether the level of some variable has changed, the fundamental strategy is to compare the current condition or level with an appropriate reference distribution. The reference distribution shows how things should be or how they used to be. Sometimes an *external reference distribution* should be created instead of simply using one of the well-known statistical reference distributions, such as the normal or t distribution, which are nicely tabulated and reasonably easy to use. Most statistical methods that rely upon these distributions require that the data are random and independent. Many sets of environmental data violate these requirements.

Non-normality in the data is a characteristic people often worry about, often unnecessarily. Many statistical procedures are robust with respect to normality. Another necessary condition for using the t distribution or the normal distribution is independence (lack of serial correlation) in the data. This requisite condition is not satisfied by most environmental data that are in the form of a time series. Data taken consecutively in time are likely to be serially correlated; high values tend to follow high values and low values follow low values. The data do not vary randomly, but instead are influenced by some inertia or memory of the system that generates the data. Averages calculated at different times drift about some long-term mean. The variation is not random because data are generally not independent of each other.

A specially constructed reference distribution will not be based on assumptions about properties of the data that may not be true. It will be based on the data themselves, whatever their properties. If serial correlation or non-normality affects the data, it will be incorporated automatically into the external reference distribution.

Making the reference distribution is conceptually and mathematically simple. No particular knowledge of statistics is needed, and the only mathematics used are counting and simple arithmetic. Despite this simplicity, the concept is statistically elegant, and valid judgements about statistical significance can be made.

CONSTRUCTING AN EXTERNAL REFERENCE DISTRIBUTION

The first 130 observations in Figure 6.1 show the natural background concentration of the pH in a stream. Table 6.1 lists the data. Suppose that a new effluent has been discharged to the stream and someone suggests it is depressing the stream pH. A survey to check this has provided ten additional consecutive measurements: 6.66, 6.63, 6.82, 6.84, 6.70, 6.74, 6.76, 6.81, 6.77, and 6.67. Their average is 6.74. We wish to judge whether this group of observations differs from past observations. These 10 values are plotted as solid circles on the right-hand side of Figure 6.1. They do not appear to be unusual, but a more careful comparison should be made with the historical data.

The obvious comparison is the 6.74 average of the 10 new values with the 6.80 average of the previous 130 pH values. One reason not to do this is that the standard procedure for comparing two averages, the t-test, is based on the data being independent of each other in time. Data that are a time series, like these pH data, usually are not independent. Adjacent values are related to each other. The data are serially correlated

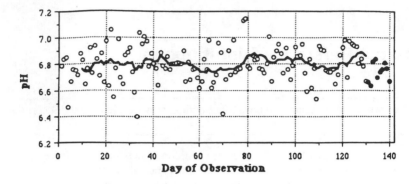

Figure 6.1 Observations with the moving average of ten consecutive values.

(autocorrelated), and the t-test is not valid unless something is done to account for this correlation. In order to avoid making any assumption about the structure of the data, the average of 6.74 should be compared with a reference distribution for averages of sets of 10 consecutive observations.

Table 6.1 gives the 121 averages of 10 consecutive observations that can be calculated from the historical data. (These are 10-day moving averages, as discussed in Chapter 3.) Figure 6.2 is a *reference distribution* for these averages. There are 11 of the 121 averages as low as 6.74. About 91% of the averages are larger than 6.74. The newest ten points are lower than usual, but with 9% of past values at this level or lower, this does not constitute strong evidence that the river pH has changed.

USING A REFERENCE DISTRIBUTION TO COMPARE TWO MEAN VALUES

Let the situation in the previous example change to the following. An experiment to evaluate the effect of an industrial discharge into a treatment process consists of making 10 observations consecutively before any addition and 10 observations afterwards. We will assume that the experiment is not affected by any transients between the two operating conditions. The average of 10 consecutive predischarge samples was 6.86, and the average of the 10 consecutive post-discharge samples was 6.74. Does the difference of $6.74 - 6.86 = -0.13$ represent a significant shift in performance?

A reference distribution for the difference between batches of 10 consecutive samples is needed. There are 112 differences that can be calculated from the data in Table 6.1. For example, the difference between the averages of the 10th and 20th batches is $6.81 - 6.76 = -0.05$. The second value is the difference between the 11th and 21st and is $6.81 - 6.74 = +0.07$. Figure 6.3 is the reference distribution of the 112 differences of batches of 10 consecutive samples. A downward difference as large as -0.13 has never been observed before. In fact, this is almost twice as large as the largest observed decrease of -0.07. The difference is so exceptional that we are prepared to conclude that the new condition is different than the recent past.

Examining the 10-day averages gave one conclusion (the effluent has not depressed the pH to an unusual level) and looking at the difference in averages gave another (the shift between 10-day averages is unusually large). How can this be? Looking at the shift focuses more on what has happened just at the intervention of the new industrial discharge. Looking at the averages brings older history more heavily into the analysis. The analyst may feel that what happened within the last month is more relevant to

Table 6.1 Data Used to Plot Figure 6.1 and the Associated External Reference Distributions

Conc.	Ave.	Conc.	Ave.	Conc.	Ave.	Conc.	Ave.	Conc.	Ave.
6.79		6.65	6.79	6.9	6.81	7.14	6.82	6.59	6.81
6.84		6.87	6.79	6.67	6.80	6.78	6.86	6.84	6.80
6.85		6.89	6.81	6.82	6.80	6.77	6.87	6.62	6.79
6.47		6.92	6.80	6.68	6.79	6.87	6.87	6.77	6.80
6.67		6.74	6.81	6.76	6.79	6.83	6.88	6.53	6.77
6.76		6.58	6.77	6.77	6.78	6.84	6.86	6.94	6.77
6.75		6.4	6.75	6.7	6.77	6.77	6.87	6.91	6.78
6.72		7.04	6.78	6.62	6.75	6.76	6.87	6.9	6.78
6.88		6.95	6.77	6.67	6.74	6.73	6.86	6.75	6.76
6.83	6.76	7.01	6.81	6.84	6.74	6.8	6.83	6.74	6.76
6.65	6.74	6.97	6.84	6.76	6.73	7.01	6.82	6.74	6.77
6.77	6.74	6.78	6.83	6.98	6.76	6.67	6.81	6.76	6.77
6.92	6.74	6.88	6.83	6.62	6.74	6.85	6.81	6.65	6.77
6.73	6.77	6.8	6.82	6.66	6.74	6.9	6.82	6.72	6.76
6.94	6.80	6.77	6.82	6.72	6.73	6.95	6.83	6.87	6.80
6.84	6.80	6.64	6.82	6.96	6.75	6.88	6.83	6.92	6.80
6.71	6.80	6.89	6.87	6.89	6.77	6.73	6.83	6.98	6.80
6.88	6.82	6.79	6.85	6.42	6.75	6.92	6.84	6.7	6.78
6.66	6.79	6.77	6.83	6.68	6.75	6.76	6.85	6.97	6.81
6.97	6.81	6.86	6.82	6.9	6.76	6.68	6.84	6.95	6.83
6.63	6.81	6.76	6.79	6.72	6.76	6.79	6.81	6.94	6.85
7.06	6.83	6.8	6.80	6.98	6.76	6.93	6.84	6.93	6.86
6.55	6.80	6.8	6.79	6.74	6.77	6.86	6.84	6.8	6.88
6.77	6.80	6.81	6.79	6.76	6.78	6.87	6.84	6.84	6.89
6.99	6.81	6.81	6.79	6.77	6.78	6.95	6.84	6.78	6.88
6.7	6.79	6.8	6.81	7.13	6.80	5.73	6.82	6.67	6.86

Note: Sequence of values is down column 1, down column, 2 etc. "Ave." is the average of the current and the nine previous values.

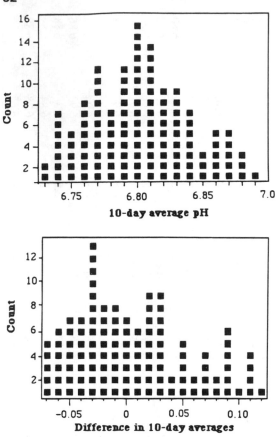

Figure 6.2 Reference distribution for averages of ten consecutive values.

Figure 6.3 Reference distribution for differences of adjacent sets of 10-day averages.

judging this situation than what happened several months ago. The results are not contradictory because they tell us different things about the data.

USING A REFERENCE DISTRIBUTION FOR MONITORING

Treatment plant effluent and water quality monitoring are usually defined in terms of 30- and 7-day averages. The effluent data themselves typically have a lognormal distribution and are serially correlated. This makes it difficult to derive the statistical properties of the 30- and 7-day averages. Fortunately, if historical data are readily available at all treatment plants, we can construct external reference distributions, not only for 30- and 7-day averages, but also for any other statistics of interest.

The data in this example are effluent BOD_5 measurements that have been made daily on 24-hr flow-weighted composite samples from an activated sludge treatment plant. We realize that BOD data are not timely for process control decisions, but they can be used to evaluate whether the plant has been performing at its normal level or whether effluent quality has changed. A more complete characterization of plant performance would include reference distributions for other variables such as suspended solids, ammonia, and phosphorus.

An available record of 1339 days (much longer than is really needed) was used to generate the open-bar histogram in Figure 6.4. The reference distribution that represents 1150 days of stable operation is shown by the shaded bars. In simple terms, "stable" is the kind of performance of which the plant is capable over long stretches of time

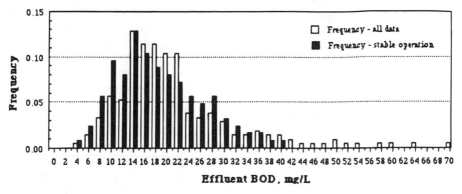

Figure 6.4 External reference distribution for the complete record and for the stable operating conditions.

(Berthouex and Fan, 1986). This is the reference distribution against which new daily effluent measurements should be compared when they become available, which unfortunately is five or six days after the event in the case of BOD data.

If the 7-day moving average is used to judge effluent quality, a reference distribution is required for this statistic. The periods of stable operation were used to calculate the 7-day moving averages (MA7) that produce the reference distribution shown in Figure 6.5 (top). Figure 6.5 (bottom) is the reference distribution of 30-day moving averages (MA30) for periods of stable operation. Plant performance can now be monitored by comparing, as they become available, new 7- or 30-day averages against these reference distributions.

SETTING CRITICAL LEVELS

The reference distribution shows at a glance which values are exceptionally high or low. What is meant by "exceptional" can be specified by setting critical decision levels

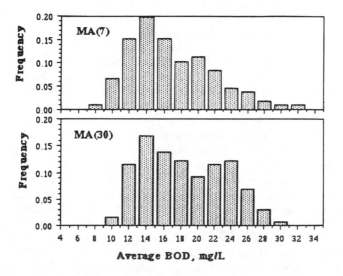

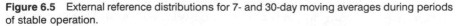

Figure 6.5 External reference distributions for 7- and 30-day moving averages during periods of stable operation.

that have a specified probability value. For example, one might specify "exceptional" as the level that is exceeded p percent of the time. The reference distribution for daily observations during stable operation (shaded bars in Figure 6.4) is based on 1152 daily values representing stable performance. The critical upper 5% level cuts are a BOD concentration of 33 mg/L. This is found by summing the probabilities, starting from the highest BOD observed during stable operation, until the accumulated percentage equals or exceeds 5%. In this case, P(BOD = 39–40) = 0.008, P(BOD = 37–38) = 0.008, P(BOD = 35–36) = 0.016, P(BOD = 33–34) = 0.016, and the sum of these probabilities is 0.048 or 4.8% percent. So, as a practical matter, we can say that the BOD exceeds 33 mg/L only about 5% of the time when operation is stable.

Upper critical levels can be set for the MA7 reference distribution as well. The upper 5% value is between 28 and 29 mg/L. The probability is less than 5% that an MA7 greater than 29 mg/L will occur when the treatment plant operation is stable. An MA7 serves warning that the process is performing poorly and may be upset. By definition, 5% of such warnings will be false alarms. A two-level warning system could be devised, for example, by using the upper 1 and 5% levels. The upper 1% level, which is about 32 mg/L, is a signal that something is almost certainly wrong; it will be false in only 1 out of 100 alerts.

There is a balance to be found between having occasional false alarms and no false alarms. Setting a warning at the 5% level or perhaps even at the 10% level means that the operator is occasionally sent to look for a problem when none exists. But, it also means that many times the operator will be warned of a problem before it becomes too serious, and on some of these occasions the operator will be able to prevent a minor upset from becoming more serious. An occasional wild goose chase is a reasonable price to pay for the early warnings.

COMMENTS

Consider why the warning levels were determined empirically instead of by calculating the mean and standard deviation and then using the normal distribution. People who know some statistics tend to think of the bell-shaped, symmetrical normal distribution when they hear that "the mean is X and the standard deviation is Y." The words "mean" and "standard deviation" create an image of approximately 95% of the values falling within two standard deviations of the mean.

A glance at Figure 6.5 reveals how this image is inappropriate for the moving average reference distributions. The distributions are not symmetrical and, furthermore, are truncated. These characteristics are especially evident in the MA30 distribution. By definition, the effluent BOD values are never very high when operation is stable, so MA cannot take on certain high values. Low values of the MA do not occur because the effluent BOD cannot be less than zero, and furthermore, values less than 4 mg/L have never been observed. The normal distribution, with its finite probability of values occurring far out on the tails of the distribution (and even into negative values), would be a terrible approximation of the reference distribution derived from the operating record.

The reference distribution for the daily values will always give a warning before the MA does. The MA is conservative. It trods upon one-day upsets, even fairly large ones, and rolls smoothly through short intervals of minor disturbances without giving much notice. The MA is like a shock absorber on a car in that it smooths out the small bumps. Also, just as a shock absorber needs to have the right stiffness, an MA needs to have the right length of memory to do its job well. An MA30 is an interesting statistic to plot only because effluent standards use a 30-day average, but it is too sluggish to usefully warn of trouble. At best, it can confirm that trouble has existed. The 7-day average is more responsive to change and serves as a better warning signal. Exponentially weighted moving averages (see Chapter 4) are also responsive, and reference distributions could be constructed for them as well.

Just as there is no reason to judge process performance on the basis of only one variable, there is no reason to select and use only one reference distribution for any particular single variable. One statistic and its reference distribution might be most useful for process control, while another is best for judging compliance. Some might give early warnings, while others provide confirmation. Since reference distributions are easy to construct and use, they should be plentiful and prominent in the control room.

REFERENCES

Berthouex, P. M. and W. G. Hunter (1983). "How to Construct a Reference Distribution to Evaluate Treatment Plant Performance," *J. Water Pollution Control Fed.*, 55, 1417–1424.

Berthouex, P. M. and R. Fan (1986). "Treatment Plant Upsets: Causes, Frequency, and Duration," *J. Water Pollution Control Fed.*, 58, 368–375.

Chapter 7

Using Transformations

Key words: transformations, logarithm, square root, variance stabilization, plankton counts, bacterial counts, Box-Cox transformation

---

There is usually no scientific reason why we should insist on analyzing data in their original scale of measurement. Instead of doing our analysis on y, it may be more appropriate to look at log(y), $\sqrt{y}$, 1/y, or some other function of y. These reexpressions of y are called *transformations*.

Some data transformations are accepted without hesitation. A pH meter reads in logarithmic units, pH $= -\log_{10}[H^+]$, and not in hydrogen ion concentration units. The instrument makes a data transformation that we accept as natural. Light absorbency is measured on a logarithmic scale by a spectrophotometer and converted with the aid of a calibration curve to a concentration. The calibration curve makes a transformation that is automatically accepted. If we are dealing with numbers of bacteria, N, we think just as well in terms of log(N) as N itself. We see that making a transformation is not cheating. It is a common scientific practice for presenting and interpreting data. When properly used, transformations eliminate distortions and give each observation equal power to inform.

There are three technical reasons for sometimes doing the calculations on a transformed scale: (1) to make the spread equal in different data sets (to make the variances uniform), (2) to make the distribution of the residuals normal, and (3) to make the effects of treatments additive (Box et al., 1978). Equal variance means having equal "spread" at the different settings of the independent variables or in the different data sets that will be compared. The requirement for a normal distribution applies to the measurement errors and not to the entire sample of data. Transforming the data makes it possible to satisfy these requirements when they are not satisfied by the original measurements.

TRANSFORMATIONS FOR LINEARIZATION

Transformations are sometimes used to obtain a straight-line relationship between two variables. This may involve, for example, using reciprocals, ratios, or logarithms. The left-hand panel of Figure 7.1 shows the exponential growth of bacteria. Notice that the variance (spread) of the counts increases as the population density increases. The right-hand panel shows that the data can be described by a straight line when plotted on a log scale. Plotting on a log scale is equivalent to making a log transformation of the data.

It is no more difficult conceptually or computationally to fit an exponential model, $y = A e^{kt}$, to the original data than to fit a straight line, $y = a + bx$, to the log-transformed data. The important characteristic of the original data is the nonconstant variance, not the nonlinearity. This is a problem when the curve or line is fitted to the data using regression. Regression tries to minimize the distance between the data points and the line described by the model. Points that are far from the line exert a strong effect because the regression mathematics want to reduce this distance. The result is that the precisely measured points at time $t = 1$ will have less influence on the position of the regression line than the poorly measured data at $t = 3$. Observations made at the different times carry different weight, and unfortunately, the weights have the undesirable property of giving too much influence to the least reliable data. We would prefer for each data point

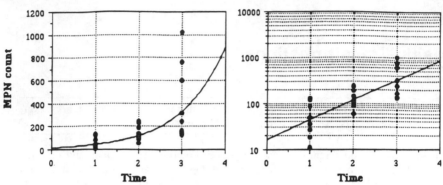

Figure 7.1 Constant variance at all levels is important so that each data point will carry equal weight in locating the position of the fitted curve.

to have about the same amount of influence on the location of the line. In this example, the log-transformed data have constant variance at the different population levels. Each data value has roughly equal weight in determining the position of the line. The log transformation is used to achieve this equal weighting and not because it gives a straight line.

A word of warning is in order about using transformations to obtain linearity. A transformation can turn a good situation into a bad one by distorting the variances and making them unequal. Figure 7.2 shows a case where the original data have constant variance which is destroyed by an inappropriate transformation. In this case, the transformation is not beneficial, even though it does linearize the data. Transformations involving reciprocals can be especially bad in distorting the variance, so use them only after checking the variances.

In the examples above, it was easy to check the variances at the different levels of the independent variables because the measurements had been replicated. If there is no replication, this check cannot be made. Replication is always a good practice, and this is one important reason to use it as part of the experimental strategy.

Lacking replication, should one assume that the variances are originally equal or unequal? Sometimes the nature of the measurement process gives a hint as to what *might* be the case. If dilutions or concentrations are part of the measurement process, if the final result is computed from the raw measurements, or if the concentration levels are widely different, it is not unusual for the variances to be unequal and to be larger

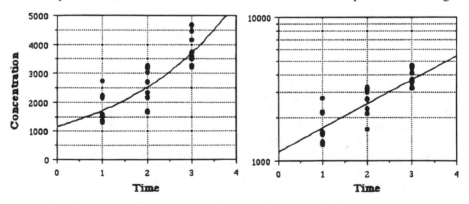

Figure 7.2 An example of how a transformation could create nonconstant variance.

at high levels of the independent variable. Biological counts frequently have nonconstant variance. These are not justifications to make transformations indiscriminately. Do not avoid making transformations, but use them wisely and with care.

VARIANCE STABILIZING TRANSFORMATIONS

When the variance changes over the range of experimental observations, the variance is said to be nonconstant or unstable. Common situations that tend to create this pattern are (1) measurements that involve making dilutions or other steps that introduce multiplicative errors; (2) using instruments that read out on a log scale, which results in low values being recorded more precisely than high values; and (3) biological counts. One of the transformations given in Table 7.1 should be suitable to obtain constant variance.

The effect of square root and logarithmic transformations is to make the larger values less important relative to the small. For example, the square root converts the values (0, 1, 4) to (0, 1, 2). The 4, which tends to dominate on the original scale, is made relatively less important by the transformation.

The log transformation is a stronger transformation than the square root transformation. "Stronger" means that the range of the transformed variables is relatively smaller for a log transformation than for the square root. When the sample contains some zero values ($\log(0)$ is undefined), the log transformation is $x = \log(y + c)$, where c is a constant. Usually the value of c is arbitrarily chosen to be 1 or 0.5. The larger the value of c, the less severe the transformation. Similarly, for square root transformations, $\sqrt{y + c}$ is less severe than $\sqrt{y}$.

> **Example 7.1.** Twenty replicate samples from five stations were counted for plankton, with the results given in Table 7.2. The computed averages and variances are in Table 7.3. The computed means and variance on the original data show that variance is not uniform; it is ten times larger at Station 5 than at Station 1. Also, the variance increases as the average increases, and s_y^2 seems to be proportional to $\bar{y}$. This indicates that a square root transformation may be suitable. Since most of the counts are small, the transformation used was $\sqrt{y + 0.5}$. Figure 7.3 shows the distribution of the original and the transformed data. The variances computed from the transformed data are uniform. Notice that the transformation also has made the distributions nearly symmetrical.

Table 7.1 **Transformations that Are Useful to Obtain Uniform Variance**

Condition	Replace y by	Notes
$\sigma \propto \bar{y}^2$	$x = 1/y$	
$\sigma \propto \bar{y}^{3/2}$	$x = 1/\sqrt{y}$	
$\sigma \propto \bar{y}\ (\sigma^2 > \bar{y})$	$x = \log(y)$ or $x = \log(y + c)$	Some $y > 0$ (some $y = 0$)
$\sigma \propto \bar{y}^{1/2}$	$x = \sqrt{y}$ (or $x = \sqrt{y + c}$)	All $y \geq 0$ (some $y < 0$)
$\sigma^2 > \bar{y}$	$x = \arcsin \sqrt{p}$	p = ratio or percentage

Table 7.2 **Twenty Replicate Values of Plankton at Five Sampling Stations**

Station 1	0	2	1	0	0	1	1	0	1	1	0	2	1	0	0	2	3	0	1	1
Station 2	3	1	1	1	4	0	1	4	3	3	5	3	2	2	1	1	2	2	2	0
Station 3	6	1	5	7	4	1	6	5	3	3	5	3	4	3	8	4	2	2	4	2
Station 4	7	2	6	9	5	2	7	6	4	3	5	3	6	4	8	5	2	3	4	1
Station 5	12	7	10	15	9	6	13	11	8	7	10	8	11	8	14	9	6	7	9	5

Table 7.3 **Statistics Computed from the Data in Table 7.2**

	Station =	1	2	3	4	5
Untransformed data:	$\bar{y}$ =	0.85	2.05	3.90	4.60	9.25
	s_y^2 =	0.77	1.84	3.67	4.78	7.57
Transformed x = $\sqrt{y + 0.5}$	$\bar{x}$ =	1.10	1.54	2.05	2.21	3.09
	s_x^2 =	0.14	0.19	0.21	0.24	0.20

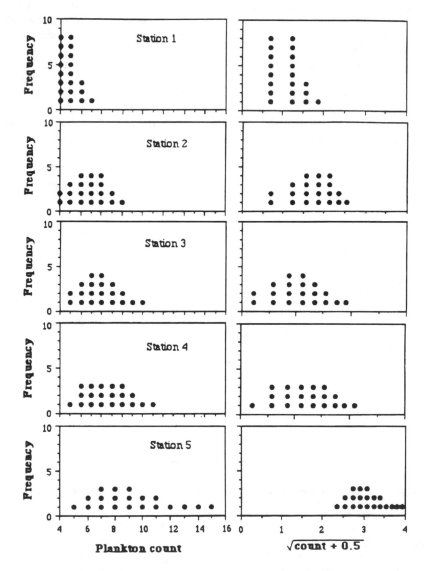

Figure 7.3 Original and transformed plankton count data.

Example 7.2. Table 7.4 shows eight replicate measurements of bacterial density were made at three locations to study the spatial pattern of contamination in an estuary. The data show that $s^2 > \bar{y}$ and that s increases in proportion to $\bar{y}$. Table 7.1 suggests that a logarithmic transformation would be suitable to stabilize the variance. The improvement due to the log transformation is shown in Table 7.4. Note that the transformation could be done using either $\log_e$ or $\log_{10}$, since they differ by only a constant multiplying factor ($\log_e = 2.303 \log_{10}$).

CONFIDENCE INTERVALS AND TRANSFORMATIONS

After summary statistics (means, standard deviations, etc.) have been calculated on the transformed scale, it is often desirable to translate the results back to the original scale of measurement. This can create some confusion. For example, if the average $\bar{x}$ has been estimated using $x = \log(y)$, the simple back-transformation of antilog $(\bar{x})$ does not give an unbiased estimate of $\bar{y}$. The correct estimate of the arithmetic mean on the original y scale is $\bar{y} = $ antilog $(\bar{x} + 0.5 \ s^2)$ (Gilbert, 1987). The antilogarithm of $\bar{x}$ is the *geometric mean* of y original data and not the arithmetic average, that is, antilog $(\bar{x}) = \bar{y}_g$.

If the transformation produced a near-normal distribution, the standard deviations and standard errors computed from the transformed data will be symmetric about the mean on the transformed scale. But, they will be asymmetric on the original scale. The options are to

1. Quote symmetric confidence limits on the transformed scale.
2. Quote asymmetric confidence limits for the original scale (recognizing that in the case of a log transformation they apply to the geometric mean and not to the arithmetic average).
3. Give two sets of results, one with standard errors and symmetric confidence limits on the transformed scale and a corresponding set of means (arithmetic and geometric) on the original scale. The reader can judge the statistical significance on the transformed scale and the practical importance on the original scale.

Two examples illustrate using log-transformed data to construct confidence limits for the geometric mean.

Table 7.4 **Eight Replicate Measurements on Bacteria at Three Sampling Stations**

	y = Bacteria/100 mL			x = $\log_{10}$(Bacteria/100 mL)		
Data Set	**1**	**2**	**3**	**1**	**2**	**3**
	27	225	1020	1.431	2.352	3.009
	11	99	136	1.041	1.996	2.134
	48	141	317	1.681	1.613	2.501
	36	60	161	1.556	1.778	2.207
	120	190	130	2.079	2.279	2.114
	85	240	601	1.929	2.380	2.779
	18	90	760	1.255	1.954	2.889
	130	112	240	2.114	2.049	2.380
$\bar{y} =$	59.4	132	421	$\bar{x} =$ 1.636	2.050	2.502
$s_y^2 =$	216	5,770	111,886	$s_x^2 =$ 0.151	0.076	0.124

Example 7.3. A sample of n = 5 observations is [95, 20, 74, 195, 71] and gives $\bar{y}$ = 91 and s_y^2 = 4,140. Clearly, $s^2 > \bar{y}$ and a log transformation should be tried. The x = $\log_{10}(y)$ data are 1.97772, 1.30103, 1.86923, 2.29003, and 1.85126. This gives $\bar{x}$ = 1.85786 and s_x^2 = 0.12780. The value of t for v = n − 1 = 4 degrees of freedom and $\alpha/2$ = 0.025 is 2.776. Therefore, the 95% confidence interval for the mean on the log-transformed scale η_x, is

$$\bar{x} \pm t \sqrt{s_x^2/n} = 1.85786 \pm 2.776 \sqrt{\frac{0.1278}{5}}$$

and

$$1.41405 < \eta_x < 2.30167$$

Transforming η_x back to the original scale gives an estimate of the *geometric mean* of the y's:

$$\bar{y}_g = \text{antilog } (\bar{x}) = \text{antilog}_{10}(1.85786) = 72.09.$$

The asymmetric 95% confidence limits for the true value of the geometric mean, η_g, are obtained by taking antilogarithms of the upper and lower confidence limits of η_x, give:

$$25.94 \le \eta_g \le 200.29$$

Note that the upper and lower confidence limits in the log metric are $\bar{x} + \beta$ and $\bar{x} - \beta$, where $\beta = t \sqrt{s^2/n}$. The upper confidence limit on the original scale is antilog $(\bar{x} + \beta)$ = antilog $(\bar{x}) \cdot$ antilog (β), which becomes $\bar{y}_g \cdot \beta'$ where β' = antilog (β). Likewise, the lower confidence limit is $\bar{y}_g/\beta'$. For this example, $\beta = 0.44381$, antilog (0.44381) = 2.778, and the 95% confidence limits for the geometric mean on the original scale are 72.09(2.7785) = 200.29 and 72.09/2.7785 = 25.94.

Example 7.4. A log transformation is needed on a sample of n = 6 that includes some zeros: y = [0, 2, 1, 0, 3, 9]. Because of the zeros, use the transformation x = log(y + 1) to find the transformed values: x = [0, 0.47712, 0.30103, 0.60206, 1.0]. This gives $\bar{x}$ = 0.39670 with standard deviation s_x^2 = 0.14730. For v = n − 1 = 5 and $\alpha/2$ = 0.025, $t_{5,\,0.025}$ = 2.571 and the 95% confidence limits on the transformed scale are

$$\text{LCL}(x) = 0.39670 - 2.571 \sqrt{0.14730/6} = -0.006 \text{ (say zero)}$$

and

$$\text{UCL}(x) = 0.39670 + 2.571 \sqrt{0.14730/6} = 0.77952$$

Transforming back to the original metric gives the geometric mean

$$\bar{y}_g = \text{antilog}_{10}(0.39670) - 1 = 2.5 - 1 = 1.5$$

The −1 is due to using x = log(y + 1). The similar inverse of the confidence limits gives

$$\text{LCL}(y) = 0 \quad \text{and} \quad \text{UCL}(y) = 5.0$$

CONFIDENCE INTERVALS ON THE ORIGINAL SCALE

Notice that the above examples are for the geometric mean and not the mean μ. The work becomes more difficult if we want the asymmetric confidence intervals on the mean on the original scale. We explain why this is so for the lognormal distribution.

A simple method of estimating the mean μ and variance σ^2 of the two-parameter lognormal distribution is to use (Gilbert, 1987)

$$\hat{\mu} = \exp\left(\bar{y} + \frac{s_y^2}{2}\right) \quad \text{and} \quad \hat{s}^2 = \hat{\mu}^2[\exp(s_y^2) - 1]$$

where $\bar{y}$ and s_y^2 are calculated on the logs of the data in the usual way shown in Chapter 2. These estimates are slightly biased upward by about 5% for n = 20 and 1% for n = 100. The importance of this when using $\hat{\mu}$ to judge compliance with environmental standards, or when comparing estimates based on unequal n, is discussed by Gilbert (1987) and Landwehr (1978).

Unfortunately, the computation of confidence limits involves much more than simply adding a multiple of the standard deviation (as we do for the normal distribution). Confidence limits for the mean, μ, are estimated using the following equations (Land, 1971).

$$UCL_{1-\alpha} = \exp\left(\bar{y} + 0.5\,s_y^2 + \frac{s_y\,H_{1-\alpha}}{\sqrt{n-1}}\right)$$

$$LCL_\alpha = \exp\left(\bar{y} + 0.5\,s_y^2 + \frac{s_y\,H_\alpha}{\sqrt{n-1}}\right)$$

The quantities $H_{1-\alpha}$ and H_α depend on s_y, n, and the significance level α. Land (1975) provides the necessary tables; a subset of these may be found in Gilbert (1987).

THE BOX-COX POWER TRANSFORMATIONS

A power transformation model developed by Box and Cox (1964) can, so far as possible, satisfy the conditions of normality and constant variance simultaneously. The method is applicable for almost any kind of statistical model and any kind of transformation. The transformed value $Y_i^{(\lambda)}$ of the original variable y_i is

$$Y_i^{(\lambda)} = \frac{y_i^\lambda - 1}{\lambda\,\bar{y}_g^{\lambda-1}}$$

where $\bar{y}_g$ is the geometric mean of the original data series and λ expresses the power of the transformation. The special case when $\lambda = 0$ is the log transformation: $Y_i^{(0)} = \bar{y}_g\ln(y_i)$. Other commonly occurring values of λ are the following: $\lambda = -1$ is a reciprocal transformation, $\lambda = 1/2$ is a square root transformation, and $\lambda = 1$ is no transformation. Examples of how this transformation is used are given in Box et al. (1978).

COMMENTS

Transformations are not tricks to reduce variation or to convert a complicated nonlinear model into a simple linear form. There are often statistical justifications for making transformations and then analyzing the transformed data. They may be needed to stabilize the variance, to make the distribution of the errors normal, or to make effects additive. The most common and important use is to stabilize (make uniform) the variance.

It can be tempting to use a transformation to make a nonlinear function linear so that it can be fitted using simple linear regression methods. Sometimes the transformation that gives a linear model will coincidentally produce uniform variance. Beware, however, because linearization can also produce the opposite effect of making constant variance become nonconstant. (See Chapter 35.)

When the analysis has been done on transformed data, the analyst must consider carefully whether to report the final results on the transformed or the original scale of measurement. Confidence intervals that are symmetrical on the transformed scale will not be symmetric when transformed back to the original scale. Care must also be taken when converting the mean on the transformed scale back to the original scale. A simple backtransformation may not give an unbiased estimate, as was demonstrated in the case of the logarithmic transformation.

REFERENCES

Box, G. E. P. and D. R. Cox (1964). "An Analysis of Transformations," *J. R. Stat. Soc., Ser. B*, 26, 211.

Box, G. E. P., W. G. Hunter, and J. S. Hunter (1978). *Statistics for Experimenters*, pp. 232–240, New York, Wiley Interscience.

Elliot, J. (1977). *Some Methods for the Statistical Analysis of Samples of Benthic Invertabrates*, 2nd ed., Ambleside, Cumbria, England, Freshwater Biological Association.

Gilbert, R. O. (1987). *Statistical Methods for Environmental Pollution Monitoring*, New York, Van Nostrand Reinhold.

Land, C. E. (1971). "Confidence Intervals for Linear Functions of the Normal Mean and Variance," *Ann. Math. Stat.*, 42, 1187–1205.

Land, C. E. (1975). "Tables of Confidence Limits for Linear Functions of the Normal Mean and Variance," in *Selected Tables in Mathematical Statistics*, Vol. 3, pp. 358–419, Providence, RI, American Mathematical Society.

Landwehr, J. M. (1978). "Some Properties of the Geometric Mean and Its Use in Water Quality Standards," *Water Resour. Res.*, 14, 467–473.

Chapter 8

Estimating Percentiles

Key words: confidence intervals, distribution free estimation, geometric mean, lognormal distribution, normal distribution, nonparametric estimation, parametric estimation, percentile, quantile, rank order statistics

---

The use of percentiles in environmental standards and regulations has grown during the past few years. England has water quality consent limits that are based on the 90th and 95th percentiles of monitoring data not exceeding specified levels. The U.S. Environmental Protection Agency (EPA) has specifications for air quality monitoring that are in effect percentile limitations. These may, for example, specify that the ambient concentration of a compound cannot be exceeded more often than once a year (the 364/365th percentile). The U.S. EPA has provided guidance for setting aquatic standards on toxic chemicals that require estimating 99th percentiles and using this statistic to make important decisions about monitoring and compliance. They have also used the 99th percentile to establish maximum daily limits for industrial effluents (e.g., pulp and paper). Specifying a 99th percentile in a decisionmaking rule gives an impression of great conservatism or of having great confidence in making the "safe" and therefore correct environmental decision. Unfortunately, the 99th percentile is a statistic that cannot be estimated precisely.

DEFINITION OF QUANTILE AND PERCENTILE

The *pth quantile*, which is a parameter, will be denoted by y_p. (In Chapter 2, we stated that parameters would be indicated with Greek letters, but this convention is violated in this chapter.) By definition, a proportion p of the population is smaller or equal to y_p, and a proportion $1 - p$ is larger than y_p. The median, $y_{0.5}$, is the 0.5 quantile, and $y_{0.25}$ and $y_{0.75}$ are the upper and lower quartiles, respectively. Quantiles expressed as a percentage are called *percentiles*. For example, the 0.5 quantile is equivalent to the 50% percentile; the 0.99 quantile is the 99th percentile. The 95th percentile will be denoted as y_{95}.

Suppose that a particular effluent (or river) produces pollutant concentration levels that give the frequency distribution represented by the solid curve in Figure 8.1. This is what would be observed if we had a "magic box" that could monitor conditions continuously without error. In other words, the solid curve in Figure 8.1 represents the true underlying pattern of the pollution level. It is the *population distribution*; it shows the properties of the population. This example has a lognormal distribution. The population 50th quantile (the median), $y_{0.50}$, and the population pth quantile, y_p, are parameters. Parameters are the true underlying values that describe populations.

In practice, the parameter, y_p, is unknown and must be estimated from data. The estimated quantile is denoted by $\hat{y}_p$. One way is to read an estimate from a probability plot (see Chapter 5 or Hahn and Shapiro, 1967). In this chapter, the parametric and nonparametric estimation methods are shown.

PARAMETRIC ESTIMATES OF QUANTILES

If we know or are willing to assume the population distribution, we can use a *parametric method*. The problem of quantile (percentile) estimation will be discussed initially in terms of the normal distribution, assuming that non-normal data can be transformed to

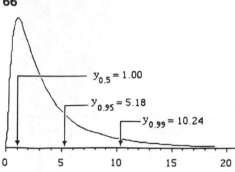

Figure 8.1 Lognormal population distribution having population percentiles $y_{0.50} = 1.00$, $y_{0.95} = 5.18$, and $y_{0.99} = 10.24$.

make them approximately normal so that normal theory may be used. This is convenient, since the properties of the normal distribution are known and accessible in tables.

The normal distribution is a bell-shaped symmetric distribution that is completely specified by two parameters, η and σ^2, which are the mean and variance of the distribution, respectively. The notation $N(\eta, \sigma^2)$ is used to denote a normal distribution with mean η and variance σ^2. Values from a $N(\eta, \sigma^2)$ distribution can be transformed to a *standard normal distribution* using

$$z = \frac{y - \eta}{\sigma}$$

z is called a *standard normal deviate*. It has mean value zero and standard deviation of unity and is denoted by $N(0,1)$.

The true *pth* quantile of the normal distribution is $y_p = \eta + \sigma z_p$. Generally, the parameters η and σ^2 are unknown, and we must estimate them by the sample average, $\bar{y}$, and sample variance, s^2. The quantile, y_p, of a normal distribution is estimated by computing

$$\hat{y}_p = \bar{y} + Z_p s$$

where z_p is the pth quantile of the standard normal distribution. The appropriate value of z_p can be found in a table of the normal distribution.

> **Example 8.1.** We have a set of data that appear to be normally distributed. The estimated mean and standard deviation are 10 and 1.2. To estimate the 99th quantile, look up $z_{0.99} = 2.326$ and compute $\hat{y}_{0.99} = 10 + 2.326(1.2) = 12.8$.

This method can be used even when a set of data indicates that the population distrtibution is not normally distributed if a transformation will make the distribution normal. For example, if a set of observations, y, appears to be from a lognormal distribution, the transformed values $x = \log(y)$ will be normally distributed. The pth quantile of y on the original measurement scale corresponds to the pth quantile of x on the log scale. Thus, $x_p = \log(y_p)$ and $y_p = \text{antilog}(x_p)$.

> **Example 8.2.** A sample of observations, y, appears to be from a lognormal distribution. A logarithmic transformation, $x = \ln(y)$, produces values that are normally distributed. The log-transformed values have an average value of 1.5 and a standard deviation of 1.0. The 99th percentile on the log scale is located at $z_{0.99} = 2.326$, which corresponds to $\hat{x}_{0.99} = 1.5 + (2.326)(1.0) = 3.826$. The 99th quantile of the lognormal distribution is found by making the transformation in reverse, $\hat{y}_{0.99} = \exp(\hat{x}_{0.99}) = 45.9$.

An upper $100(1 - \alpha)\%$ confidence limit for the true pth quantile, y_p, can be easily obtained if the underlying distribution is normal (or has been transformed to become normal). This upper confidence limit is

$$UCL_{1-\alpha}(y_p) = \bar{y} + K_{1-\alpha,p} \, s$$

where $K_{1-\alpha,p}$ is obtained from a table by Owen (1972), which is reprinted in Gilbert (1987).

Example 8.3. From $n = 300$ normally distributed observations, we have estimated $\bar{y}$ $= 10$ and $s = 1.2$. The estimated 99th quantile is $\hat{y}_{0.99} = 10 + (2.326)1.2 = 12.79$. We wish to state the precision of this estimate in the form of a $100(1 - \alpha)\% = 95\%$ upper confidence interval. For $n = 300$, $1 - \alpha = 0.95$ and $p = 0.99$, we find (in Gilbert, 1987) that $K_{0.95,\,0.99} = 2.522$. Using the formula given above,

$$UCL_{0.95}(y_{0.99}) = 10 + (2.522)(1.2) = 13.02.$$

In summary, the best estimate of the 99th quantile is 12.79, and we can state with 95% confidence that its true value is less than 13.02.

NONPARAMETRIC ESTIMATES OF QUANTILES

In the previous section, a *parametric method* was used to estimate quantiles. The parameters were the mean and variance of the assumed normal distribution. The alternate approach is to use *nonparametric* estimation methods. The methods are called nonparametric (also known as distribution-free methods) because their validity does not depend on the data being drawn from any particular distribution. The estimates obtained by these methods are not as precise as we could obtain with a parametric method. The advantage of the nonparametric methods is that they can be used with any data set. The price paid for this convenience is a loss in precision of the estimated values because we are unable (or unwilling) to make a constraining assumption regarding the population distribution. The nonparametric method should be used only when the underlying distribution is unknown or cannot be transformed to make it become normal.

The data are ranked just as would be done to construct a probability plot (Chapter 5). Percentiles could be estimated by reading them from a probability plot, but the method used here skips the plotting. The estimated pth percentile, $\hat{y}_p$, is simply the kth largest datum in the set, where $k = p(n + 1)$, n is the number of data points, and p is the percentile level of interest. If k is not an integer, $\hat{y}_p$ is obtained by linear interpolation between the two closest order statistics.

Example 8.4. A sample of $n = 575$ daily biochemical oxygen demand (BOD) observations is available to estimate the 99th percentile by the nonparametric method for the purpose of setting a maximum limit in a paper mill's discharge permit. The 11 largest ranked observations are

Rank	575	574	573	572	571	570	569	568	567	566	565
BOD	10565	10385	7820	7580	7322	7123	6627	6289	6261	6079	5977

The 99th percentile is located at observation number $p(n + 1) = 0.99(575 + 1) = 570.24$. Since this is not an integer, interpolate between the 570th and 571st largest observations to estimate $\hat{y}_{0.99} = 7171$.

An advantage of this method is that we do not have to specify or assume a distribution for the data. The disadvantage is that only the few largest observed values are used to estimate the percentile. The lower values are not used, except as they contribute to ranking the large values. Discarding these lower values throws away information that could be used get more precise parameter estimates if the shape of the population distribution could be identified and used to make a parametric estimate.

In order to use this method, the data set must be large enough that extrapolation is unnecessary. That is, a 95th percentile could be estimated from 20 observations, but a 99th percentile could not unless the data set contains at least 100 observations. The data set should actually be much larger than the minimum if the estimates are to be much good. The advisability of this is obvious from a probability plot, which shows clearly that the greatest uncertainty is in the location of the extreme quantiles. This uncertainty can be expressed as confidence limits.

The confidence limits for quantiles that have been estimated using the nonparametric method can be determined with the following method if n > 20 observations. Compute the rank order of the lower and upper confidence limits (LCL and UCL):

$$\text{Rank(LCL)} = p(n+1) - z_{\alpha/2} [np(1-p)]^{1/2}$$

$$\text{Rank(UCL)} = p(n+1) + z_{\alpha/2} [np(1-p)]^{1/2}$$

Since Rank(UCL) and Rank(LCL) are usually not integers, the limits are obtained by linear interpolation between the closest order statistics.

A rank of the one-sided $1 - \alpha$ UCL is obtained by computing

$$\text{Rank(UCL)} = p(n+1) + z_{\alpha} [np(1-p)]^{1/2}$$

and, if necessary, interpolating between the two closest order statistics. Once again, this computation is for the case where n > 20.

Example 8.5. We continue Example 8.4 to establish 95% confidence limits on the estimated 99th percentile value of 7171, which was located by interpolating between the 570th and 571st largest observed values. For n = 575 observations, and $\alpha/2 = 0.025$, $z_{.025} = 1.96$ and

$$\text{Rank(LCL)} = 0.99(576) - 1.96[575(0.99)(0.01)]^{1/2} = 565.6$$
$$\text{Rank(UCL)} = 0.99(576) + 1.96[575(0.99)(0.01)]^{1/2} = 574.9$$

Interpolating between observations 563 and 564 and between observations 574 and 575 gives LCL = 6038 and UCL = 10,457.

COMMENTS

Percentiles can be estimated using parametric or nonparametric methods. The nonparametric method is simple, but the sample must contain more than p observations to estimate the pth percentile. Use the nonparametric method whenever you are unwilling or unable to specify a plausible distribution for the sample. Parametric estimates should be made whenever the distribution can be identified because the estimates will be more precise than those obtained from the nonparametric method.

The 50th percentile can be estimated with greater precision than any other, and precision decreases rapidly as the estimates move toward the extreme tails of the distribution. Neither estimation method produces very precise estimates of extreme percentiles, even with large data sets.

REFERENCES

Bisgaard, S. and W. G. Hunter (1986). "Studies in Quality Improvement: Designing Environmental Regulations," Tech. Report No. 7, Center for Quality and Productivity Improvement, University of Wisconsin-Madison.

Crabtree, R. W., I. D. Cluckie, and C. F. Forster (1987). "Percentile Estimation for Water Quality Data," *Water Res.*, 23, 583–590.

Gilbert, R. O. (1987). *Statistical Methods for Environmental Pollution Monitoring*, New York, Van Nostrand Reinhold.

Hahn, G. J. and S. S. Shapiro (1967). *Statistical Models in Engineering*, New York, Wiley.

Owen, D. B. (1962). *Handbook of Statistical Tables*, Palo Alto, CA, Addison Wesley.

Chapter 9

The Limit of Detection

Key words: Limit of detection, measurement error, method limit of detection, percentile, variance, standard deviation

---

The *method limit of detection* or *method detection limit* (MDL) is based on the ability of a measurement method to determine an analyte in a sample matrix, regardless of its source of origin. Processing the specimen by dilution, extraction, drying, etc. introduces variability, and it is essential that the MDL include this variability.

The MDL is a statistical concept, although it is often thought of as a chemical concept, because it varies from substance to substance and it becomes possible to measure progressively smaller quantities as analytical methods improve. Nevertheless, the MDL is a statistic that is estimated from data. As such, it has no scientific meaning until it is operationally defined in terms of a measurement process and a statistical method for analyzing the measurements that are produced. Without a precise statistical definition, one cannot determine a numerical value for the limit of detection or expect different laboratories to be consistent in how they determine the limit of detection.

Many definitions have been published. They may differ in detail, but broadly speaking they are all defined in terms of a multiple of the standard deviation of measurements on blank specimens or, alternately, on specimens that have a very low concentration of the analyte of interest. All definitions exhibit the same difficulty with regard as to how the standard deviation of blank specimens is to be estimated.

The U.S. EPA's definition and suggested method of estimation is reviewed, and then we examine an alternative approach to understanding the precision of measurements at low concentrations, which has been proposed by Pallesen (1985).

CASE STUDY — LEAD MEASUREMENTS

Lead is a toxic metal that is regulated in wastewater effluents and in drinking water. Five laboratories are going to be sharing samples from time to time as a check on the quality of their work. They want to know whether the MDL for lead is the same at each lab. As a first step in this evaluation, each laboratory analyzed 50 test specimens containing lead. They did not know how many levels of lead had been prepared (by spiking known amounts of lead into a large quantity of common solution) nor did they know what concentrations they were given, except that they were low, near the expected MDL. The background matrix of all the test specimens was filtered effluent from a well-operated activated sludge plant. A typical data set is shown in Table 9.1 (Berthouex, 1993).

METHOD DETECTION LIMIT — GENERAL CONCEPTS

The *method detection limit*, or MDL, is much more a statistical than a chemical concept. The term has no scientific meaning until it is operationally defined in terms of a measurement process and a statistical method for analyzing the measurements that are produced. Without a precise statistical definition, one cannot determine a scientifically defensible value for the limit of detection, expect different laboratories to be consistent in how they determine the limit of detection, or be scientifically honest about declaring that a substance has (or has not) been detected. Beyond the statistical definition, there must be a clear set of operational rules for how this measurement error is to be determined

Table 9.1 **Typical Lead Data (Laboratory B)**

	Spiked Lead Concentration			
Zero	**1.25 μg/L**	**2.5 μg/L**	**5 μg/L**	**10 μg/L**
2.5	2.8	4.5	3.9	12.2
3.0	2.7	3.7	5.0	13.8
2.2	3.4	3.8	5.4	9.9
2.2	2.4	4.4	4.9	9.5
3.1	3.0	5.4	6.2	8.9
2.6	3.7	3.9		
	4.6	4.1		
	3.1	3.7		
	3.6	3.0		
	3.1	4.5		
	4.3	4.8		
	4.0	3.3		
	1.7	4.7		
	2.2	4.4		
	2.4			
	3.5			
	2.2			
	2.7			
	3.2			
	2.8			

in the laboratory. Most published definitions are weak with respect to these instructions, which must explain how to estimate the variances and what kind and number of blanks to be used.

The U.S. EPA[1] says, "The method detection limit (MDL) is defined as the minimum concentration of a substance that can be measured and reported with 99% confidence that the analyte concentration is greater than zero and is determined from analysis of a sample in a given matrix containing the analyte. . . . It is essential that all sample processing steps of the analytical method be included in the determination of the method detection limit." The mention of "a sample in a given matrix" indicates that the MDL may vary as a function of specimen composition. The phrase ". . . containing the analyte" may be confusing since the procedure was designed to apply to a wide range of samples, including reagent blanks that would not include the analyte.

The U.S. EPA gives a procedure for establishing an initial estimate of the MDL for the purpose of making samples to be used to determine the MDL. They point out that the variance of the analytical method may change with concentration and that the MDL determined using their procedure may not truly reflect method variance at lower analyte concentrations. A minimum of seven aliquots of the prepared solution shall be used to calculate the MDL. If a blank measurement is required in the analytical procedure, obtain a separate blank measurement for each aliquot and subtract the average of the blank measurements from each specimen measurement.

Taylor (1987) pointed out that the term "limit of detection" is used in several different situations, which should be clearly distinguished. One of these is the *instrument limit*

---

[1] Appendix B to Part 136, 40 CFR Ch. 1 (7-1-89 edition).

of detection (IDL). This limit is based on the ability to detect the difference between "signal" and "noise." If the measurement method is entirely instrumental (no chemical or procedural steps are used), the signal-to-noise ratio has some relation to MDL. Whenever other procedural steps are used, the IDL relates only to instrumental measurement error, which may have a small variance in comparison with variance contributed by other steps. In such a case, improving the signal-to-noise ratio by instrumental modifications would not necessarily lower the MDL.

As a practical matter, we are interested in the MDL. Processing the specimen by dilution, extraction, drying, etc. introduces variability, and it is essential that the MDL reflect this variability. The MDL is based on a method's ability to determine an analyte in a sample matrix, regardless of its source of origin.

It should be remembered that the various limits of detection (MDL, IDL, etc.) are not unique constants of methodology (for example, see "Standard Methods," APHA et al., 1989, Section 1030E). They depend on the statistical definition and how measurement variability at low concentrations is estimated. They also depend on the expertise of the analyst, the quality control procedures used in the laboratory, and on the sample matrix measured. Thus, two analysts in the same laboratory, using the same method, can show significant differences in precision, and hence their MDLs will differ. From this, it follows that published values for an MDL have no application in a specific case, except possibly to provide a rough reference datum against which a laboratory or analyst could check a specifically derived MDL. The analyst's value and the published MDL could show poor agreement because the analyst's value includes specimen matrix effects and interferences that could substantially shift the MDL.

THE U.S. EPA APPROACH TO ESTIMATING THE MDL

The U.S. EPA defines the MDL as the minimum concentration of a substance that can be measured and reported with 99% confidence that the analyte concentration is greater than zero and is determined from analysis of a sample in a given matrix containing the analyte. Similar definitions have been given by Glaser et al. (1981), Hunt and Wilson (1986), the American Chemical Society (1983), Kaiser and Menzes (1968), Kaiser (1970), Holland and McElroy (1986), and Porter et al. (1988).

The U.S. EPA gives a procedure for establishing an initial estimate of the MDL for the purpose of making samples to be used to determine the MDL. Measurements are made on a minimum of seven aliquots of the prepared solution, and these are used to calculate the variance of the replicate measurements:

$$s^2 = \frac{1}{n-1} \left[\sum_{i=1}^{n} (x_i - \bar{x})^2 \right]$$

where x_i, $i = 1$ to n are the analytical results in the final method reporting units obtained from the n aliquots. The MDL is

$$MDL = t_{u, \alpha = .01} \cdot s$$

where $s = \sqrt{s^2}$ and $t_{v, \alpha = 0.01}$ is the Student's t value appropriate for a 99% confidence level and a standard deviation estimate with $v = n - 1$ degrees of freedom. Note that this is the t value that cuts off the upper 1% of the t distribution. For $n = 7$, $t = 3.143$ and the estimated MDL = 3.143 s.

The method points out that the variance of the analytical method may change with concentration. If this happens, the estimated MDL will also vary depending on the concentration of the prepared solution on which the replicate measurements were made. The U.S. EPA suggests that the analyst check this by analyzing seven replicate aliquots at a slightly different concentration in order to "verify the reasonableness" of the estimate of the MDL. If the difference between the variance of the set of seven aliquots measured at concentration 1, s_1^2, and at a second concentration, s_2^2, can be considered statistically insignificant (based on the F statistic of their ratio) these two variances are combined (pooled) to obtain a single estimate for s^2.

$$s_{pooled}^2 = \frac{6s_1^2 + 6 s_2^2}{6 + 6} = \frac{s_1^2 + s_2^2}{2}$$

The 6's are the degrees of freedom for the two sets of seven aliquot observations. In general, the degrees of freedom are n − 1. If the two batches of specimens have identical degrees of freedom, then "pooling" is just averaging the variances of the two batches.

Based on the 12 degrees of freedom for the pooled standard deviation, $t_{12, 0.01} = 2.681$, and the MDL is computed as

$$\text{MDL} = 2.681 \; s_{pooled}$$

If more than seven aliquots were used, the appropriate Student's t value at the 1% significance limit would be taken from Table 9.2.

The U.S. EPA's definition of the MDL is illustrated by Figure 9.1. Measurements on test specimens are pictured as being normally distributed about the true concentration. A small probability, α, exists that a measurement on a sample specimen will exceed

Table 9.2 Values of the t Statistic for Use in the U.S. EPA's Procedure to Set the MDL

For s^2

n	7	8	9	10	11	12	16	21	∞
df = n − 1	6	7	8	9	10	11	15	20	∞
$t_{n-1, 0.01}$	3.143	2.998	2.896	2.821	2.764	2.718	2.602	2.528	2.326

For s_{pooled}^2

n + n	14	16	18	20	22	24	32	42	∞
df = 2n − 2	12	14	16	18	20	22	30	40	∞
$t_{2n-2, 0.01}$	2.681	2.624	2.583	2.552	2.528	2.508	2.457	2.423	2.326

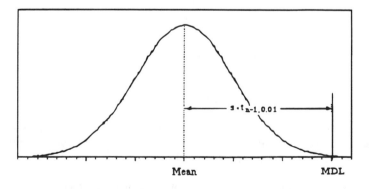

Figure 9.1 The U.S. EPA's definition of the MDL.

the MDL, which is located a distance $t_{n-1,\alpha=0.01}$ s to the right of the mean concentration. If the specimens being measured were true blanks (concentration equal zero), this definition of the MDL would provide a high degree of protection $(1 - \alpha)$ against declaring that the analyte was detected in a specimen in which it was not present.

Example 9.1. A set of seven replicate specimens were measured to obtain the following results: 2.5, 2.7, 2.2, 2.2, 3.1, 2.6, and 2.8. The estimated variance is $s^2 = 0.105$, s = 0.324, and the MDL = 3.143(0.324) = 1.01. As a check on the reasonability of this variance estimate, another set of seven replicate specimens was analyzed and the data were 1.6, 1.9, 1.3, 1.7, 2.1, 0.9, and 1.8. These data give $s^2 = 0.161$, s = 0.402, and the MDL = 3.143(0.402) = 1.26. A statistical F test shows that the two estimates of the variance are not significantly different. Therefore, the two samples are pooled to give an improved estimate of the variance:

$$s^2_{pooled} = \frac{0.105 + 0.161}{2} = 0.133 \quad \text{and} \quad s_{pooled} = 0.365$$

s_{pooled} has 12 degrees of freedom. The improved estimate of the MDL is

$$MDL = 2.681 \; s_{pooled} = 2.681(0.365) = 0.98$$

Example 9.2. Calculate, using the U.S. EPA's method, the MDL for the lead data given in Table 9.1 for Laboratory B. By definition, a minimum of seven replicates containing the analyte are needed to compute the MDL. Therefore, we can use only the 1.25 and 2.50 μg/L spiked specimens. The summary statistics are

	1.25 μg/L	2.50 μg/L
$\bar{y}$	3.07	4.16
s^2	0.548	0.407
s	0.746	0.638
n	20	14

Pooling the sample variances yields

$$s^2_{pooled} = \frac{(19)(0.548) + (13)(0.407)}{(19 + 13)} = 0.491$$

with $\nu = 19 + 13 = 32$ degrees of freedom. The pooled standard deviation is

$$s_{pooled} = 0.701 \; \mu g/L$$

The appropriate t statistic to compute the MDL is $t_{32,0.01} = 2.45$, and

$$MDL = 2.45 \; (s_{pooled}) = 2.45 \; (0.701) = 1.72 \; \mu g/L$$

AN ALTERNATE MODEL FOR THE MDL

Pallesen (1985) defines the limit of detection as "the smallest value of analyte that can be reliably detected above the random background noise present in the analysis of blanks." The MDL is defined in terms of the background noise of the analytical procedure for the matrix being analyzed. He proposed an alternate model for the MDL that separately considers analytical error and background noise and provides a clear method for estimating the variance of the background noise and the analytical error.

The error structure in the neighborhood of the true mean where limit of detection considerations are relevant is assumed to be

$$y_i = \eta + e_i = \eta + a_i + b_i$$

The a_i and b_i are two kinds of random error that affect the measurements. a_i is random analytical error and b_i is background noise. The total random error is the sum of these two component errors: $e_i = a_i + b_i$. Both errors are assumed to be random and normally distributed with mean zero.

Background noise, b_i, exists even in blank measurements and has constant variance, σ_b^2. The measurement error in the analytical signal, a_i, is assumed to be proportional to the measurement signal, η. That is, $\sigma_a = \kappa\eta$ and the variance $\sigma_a^2 = \kappa^2\eta^2$. Under this assumption, the total error variance, σ_e^2, of any measurement is

$$\sigma_e^2 = \sigma_b^2 + \sigma_a^2 = \sigma_b^2 + \kappa^2\eta^2$$

According to this model, both σ_a^2 and σ_e^2 decrease as the signal decreases. When the specimen is blank (i.e., when $\eta = 0$), σ_a^2 decreases to zero and $\sigma_e^2 = \sigma_b^2$, the variance of the blank. Also, for low η, the analytical error variance, $\sigma_a^2 = \kappa^2\eta^2$, becomes small compared to the background variance σ_b^2. Figure 9.2 shows this relation in terms of the variances and the standard deviations.

The MDL is the smallest value of y for which the hypothesis $\eta = 0$ cannot be rejected with some stated level of confidence. If y is larger than the limit of detection, it is very improbable that $\eta = 0$, and such a measurement would indicate that the analyte of interest has been detected. If y is less than the MDL, we cannot say with confidence that the analyte of interest has been detected.

Pallesen (1985) defines the MDL as a multiple of the background noise, σ_b:

$$MDL = k_d\sigma_b$$

and assumes that y, in the absence of analyte, is normally distributed with mean zero and variance σ_b^2. k_d is chosen such that the probability of y being greater than MDL is kept suitably low. The probability that a blank specimen gives y > MDL is α. The k_d values are z values of the standardized normal distribution. k_d corresponding to any given α value can be found in a table of the normal distribution, where α is the area under the tail that lies above the MDL. For example,

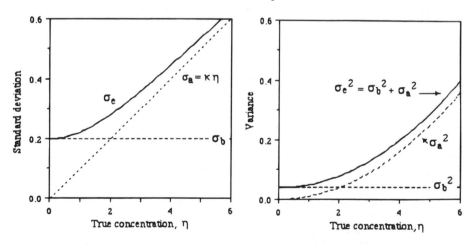

Figure 9.2 Error structure of the alternate model of the MDL. (Adapted from Pallesen [1985].)

$$k_d = 1.63 \qquad \alpha = 0.05 \ (5\%)$$
$$k_d = 2.00 \qquad \alpha = 0.023 \ (2.3\%)$$
$$k_d = 2.33 \qquad \alpha = 0.01 \ (1\%)$$
$$k_d = 3.00 \qquad \alpha = 0.0013 \ (0.13\%)$$

Using $k_d = 3.00$ means that observing $y > MDL$ justifies a conclusion that η is not zero at a significance level of $\alpha = 0.0013$ (or confidence level of 99.87%). Using this MDL repeatedly, just a little over 1 in 1000 (0.13%) of true blanks specimens will be misjudged as positive determinations. Note that the U.S. EPA's definition chooses k_d so $\alpha = 0.01$ (1%).

CASE STUDY SOLUTION

The replicate lead measurements in Table 9.1 will be used to estimate the parameters σ_b^2 and κ^2 as in the model

$$\sigma_e^2 = \sigma_b^2 + \sigma_a^2 = \sigma_b^2 + \kappa^2 \eta^2$$

The values of σ_e^2 and η^2 are estimated from the data by computing s_e^2 and $\bar{y}^2$. The average and variance for the five laboratories are listed in Table 9.3.

The parameters σ_b^2 and κ^2 are estimated by regression, as shown in Figure 9.3. The intercept is an estimate of σ_b^2, and the slope is estimate of κ^2. Table 9.4 gives the estimated

Table 9.3 **Estimated Averages ($\bar{y}$) and Variances (s_e^2) at Each Concentration Level for Five Laboratories**

Spike	Laboratory A $\bar{y}$	s_e^2	Laboratory B $\bar{y}$	s_e^2	Laboratory C $\bar{y}$	s_e^2	Laboratory D $\bar{y}$	s_e^2	Laboratory E $\bar{y}$	s_e^2
0	2.30	0.45	2.60	0.15	1.90	0.12	1.09	0.22	1.50	0.06
1.25	2.55	0.67	3.07	0.55	2.76	0.22	1.82	0.37	1.65	0.68
2.5	4.55	1.24	4.16	0.41	4.53	0.36	3.08	0.43	3.21	0.66
5.0	8.14	2.31	5.08	0.70	6.90	0.44	4.59	0.44	5.16	0.75
10.0	11.90	2.64	10.86	4.266	11.23	1.26	9.12	1.15	10.66	1.51

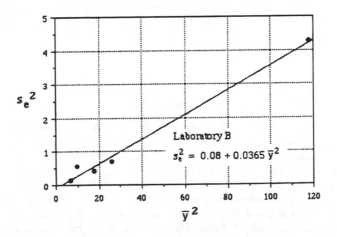

Laboratory B
$$s_e^2 = 0.08 + 0.0365 \, \bar{y}^2$$

Figure 9.3 Plots of σ_e^2 vs $\bar{y}^2$ used to estimate the parameters κ and σ_b^2.

Table 9.4 **Estimates of σ_b^2, κ, and the MDL**

Laboratory	σ_b^2	σ_b	κ^2	κ	MDL
A	0.72	0.85	0.0155	0.12	2.55
B	0.08	0.28	0.0365	0.19	0.84
C	0.12	0.34	0.0089	0.09	1.02
D	0.27	0.52	0.0104	0.10	1.56
E	0.43	0.66	0.0097	0.10	1.98

values of σ_b^2 and κ^2 for each of the five laboratories. From these we can compute the limit of detection (MDL) that are given in Table 9.4. We conclude that the MDLs of the five laboratories are fairly consistent for the analytical procedure used to measure lead.

Consider the differences between the U.S. EPA's approach and Pallesen's (1985) alternate method for determining the MDL. We will use Example 9.2 (Laboratory B) as a basis for discussion. In Example 9.2, the U.S. EPA's method gave MDL = 1.72 μg/L. Pallesen's method gave MDL = 0.84 μg/L. There are important differences in the two methods.

1. The z or t statistic multipliers are different: $z(\alpha = 0.01) = 2.33$ for the U.S. EPA's method and $z(\alpha = 0.0013) = 3.00$ for Pallesen's method. On purely statistical grounds, the U.S. EPA's arbitrary specification of $\alpha = 0.01$ is no more justified than Pallesen's recommended multiplier of 3.00.

2. Users of the U.S. EPA's method will often estimate the MDL using 7, or perhaps 14, replicate measurements, since the procedure requires "a minimum of seven replicates." In Example 9.2, the MDL was based on 34 measurements. Pallesen's method requires replicated measurements at several concentration levels (how many levels or replicates is not specified). The case study used 40 measured values, divided among five concentration levels. Each of these values contributes to the estimation of the MDL.

3. The U.S. EPA's method estimates the MDL from measurements "in a given matrix containing the analyte" and then assumes that the background noise observed is applicable to specimens that are blank (zero concentration). The alternate method extrapolates to zero analyte concentration to estimate the background noise, σ_b, that serves as the basis for computing the MDL. As shown by Table 9.4 and Figure 9.3, the total error variance increases as the concentration increases, and Pallesen's method accounts for this property of the data. We note that if the data have this characteristic, the U.S. EPA's approach of pooling replicates from two different concentrations will tend to inflate the MDL over that estimated using the alternate method.

COMMENTS

The limit of detection is a troublesome concept. It causes difficulties for the chemist who must determine its value, for the analyst who must work with data that are censored by being reported as "<MDL," and for the regulator and discharger who must make important decisions with incomplete information. Some statisticians and scientists think we would do better without it (Rhodes, 1981; Gilbert, 1987). Nevertheless, it is an idea that seems firmly fixed in environmental measurement work, and we need to understand what it is and what it is not. It is an attempt to prevent deciding that a blank sample is contaminated with an analyte. Pallesen's model is appealing because it distinguishes between error contributions from both analytical method and background noise.

By the U.S. EPA's definition, if measurements were made on blanks, the odds of finding any one particular blank "contaminated" is 1%. The collective odds, however, become unfavorable if we make this test on many blanks. If, for example, we made a measurement on blank specimens every day for one month, the probability of having *at least one* value declared ">MDL" approaches 100%.

The MDL is based directly on an estimate of the measurement error at low concentrations. This quantity is not easy to estimate, which means that the MDL is not a well-known, fixed value. Different chemists and different laboratories would estimate different MDLs working from the same specimens. This was shown by example with real data. A result that is declared "≤MDL" can come from a sample that has a concentration actually greater than the MDL.

REFERENCES

American Chemical Society, Committee on Environmental Improvement (1983). "Principles of Environmental Analysis," *Anal. Chem.*, 55, 14, 2210–2218.

APHA, AWWA, and WEF (1989). *Standard Methods for the Examination of Water and Wastewater*, 17th ed., Eds. Clesceri, L. S., A. E. Greenberg, and R. R. Trussell.

Berthouex, P. M. (1993). "A Study of the Precision of Lead Measurements at Concentrations Near the Limit of Detection," *Water Environ. Res.*, 65, 5, 620–629.

Berthouex, P. M. and D. R. Gan (1993). "A Model for the Precision of Measurements at Low Concentrations," *Water Environ. Res.*, 65, 6, 759–763.

Currie, L. A. (1984). "Chemometrics and Analytical Chemistry," in *Chemometrics: Mathematics and Statistics in Chemistry*, Vol. 138, pp. 115–146, NATO ASI Series C, Dordrect, Germany, D. Reidel.

Gilbert, R. O. (1987). *Statistical Methods for Environmental Pollution Monitoring*, New York, Van Nostrand Reinhold.

Glaser, J. A., D. L. Foerst, G. D. McKee, S. A. Quave, and W. L. Budde (1981). "Trace Analyses for Wastewaters," *Environ. Sci. Technol.*, 15, 12, 1426–1435.

Holland, D. M. and F. F. McElroy (1986). "Analytical Method Comparison by Estimates of Precision and Lower Detection Limit," *Environ. Sci. Technol.*, 20, 1157–1161.

Hunt, D. T. E. and A. L. Wilson (1986). *The Chemical Analysis of Water*, 2nd ed., London, The Royal Society of Chemistry.

Kaiser, H. H. (1970). "Quantitation in Elemental Analysis: Part II," *Anal. Chem.*, 42, 4, 26A–59A.

Kaiser, H. H. and A. C. Menzes (1968). *Two Papers on the Limit of Detection of a Complete Analytical Procedure*, London, Adam Hilger Ltd.

Pallesen, L. (1985). "The Interpretation of Analytical Measurements Made Near the Limit of Detection", Technical Report, IMSOR, Technical University of Denmark.

Porter, P. S., R. C. Ward, and H. F. Bell (1988). "The Detection Limit," *Environ. Sci. Technol.*, 22, 8, 856–861.

Rhodes, R. C. (1981). "Much Ado About Next to Nothing, Or What to Do with Measurements Below the Detection Limit," in *Environmetrics 81 — Selected Papers*, pp. 174–175, SIAM-SIMS Conference Series No. 8, Philadelphia, Society for Industrial and Applied Mathematics.

Taylor, J. K. (1987). *Quality Assurance of Chemical Measurements*, Chelsea, MI, Lewis Publishers.

Chapter 10

Simple Methods for Analyzing Data That Are Below the Limit of Detection

---

Key words: censored data, median, limit of detection, trimmed mean, Winsorized mean, probability plot, rankit, regression

---

In the past, most environmental measurements were made on chemicals that were expected to exist at easily measurable concentrations. Today the most important scientific problems in environmental quality focus on chemicals that are expected either to be absent or to exist at very low concentrations. Under these conditions, it can be expected that a set of data on trace chemicals will include some observations for which the chemical analyst has reported that the analyte concentrations are "below the limit of detection (MDL)." Such a data set is said to be *censored*. This chapter considers simple ways of estimating the mean and variance of a censored data set.

Arbitrarily assigning a numerical value to censored values will give biased estimates of the mean value. The median, trimmed mean, and Winsorized mean are three unbiased estimates of the mean for normal or other symmetrical distributions. They are insensitive to information from the extremes of the distribution and can be used when the extent of censoring is moderate (say, not more than 15–25% censored). Graphical interpretations, especially probability plots, are are demonstrated as useful methods.

REPLACEMENT OR DELETION METHODS

Censored data, in essence, are missing values. Missing values are not always a serious problem. If we collected 50 specimens and 5 of them, selected at random, were damaged or lost, we could do the data analysis as though there were only 45 specimens. If a few values were missing from a time series, we could probably find some way to fill in the missing values without distorting the pattern of the series. The difficulty with censored data is that they are not missing in a random pattern, but they are all missing at one end of the distribution. We cannot go ahead as if they never existed, we cannot pretend they are zeros (if they are missing at the low end), and we will see that the possibilities for replacing the censored observations with some arbitrary value are not entirely satisfactory either.

An example will show that all these methods give biased estimates of the mean and variance. In order to show this, we need to start with a sample of fully measured data from which a censored sample can be constructed. We have used the nitrate data from Chapter 2 as the "full" data set shown in Table 10.1. (Since nitrate data are almost never affected by censoring, imagine that they are something more exotic, and that the units are in micrograms per liter.) Assume the MDL = 6.0, which censors 4 out of 27 values and produces the censored data sample in the second column of Table 10.1.

The last four columns of Table 10.1 were constructed by using these four ad hoc methods to replace the censored values:

1. Replace all "<MDL" values with the MDL.
2. Replace all "<MDL" values with the zeros.
3. Replace all "<MDL" values with one half the MDL.
4. Use only the ">MDL" values.

Table 10.1 **Illustration of Bias in ad hoc Methods for Replacing Censored Observations**

		Method Used to Construct Deal with Censored Data			
Full Data Set	**Censored Data Set**	**<MDL = MDL**	**<MDL = 0**	**<MDL = 0.5 MDL**	**Delete <MDL**
6.9	6.9	6.9	6.9	6.9	6.9
7.8	7.8	7.8	7.8	7.8	7.8
8.9	8.9	8.9	8.9	8.9	8.9
5.2	<MDL	6.0	0	3.0	—
7.7	7.7	7.7	7.7	7.7	7.7
9.6	9.6	9.6	9.6	9.6	9.6
8.7	8.7	8.7	8.7	8.7	8.7
6.7	6.7	6.7	6.7	6.7	6.7
4.8	<MDL	6.0	0	3.0	—
8.0	8.0	8.0	8.0	8.0	8.0
10.1	10.1	10.1	10.1	10.1	10.1
8.5	8.5	8.5	8.5	8.5	8.5
6.5	6.5	6.5	6.5	6.5	6.5
9.2	9.2	9.2	9.2	9.2	9.2
7.4	7.4	7.4	7.4	7.4	7.4
6.3	6.3	6.3	6.3	6.3	6.3
5.6	<MDL	6.0	0	3.0	—
7.3	7.3	7.3	7.3	7.3	7.3
8.3	8.3	8.3	8.3	8.3	8.3
7.2	7.2	7.2	7.2	7.2	7.2
7.5	7.5	7.5	7.5	7.5	7.5
6.1	6.1	6.1	6.1	6.1	6.1
9.4	9.4	9.4	9.4	9.4	9.4
5.4	<MDL	6.0	0	3.0	—
7.6	7.6	7.6	7.6	7.6	7.6
8.1	8.1	8.1	8.1	8.1	8.1
7.9	7.9	7.9	7.9	7.9	7.9
Average =	7.51	7.62	6.73	7.17	7.90
Variance =	1.914	1.464	9.181	4.121	1.166

Example 10.1. The mean of the full sample of 27 values is 7.51 μg/L, and the standard deviation is 1.38 μg/L. The averages and variances resulting from the four ad hoc methods are biased. Replacing the censored observations with zero or 0.5 MDL gives estimates of the mean that are biased low. Replacing the censored values with the MDL, or omitting the censored observations, gives estimates that are biased high. The variances are also biased and will have, in general, a pattern opposite that of the estimates of the mean; replacement with zero or 0.5 MDL leads to estimates of the variance and standard deviation that are too large, while deletion or replacement with the MDL underestimates the variance. The bias of both the mean and variance would increase as (1) the fraction of observations censored increases or (2) the MDL increases.

THE MEDIAN

If the data are normally distributed, the true mean and true median are the same value. Therefore, the median is an estimate of the mean. The median gives an unbiased estimate of the mean if the underlying distribution is believed to be symmetric, but not necessarily normal. The median is unaffected by the magnitude of observations on the tails of the distribution. It is also unaffected by censoring, so long as more than half of the observations have been quantified. If more than half the values are censored, the median itself is below the MDL and cannot be estimated.

Example 10.2. The median is the middle value in a ranked data set if the number of observations is odd. If the number of observations is even, the two middle values are averaged to estimate the median. For the censored data in Table 10.1, the median is 7.6, the 14th largest of the 27 ranked observations shown below.

<MDL	<MDL	<MDL	<MDL	6.1	6.3	6.5	6.7	6.9
7.2	7.3	7.4	7.5	[7.6]	7.7	7.8	7.9	8.0
8.1	8.3	8.5	8.7	8.9	9.2	9.4	9.6	10.1

THE TRIMMED MEAN

The *trimmed mean* and the *Winsorized mean* can be used to estimate the mean if the underlying distribution is believed to be symmetric, but not necessarily normal. In this case, they are unbiased estimators, but they do not have minimum variance.

Censoring is in effect trimming. It trims away part of the lower tail of the distribution and creates an unsymmetric data set, one with more known values above than below the median. Symmetry could be returned by trimming away some values on the upper tail of the distribution.

A trimmed mean is computed from a set of n observations by trimming away (eliminating) the largest np values and the smallest np values. The average is computed using the remaining n – 2np values. This is called a 100 p% trimmed mean, where 0 < p < 0.5. The degree of trimming (i.e., the value of p) that can be used depends on the size of the sample. p does not have to equal the percentage of observations that have been censored. It could be a higher, but obviously not lower, percentage.

Hoaglin et al. (1983) suggest that a 25% trimmed mean is a good estimator of the mean for symmetric distributions. This uses the middle 50% of the data. Hill and Dixon (1982) considered asymmetric distributions and found that a 15% trimmed mean was a "safe" estimator, in the sense that its performance did not vary markedly from one situation to another. It is obvious that if more than 15% of the observations in the sample have been censored, the 15% trimmed mean cannot be computed.

Example 10.3. Compute the trimmed mean for the censored sample in Table 10.1 which has four censored values. To create symmetry, trim away the largest four values, leaving the 19 observations listed below.

6.1	6.3	6.5	6.7	6.9	7.2	7.3	7.4	7.5	7.6	7.7
7.8	7.9	8.0	8.1	8.3	8.5	8.7	8.9			

The trimmed mean is computed as

$$\bar{y}_t = \frac{\Sigma y_i}{n - 2 np} = \frac{143.4}{27 - 2(4)} = 7.55 \ \mu g/L$$

The percent trimming is 100(4/27) = 15%, and $\bar{y}_t = 7.55$ is a 15% trimmed mean.

THE WINSORIZED MEAN

Winsorization can be used to estimate the mean and standard deviation of a distribution even though the data set has a few missing or unreliable values at either or both ends of the distribution. The method is most easily explained by example, once again using the data from Table 10.1.

> **Example 10.4.** Compute the Winsorized mean for the censored samples in Table 10.1. The procedure is to (1) replace the four "<MDL" values by the next largest datum, that is, by 6.1; (2) replace the four largest values by the next smallest datum, that is, by 8.9; and (3) compute the sample mean, $\bar{y}$, and standard deviation, s, of the resulting sample.
>
> Completing the first two steps gives the Winsorized sample below:
>
> 6.1 6.1 6.1 6.1 6.1 6.3 6.5 6.7 6.9 7.2 7.3 7.4 7.5 7.6
> 7.7 7.8 7.9 8.0 8.1 8.3 8.5 8.7 8.9 8.9 8.9 8.9 8.9

The mean $\bar{y}$ and standard deviation s of the Winsorized sample are computed in the usual way.

$$\bar{y} = 7.53 \ \mu g/L \quad \text{and} \quad s = 1.022 \ \mu g/L$$

The Winsorized mean is the same as the mean of the Winsorized sample.

$$\bar{y} = \bar{y}_w = 7.53 \ \mu g/L.$$

The Winsorized standard deviation, s_w, is an approximately unbiased estimator of σ and is computed from the standard deviation of the Winsorized sample as

$$s_w = \frac{s(n-1)}{v-1} = 1.02 \ (27-1)/(19-1) = 1.398 \ \mu g/L$$

where n is the total number of data values and v is the number of data not replaced during the Winsorization. In this example, $v = 27 - 4 - 4 = 19$ because four "<MDL" values and the four largest values have been replaced.

If the data are from a normal distribution, the upper and lower limits of a two-sided $100(1 - \alpha)\%$ confidence interval about μ are

$$\bar{y}_w \pm t_{v-1,\alpha/2} \frac{s_w}{\sqrt{n}}$$

where $t_{v-1,\alpha/2}$ is the value of t that cuts off $(100\alpha/2)\%$ tail of the t distribution with $v - 1$ degrees of freedom. Note that this equation is identical to the usual limits, except the degrees of freedom are $v - 1$ instead of the usual $n - 1$, and s_w replaces s.

Both the trimmed mean and the Winsorized mean are best used on symmetric, but not necessarily normally distributed data sets that have either missing (censored) or unreliable data on the tail of the distribution. If the distribution is symmetric, they give unbiased estimates of the true mean. The distinction between the two methods is that trimming discards data on both ends of the distribution and the trimmed mean is computed with the remaining data. Winsorizing replaces data on both tails with the next most extreme datum in each tail and computes the mean and standard deviation using the newly constructed data set.

GRAPHICAL METHODS

The various ad hoc methods for replacing or deleting censored data give biased estimates. The median, trimmed mean, and Winsorized mean give unbiased estimates of the mean,

Table 10.2 **Measurable Mercury Concentrations**

Time (days)	2	4	10	12	13	16	17	22	26	27	35	38	43	44
Conc. (μg/L)	0.5	0.6	0.8	1.0	3.4	5.5	1.0	1.0	3.5	2.8	0.3	0.5	0.5	1.0

but are only suitable when the distribution is symmetrical. Many sets of environmental data are not symmetrical, so other approaches are needed. Even if the distribution is symmetric, it may happen that the proportion of censored observations is too great (say, more than 15–25%), and these methods are inappropriate. In such a case, there is no way to get a precise estimate of the mean, and about all that can be hoped for is to do something helpful that is not misleading or scientifically dishonest.

Graphical methods, when properly used, are clear, and they reveal all the important features of the data. For this reason, important data should always be displayed graphically, regardless of any numerical data analysis to which they may be subjected. Graphical methods take on a special importance when one is unsure about using a particular statistical method. A wise policy often will be to display the data rather than to compute a summarizing statistic that hides the data structure and, therefore, may be misleading.

Suitable plots convey a great deal of information and very often will enable a decision to be reached without further analysis. Two useful graphs, the time series plot and the cumulative probability plot, are illustrated using a highly censored sample.

Forty-five grab samples were taken on consecutive days from a large river and the mercury concentration of each sample specimen was measured. The MDL of the measurement method was 0.2 μg/L, and only 14 samples had a concentration above this level. Table 10.2 gives the measured concentrations and the days on which they were observed.

With 31 out of 45 observations censored, it is impossible to compute any meaningful statistics for this data set. The median is below the MDL, and since more than half the data are censored, neither the trimmed or Winsorized mean can be computed. One approach is plot the data and see what can be learned. Figure 10.1 shows that high concentrations seem to occur at random, and they last one day (or less). Values reported as "≤MDL" are plotted at the MDL. The plot shows that the occurrence of measureable concentrations is random, and it reveals how difficult it is to say much about the mean or median mercury level.

The probability plot (see also Chapter 5) is prepared by counting the number of samples in which the concentration is below a specified concentration and presenting this

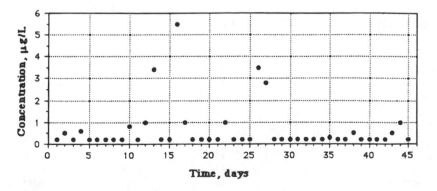

Figure 10.1 Time series plot of the data.

as the probability of the particular value being exceeded. When making this probability statement, we need to make some provision for having a finite number of observations. One way of doing this is to calculate the probability as $p = (i - 1/2)/n$, where n is the sample size. This method is preferred by statisticians, but in many engineering texts the plotting positions are computed as $p = i/(n + 1)$. Typically, any difference between the two plotting probabilities will be at the most extreme points and not in the center of the distribution. Such a discrepancy has little practical importance in a graphical analysis, since the extreme points are always less reliable than the more centrally located points and should not be given too strong an influence on the position of the plotted line.

Figure 10.2 is the cumulative probability plot for the mercury concentration data. A logarithmic scale was used so that the data would plot as a straight line. This plot reveals a great deal about the data. About 65% of the observations were below the MDL, and the median value was estimated as 0.1 μg/L by extrapolating the straight line below the MDL. We cannot be certain, based on these data alone, that the measurements below the MDL would fall on the extrapolated straight line, but at least the data do not preclude this possibility. The line was drawn by eye. In the next section, a regression method is used to fit the probability plot.

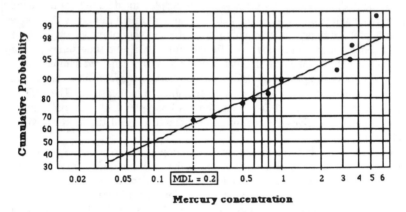

Figure 10.2 Cumulative probability plot.

Table 10.3 Construction of the Cumulative Probability Distribution Shown in Figure 10.2

Concentration y_i	No. of Obs. $\leq y_i$	Rank i	Probability Value $\leq y_i$	
			$p = (i - 0.5)/n$	$p = i/(n + 1)$
<0.200	31	31	0.678	0.674
0.300	1	32	0.700	0.696
0.500	3	35	0.767	0.761
0.600	1	36	0.789	0.783
0.800	1	37	0.811	0.804
1.000	4	41	0.900	0.891
2.800	1	42	0.922	0.913
3.400	1	43	0.944	0.935
3.500	1	44	0.967	0.957
5.500	1	45	0.989	0.978

REGRESSION ON ORDER STATISTICS

It is possible to replace the probabilities with *normal order scores* (also called *order statistics*) and then to use regression to fit a line to the probability plot (Gilliom and Helsel, 1986; Hashimoto and Trussell, 1983; Travis and Land, 1990). This is equivalent to rescaling the graph in terms of standard deviations instead of probabilities.

If the data are normally distributed, or have been transformed to make them normal, the probabilities, p_i, are converted to *normal order scores*, $R_i = F^{-1}(p_i)$, where F^{-1} is the inverse cumulative normal probability distribution and p_i is the plotting position of the ith ranked observation $x_i = log(y_i)$. The normal order statistics are also known as *rankits*. Rankits are tabulated in standard statistical tables for up to n = 50. (The analysis could also be done using probits, which are obtained by adding five to the rankits to eliminate the negative values.) In the previous graphical example, the plotting positions were p = i/(n + 1). Blom (1958), Mandel (1964), and Press et al. (1986) show that plotting positions closely approximating the order statistics of the normal distribution are given by p = (i − 0.375)/(n + 0.25).

The regression model $y_i = \beta_1 + \beta_2 R_i + e_i$ is used with the normal order scores of the noncensored portion of the data to estimate β_1 and β_2. The e_i are the deviations of the fitted line and the observed values. For normally distributed data, the coefficients β_1 and β_2 are, respectively, the mean and standard deviation of the noncensored distribution. If the data have been transformed to obtain normality, then β_1 and β_2 estimate the mean and standard deviation of the transformed data, but not of the data on the original measurement scale. This is because rankits are symmetrical about zero, and therefore, the 50th percentile corresponds to $R_i = 0$. For the normal distribution, the 50th percentile is the mean and also the median. For the lognormal distribution, the median is the geometric mean.

The method is using the 45 values given in Table 10.4 (these are simulated random normal values). The concentrations have been ranked, and the rankits are from standard statistical tables (Rohlf and Sokal, 1981). Figure 10.3 shows the probability plot constructed from the rankits. The top panel shows 45 normally distributed observations. The linear regression of concentration on rankits estimates a mean of 33.6 and a standard deviation of 3.6. The middle panel shows the result assuming censoring of the 10 smallest observations (22% censoring), and the bottom panel has the lowest 20 values censored (44% censoring). The censoring has not greatly changed the estimates of the mean and standard deviation. It is apparent, however, that more censoring increases the relative weight given to values in the upper tail of the distribution.

SUMMARY

Replacement and deletion methods, though simple, introduce bias. The median, trimmed mean, and Winsorized mean are unbiased estimates of the mean when the distribution is symmetric, but not necessarily normal. The trimmed mean should be useful for up to 25% censoring and the Winsorized mean for up to 15% censoring. The median can be estimated if less than half the observations are censored.

Special problems arise when more than half the observations are censored. In such cases, the best approach is to display the data graphically. Simple time series plots and probability plots will reveal a great deal about the data and will never mislead, whereas presenting any single numerical value may be misleading.

The time series plot gives a good impression about variability and randomness. The probability plot shows how frequently any particular value has occurred in the past. If the system is not changed, one might expect the same frequencies in the future. The probability plot also can be used to estimate the median value. If the median is above

Table 10.4 The Probability Analysis in Terms of Rankits or Normal Order Statistics

Observ.	Conc.	Rankit	Observ.	Conc.	Rankit	Observ.	Conc.	Rankit
1	23.8	-2.21	16	31.0	-0.4	31	36.4	0.46
2	24.4	-1.81	17	31.6	-0.3	32	37.0	0.52
3	25.0	-1.58	18	31.6	-0.3	33	37.0	0.59
4	25.0	-1.41	19	31.6	-0.2	34	37.0	0.65
5	25.6	-1.27	20	32.2	-0.2	35	37.0	0.72
6	25.6	-1.16	21	32.8	-0.1	36	37.0	0.80
7	28.0	-1.05	22	33.4	-0.1	37	37.6	0.88
8	28.0	-0.96	23	34.1	0.0	38	37.6	0.96
9	28.0	-0.88	24	34.6	0.1	39	38.2	1.05
10	28.6	-0.80	25	35.2	0.1	40	38.2	1.16
11	28.6	-0.72	26	35.2	0.2	41	39.4	1.27
12	29.2	-0.65	27	35.2	0.2	42	39.4	1.41
13	29.8	-0.59	28	35.8	0.3	43	40.6	1.58
14	29.8	-0.52	29	35.8	0.3	44	43.6	1.81
15	31.0	-0.46	30	35.8	0.4	45	47.8	2.21

Note: Rankits are from Table 27 in Rohlf and Sokal (1981).

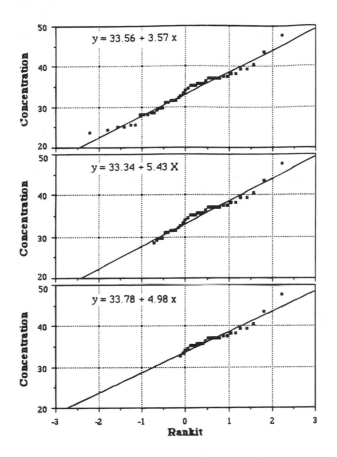

Figure 10.3 The top panel shows a sample of 45 normally distributed values. In the middle panel, 10 values have been censored. The bottom panel has 20 censored values.

the MDL, draw a smooth curve through the plotted points and estimate the median directly. If the median is below the MDL, extrapolation often will be justified on the basis of experience with similar data sets. If the data are distributed normally, the median is also the arithmetic mean. If the distribution is lognormal, the median is the geometric mean.

REFERENCES

Blom, G. (1958). *Statistical Estimates and Transformed Beta Variables*, New York, John Wiley & Sons.

Gilliom, R. J. and D. R. Helsel (1986). "Estimation of Distribution Parameters for Censored Trace Level Water Quality Data: Estimation Techniques," *Water Resour. Res.*, 22, 135–146.

Hashimoto, L. K. and R. R. Trussell (1983). *Proc. Annual Conf. of the American Water Works Association*, Denver, AWWA, p. 1021.

Hill, M. and W. J. Dixon (1982). "Robustness in Real Life: A Study of Clinical Laboratory Data," *Biometrics*, 38, 377–396.

Hoaglin, D. C., F. Mosteller, and J. W. Tukey (1983). *Understanding Robust and Exploratory Data Analysis*, New York, Wiley.

Travis, C. C. and M. L. Land (1990). "Estimating the Mean of Data Sets with Nondetectable Values," *Environ. Sci. Tech.*, 24, 7, 961–962.

Mandel, J. (1964). *The Statistical Analysis of Experimental Data*, New York, Interscience Publishers.

NCASI (1991). "Estimating the Mean of Data Sets that Include Measurements Below the Limit of Detection," Technical Bulletin No. 621 (December 1991).

Press, W. H., B. P. Flannery, S. A. Teukolsky, and W. T. Vetterling (1986). *Numerical Recipes: The Art of Scientific Computing*, Cambridge, Cambridge University Press.

Rohlf, F. J. and R. R. Sokal (1981). *Statistical Tables*, 2nd ed., San Francisco, W. H. Freeman and Co.

Chapter 11

Estimating the Mean of Censored Samples

Key words: censored data, Cohen's method, maximum likelihood estimates, transformations

---

Here we present more formal answers to the questions, "How should a reasonable average be calculated for a censored sample?" and "What would be a meaningful measure of variability?"

Several ways of doing this are referenced at the end of this chapter. None is always superior, so we have chosen to present only one, one that is easy to compute and that has fared well in comparisons with other methods. This is Cohen's maximum likelihood estimator method. The "maximum likelihood method" estimates the parameters of a population by finding the parameter values that maximize the probability of obtaining the observed sample.

These methods are for samples having a normally distributed parent population or that can be made normal by a suitable transformation. Having to make a transformation is not really a limitation of the method since, in theory, any set of data can be transformed to obtain normality. The questions of deciding whether to make a transformation and, if so, how to make the transformation are discussed.

COHEN'S MAXIMUM LIKELIHOOD ESTIMATOR METHOD

A sample of n observations is subject to the restriction that full measurement of the variable is possible only if $y \geq y_c$, where y_c is a known and fixed point of censoring. In our application, y_c is the method detection limit (MDL), and it is assumed that the same MDL applies to each observation. Of the total of n observations in the sample, n_c observations have $y \leq y_c$ and are censored. The number of observations with $y > y_c$ is $k = n - n_c$. The fraction of censored data is $h = n_c/n$. Cohen's method begins by using the n fully measured observations (i.e., the k non-censored data values) to calculate a crude estimate of the mean and variance using the noncensored data using

$$\bar{y} = \Sigma y_i / k$$

and

$$s^2 = \Sigma(y_i - \bar{y})^2 / k$$

Note that s^2 is calculated using a denominator of k and not $k - 1$. For the case where the left tail of the distribution is truncated, this estimate of the mean will be too high, and the estimate of the variance will be too low.

These estimates are adjusted using the factor λ. The maximum likelihood estimates (MLE) for the mean, η, and the variance, σ^2, are

$$\hat{\eta} = \bar{y} - \hat{\lambda}(\bar{y} - y_c)$$
$$\hat{\sigma}^2 = s^2 + \hat{\lambda}(\bar{y} - y_c)^2$$

The ^ symbol indicates that the value is an estimate; for example, $\hat{\eta}$ is an estimate of the true mean η.

The standardized distance between the mean and the censoring point (also called the "terminus") y_c is defined as $\xi = (y_c - \eta)/\sigma$. Cohen (1961) provided tables of $\hat{\lambda}$ as a function of h and $\gamma = s^2/(\bar{y} - y_c)^2$. A portion of his table is reproduced in Table 11.1. Haas and Scheff (1990) developed the following power series expansion, which they say can estimate $\hat{\lambda}$ to within 6% relative error of the tabulated values.

$$\ln(\hat{\lambda}) = 0.182344 - 0.3756/(\gamma+1) + 0.10017\gamma + 0.78079\omega - 0.00581\gamma^2 - 0.06642\omega^2$$
$$- 0.0234\ \gamma\omega + 0.000174\ \gamma^3 + 0.001663\gamma^2\omega - 0.00086\gamma\omega^2 - 0.00653\omega^3$$

where γ defined above and $\omega = \ln [h/(1 - h)]$.

Example 11.1. Using Cohen's method, estimate the mean and variance of the censored sample from Chapter 10, Table 10.1. The sample of n = 27 observations includes 23 measured and 4 censored values. The censoring limit is 6 μg/L. The 23 measured (uncensored) values are 6.9, 7.8, 8.9, 7.7, 9.6, 8.7, 6.7, 8.0, 8.5, 6.5, 9.2, 7.4, 6.3, 7.3, 8.3, 7.2, 7.5, 6.1, 9.4, 7.6, 8.1, 7.9, and 10.1. The four censored values are, of course, unknown. The average and variance of the k = 23 observations with measurable values are

$$\bar{y} = \Sigma y_i /k = 7.9\ \mu g/L$$

and

$$s^2 = \Sigma(y_i - \bar{y})^2/k = 1.1078\ (\mu g/L)^2$$

The limit of censoring is at $y_c = 6$, and the proportion of censored data is

$$h = (27 - 23)/27 = 0.1481,$$

which gives

$$\gamma = s^2/(\bar{y} - y_c)^2 = 1.1078/(7.9 - 6.0)^2 = 0.30687$$

The value of the adjustment factor $\hat{\lambda}$ is found by interpolating Table 11.1 to find $\hat{\lambda} = 0.20392$. The estimates of the mean and variance are

$$\hat{\eta} = \bar{y} - \hat{\lambda}(\bar{y} - y_c) = 7.9 - 0.20392\ (7.9 - 6.0) = 7.52\ \mu g/L$$
$$\hat{\sigma}^2 = s^2 + \hat{\lambda}(\bar{y} - y_c)^2 = 1.1078 + 0.20392\ (7.9 - 6.0)^2 = 1.844.$$
$$\sigma = 1.36\ \mu g/L$$

Table 11.1   **Cohen's $\hat{\lambda}$, as a Function of $h = n_c/n$ and $\gamma = s^2/(\bar{y} - y_c)^2$**

γ\h	0.1	0.2	0.3	0.4	0.5
0.2	0.12469	0.27031	0.4422	0.6483	0.9012
0.3	0.13059	0.28193	0.4595	0.6713	0.9300
0.4	0.13595	0.29260	0.4755	0.6927	0.9570
0.5	0.14090	0.30253	0.4904	0.7129	0.9826
0.6	0.14552	0.31184	0.5045	0.7320	1.007

Data from Cohen (1961).

Notice that the censoring affects the estimates of both $\hat{\eta}$ and $\hat{\sigma}^2$. This means that these two statistics are not estimated independently (as they would be in the case without censoring). The result is that we cannot compute confidence intervals in the usual way. Cohen (1961) provides tables and formulas to correct for this dependency, and describes how to compute confidence intervals. See Cohen's paper for example computations.

The method described above can be used to analyze lognormally distributed data by transforming the data to make them normal. The calculations are done on the transformed values. That is, if x is distributed lognormally, then $y = \ln(x)$ has a normal distribution. The computations are the same as illustrated above, except that they are done on the log-transformed values instead of on the original measured values.

Example 11.2. Use Cohen's method to estimate the mean and variance of a censored sample of n = 30 that is lognormally distributed. The 30 individual values ($x_1 \; x_2, \ldots, x_{30}$) are not given, but assume that a logarithmic transformation, $y = \ln(x)$, will make the distribution normal with mean η_y and variance σ_y^2.

Of the 30 measurements, 12 fall below MDL of 18 $\mu g/L$, so the fraction censored is h = 12/30 = 0.40. The mean and variance computed from the logarithms of the noncensored values are

$$\bar{y} = 3.2722 \quad \text{and} \quad s^2 = 0.03904$$

The limit of censoring is also transformed:

$$y_c = \ln(x_c) = \ln(18) = 2.8904$$

Using these values, we compute

$$\gamma = s^2/(\bar{y} - y_c)^2 = 0.03904/(3.2722 - 2.8904)^2 = 0.2678$$

which is used with h = 0.4 to look up λ = 0.664 in Table 11.1. The estimated mean and variance of the log-transformed values are

$$\hat{\eta}_y = 3.2722 - 0.664 \, (3.2722 - 2.8904) = 3.0187$$
$$\hat{\sigma}_y^2 = 0.03904 + 0.664 \, (3.2722 - 2.8904)^2 = 0.1358, \text{ giving } \hat{\sigma}_y = 0.37.$$

We wish to convert these into estimates of the mean and variance of the original data (the untransformed x's). The appropriate transformation equations are

$$\hat{\eta}_x = \exp(\hat{\eta}_y + 0.5 \, \hat{\sigma}_y^2)$$
$$\hat{\sigma}_x^2 = \hat{\eta}_y^2 \exp(\hat{\sigma}_y^2 - 1)$$

Substituting the parameter estimates $\hat{\eta}_y$ and $\hat{\sigma}_y^2$,

$$\hat{\eta}_x = \exp[3.0187 + 0.5 \, (0.1358)] = \exp(3.0866) = 21.9 \mu g/L$$
$$\hat{\sigma}_x^2 = (21.90)^2 \exp(0.1358 - 1) = 479.2 \, (0.4214) = 202.1, \text{ giving } \hat{\sigma}_x = 14.7 \mu g/L$$

COMMENTS

The problem of censored data arises because a limit of detection is established for analytical measurements. This practice is understandable from the point of view of the chemist, but for the data analyst it would be preferable to have numerical values always reported, even if the measurement error is large relative to the value itself. Even if this practice could be encouraged in the future, there are many important data sets that have already been censored and which must be analyzed.

The precision of the estimated mean and variances become progressively worse as the fraction of observations censored is increased. Judging from the results of several comparative studies on simulated data, Cohen's method seems to work quite well up to 20% censoring. As the extent of censoring reaches 20–50%, the estimates suffer in terms of increased bias and variability.

There have been some well-organized studies to compare the various methods of estimating means of censored data sets (Gilliom and Helsel, 1986; Haas and Scheff, 1990; Newman et al., 1989). They showed that Cohen's MLE method was satisfactory, and regression of normal scores (Chapter 10) was effective. Several other methods were investigated, and on some samples these methods performed better than the method discussed in this chapter.

The method of Cohen (maximum likelihood estimation) is easy to use for data that have a normal or lognormal distribution. Many sets of environmental samples are lognormal, or at least closely approximated by the lognormal, and the log transformation frequently will be needed. Failing to transform the data when they are skewed causes serious bias in the estimates of the mean.

The normal and lognormal distributions have been used often because of (1) limited data which cannot justify any other model, (2) traditional faith in the normal and lognormal models, and (3) the lack of a readily available method to test various models. It is not always possible to correctly identify the true distribution of a small sample ($n = 20$–50), which is the size of many data sets. Hahn and Shapiro (1967) showed this graphically, and Shumway et al. (1989) have shown it using simulated data sets. They have also shown that when we are unsure of the correct distribution, making the log transformation is usually beneficial or, at worst, harmless.

REFERENCES

Cohen, A. C., Jr. (1959). "Simplified Estimators for the Normal Distribution when Samples are Singly Censored or Truncated," *Technometrics*, 1, 217–237.

Cohen, A. C., Jr. (1961). "Tables for Maximum Likelihood Estimates: Singly Truncated and Singly Censored Samples," *Technometrics*, 3, 535–541.

Cohen, A. C. (1979). "Progressively Censored Sampling in the Three Parameter Log-Normal Distribution," *Technometrics*, 18, 99–103.

Gilbert, R. O. (1987). *Statistical Methods for Environmental Pollution Monitoring*, New York, Van Nostrand Reinhold.

Gilliom, R. J. and D. R. Helsel (1986). "Estimation of Distribution Parameters for Censored Trace Level Water Quality Data: 1. Estimation Techniques," *Water Resour. Res.*, 22, 135–146.

Haas, C. N. and P. A. Scheff (1990). "Estimation of Averages in Truncated Samples," *Environ. Sci. Technol.*, 24, 912–919.

Hahn, G. A. and S. S. Shapiro (1967). *Statistical Methods for Engineers*, New York, John Wiley & Sons.

Helsel, D. R. and R. J. Gilliom (1986). "Estimation of Distribution Parameters for Censored Trace Level Water Quality Data: 2. Verification and Applications," *Water Resour. Res.*, 22, 146–155.

NCASI (1991). "Estimating the Mean of Data Sets that Include Measurements Below the Limit of Detection," Technical Bulletin No. 621.

Newman, M. C. and P. M. Dixon (1990). "UNCENSOR: A Program to Estimate Means and Standard Deviations for Data Sets with Below Detection Limit Observations," *Anal. Chem.*, 26, 4, 26–30.

Newman, M. C., P. M. Dixon, B. B. Looney, and J. E. Pinder (1989). "Estimating Means and Variance for Environmental Samples with Below Detection Limit Observations," *Water Resour. Bull.*, 25, 4, 905–916.

Shumway, R. H., A. S. Azari, and P. Johnson (1989). "Estimating Mean Concentrations Under Transformation for Environmental Data with Detection Limits," *Technometrics*, 31, 3, 347–356.

Chapter 12

Assessing Conformance with a Standard

Key words: t test, hypothesis test, confidence interval, dissolved oxygen, lead

---

We are sometimes concerned with comparing measured values with a known value. The application might be a quality control check on laboratory work in which several laboratories have been sent specimens that contain a substance in a precisely known amount. A prescribed analytical method is used, and the results are analyzed to determine whether the measured values and the known concentration of the standard specimens are in agreement. This kind of checking on laboratory work must be done regularly. A related question may be whether the average of some measurements on groundwater quality is greater than a standard value established by a regulatory agency.

How can a value be "known"? It may be known because it is a primary standard provided by a certified laboratory. The government has primary standards for the kilogram, the meter, and other physical measurements, and we can obtain weights and measures that have been carefully calibrated against these primary standards in a certified way that is supposed to eliminate bias. It is possible in a chemical laboratory to prepare solutions according to precise recipes and procedures that are known as *primary standards*. The U.S. Environmental Protection Agency (EPA) can provide such standards for almost all quantities that are measured by environmental laboratories. A value may be known because of calibration against a standard of known properties. For example, a spectrophotometer wavelength scale can be calibrated with the aid of standard light sources, and the absorbance scales can be calibrated with standard filters. This kind of calibration against primary standards gives one great confidence in the stated values and as a practical matter we tend to say that the quantities are "known."

The difference between the measured values and the standard value is examined to decide whether we can confidently declare the difference to be positive, negative, or whether the difference is so small that we are uncertain about the direction of the difference. The standard procedure for making such comparisons is to construct a *null hypothesis* which is tested statistically using a t-test. The classical null hypothesis is "the difference is zero."

No scientist or engineer ever believes this hypothesis to be strictly true. There will always be a difference at some decimal place. Why propose what we do not believe is true? The answer is a philosophical one. We cannot prove equality, but we may collect data that shows a difference so large that it is unlikely to arise from chance. The null hypothesis, therefore, is an artifice for letting us conclude, at some stated level of confidence, that there is a difference. If no difference is evident, we should state that, "The evidence at hand does not permit me to state with a high degree of confidence that the measurements and the standard are different."

An alternate, but equivalent, approach to constructing a null hypothesis is to compute the difference and also the interval in which the difference is expected to fall if the experiment were repeated many, many times. This interval is called the *confidence interval*. For example, the value of a primary standard is 7.0 and the average of several measurements is 7.2, giving a difference of 0.20. Suppose that the results further show that the 95% confidence interval of the true difference is 0.12 to 0.28. This is what we want to know. A confidence interval is more direct and often less confusing than null hypotheses and significance tests. In this book, we often prefer to compute the confidence interval instead of making significance tests. In this chapter, we will try to illustrate the equivalence of the two approaches to decision-making.

CASE STUDY — INTERLABORATORY STUDY OF DO MEASUREMENTS

This example is loosely based on a study by Wilcock et al. (1981). Fourteen laboratories were sent samples that contained 1.2 mg/L dissolved oxygen (DO) and were asked to measure the DO concentration using the Winkler titration method. The concentrations, as milligrams per liter per DO, reported by the participating laboratories are 1.2, 1.4, 1.4, 1.3, 1.2, 1.35, 1.4, 2.0, 1.95, 1.1, 1.75, 1.05, 1.05, and 1.4. Do the laboratories, on average, measure the 1.2 mg/L or is there some bias?

THEORY: t-TEST TO ASSESS AGREEMENT WITH A KNOWN VALUE

The known value of a population mean is denoted by η_0. If the true mean value of a set of tested specimens is η, then the expected value of measured concentrations will estimate η. We wish to test whether η, the mean estimated by $\bar{y}$ is equal to the known value, η_0. This comparison can be expressed as the null hypothesis

$$H_0 : \eta - \eta_0 = 0$$

which is read "the expected difference between η and η_0 is zero." The "null" is the zero. The extent to which $\bar{y}$ differs from η will be due only to random measurement error and not to bias. The extent to which $\bar{y}$ differs from η_0 will be due to both random measurement error and to bias (i.e., $\eta - \eta_0$). We hypothesize the bias to be zero, and test for evidence to the contrary.

The sample average is

$$\bar{y} = \frac{\Sigma y_i}{n}$$

the sample variance is

$$s^2 = \frac{\Sigma(y_i - \bar{y})^2}{n - 1}$$

and the standard error of the mean is estimated by

$$s_{\bar{y}} = \frac{s}{\sqrt{n}}$$

The t statistic is constructed, assuming the null hypothesis to be true, i.e., $\eta = \eta_0$.

$$t_0 = \frac{\bar{y} - \eta_0}{s_{\bar{y}}}$$

On the assumption of random sampling from a normal distribution, t_0 will have a t distribution with $\nu = n - 1$ degrees of freedom. Notice that t_0 may be positive or negative, depending upon whether $\bar{y}$ is greater or less than η_0.

If we are interested in a one-sided test — that is, $\eta > \eta_0$ (or $\eta < \eta_0$) — the null hypothesis is rejected if the absolute value of the calculated t_0 is greater than (or less than) the value of Student's t statistic, $t_{\nu,\alpha}$, where α is the selected probability point of the t distribution and $\nu = n - 1$ degrees of freedom. If we are interested in the case η may be either greater than or less than $\eta_{\alpha0}$, the null hypothesis is rejected if the absolute value of the calculated t_0 is greater than the value of Student's t statistic, $t_{v\alpha/2}$, where the total probability α is divided between the two ends of the t distribution and $v = n - 1$ degrees of freedom.

The $(1 - \alpha)100\%$ confidence interval for the difference $\bar{y} - \eta_0$ is constructed using Student's t distribution as follows

$$t_{v,\alpha/2} \cdot S_{\bar{y}} < \bar{y} - \eta_0 < + t_{v\alpha/2} \cdot S_{\bar{y}}$$

If this confidence interval does not include zero, the difference between the known and measured values is large enough that it is unlikely to arise from chance, and it is concluded that there is a difference between the measured mean $\bar{y}$ and the known value η_0.

A similar confidence interval can be defined for the mean population

$$\bar{y} - t_{v\alpha/2} \cdot S_{\bar{y}} < \eta < \bar{y} + t_{v\alpha/2} \cdot S_{\bar{y}}$$

If the known value, η_0 falls outside this interval it is declared to be different from the population mean, η, as estimated by $\bar{y}$.

SOLUTION

The concentration of the standard specimens that were analyzed by the participating laboratories was 1.2 mg/L. This value presumably was known with such accuracy that it was considered to be the true mean value η_0. The average concentration of the 14 measured DO concentrations is $\bar{y} = 1.4$ mg/L, the standard deviation is s = 0.30 mg/L, and the standard error of the mean is $s_{\bar{y}} = 0.08$ mg/L. The difference between the known concentration of 1.2 mg/L and the average measured concentration of 1.4 mg/L seems large, but the variability in the measured values is also large. A t-test can be used to assess whether this difference is so exceptionally large as to be unlikely to occur through chance.

The test t statistic is $t_0 = (1.4 - 1.2)/0.08 = 2.5$. This is compared with the t distribution with 13 degrees of freedom, which is shown in Figure 12.1a. The values t = -2.16 and t = +2.16 that cut off 5% of the area under the bell-shaped curve are shaded. Notice that the 5% comprises 2.5% on the upper tail and 2.5% on the lower tail of the distribution. The test value of $t_0 = 2.5$, located by the arrow, falls outside this range and therefore is considered to be exceptionally large. It is highly unlikely (less than 5%) that such a result would occur by chance. In statistical jargon, this means that "the null hypothesis is rejected." In engineering jargon, this means that "there is strong evidence that the measurement method used in these laboratories gives results that are too high."

Now we look at the equivalent interpretation using a 95% confidence interval, which is constructed using t = 2.16 for $\alpha = 0.05/2$ and $v = 13$. The 95% confidence interval for the difference, which has expected value zero under the null hypothesis, is $0 \pm 2.16(0.08) = \pm0.17$ mg/L. The portion of the reference distribution for the difference that falls outside this range is shaded in Figure 12.1b. The observed difference $(\bar{y} - \eta_0)$ of 0.2 mg/L falls in the shaded area, which means it is beyond the 95% confidence limits. The difference is so large that it is unlikely to occur due to pure random variation in the measurement process. "Unlikely" in this case means "a probability of 5% that a difference this large could occur due to random measurement variation."

Figure 12.1c is the reference distribution that shows the expected variation of the sample average about the true mean. It also shows the 95% confidence interval for the average. If the average is estimating the true concentration of 1.2, it is expected to vary within the range of $1.2 \pm 2.16(0.08) = 1.2 \pm 0.17$. The lower bound of the 95% confidence interval is 1.03, and the upper bound is 1.37. The measured average value

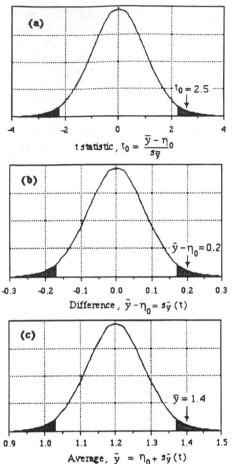

$$t \text{ statistic}, t_0 = \frac{\bar{y} - \eta_0}{s_{\bar{y}}}$$

$$\text{Difference}, \bar{y} - \eta_0 = s_{\bar{y}} (t)$$

$$\text{Average}, \bar{y} = \eta_0 + s_{\bar{y}} (t)$$

Figure 12.1 Three equivalent reference distributions scaled to compare the observed average with the known value on the basis of the distribution of the (a) t statistic, (b) difference between the observed average and the known level, and (c) the sample average. The distributions were constructed using $\eta_0 = 1.2$ mg/L, $\bar{y} = 1.4$ mg/L, $t_{13,0.029} = 2.16$, and $s_{\bar{y}} = 0.08$ mg/L.

of 1.4 mg/L does not fall within the 95% confidence interval, which leads us to conclude that the concentration is higher than the true mean of the known value of 1.2 mg/L.

Alternatively, a 95% confidence interval for the true mean η can be computed from $\bar{y} \pm t_{\nu,\alpha/2} \, s_{\bar{y}} = 1.4 \pm 2.16(0.08) = 1.4 \pm 0.17$ or 1.23 to 1.57. The known value $\eta_0 = 1.2$ mg/L, is not in this interval, again leading to the conclusion that the true concentration is greater than 1.2 mg/L.

By now the reader will have noticed that the shapes of the three reference distributions are identical. The only difference is the scaling of the horizontal axis, whether we choose to consider the difference in terms of the t statistic, the difference, or the concentration scale. Many engineers will prefer to make this judgment on the basis of a value scaled as the measured values are scaled, for example, as milligrams per liter instead of on the dimensionless scale of the t statistic. This is easily done by computing the confidence intervals for the difference or for the average.

COMMENTS

The conclusion that the measured concentrations are higher than the "known" concentration of 1.2 mg/L could be viewed in two ways. If the concentration in the standard specimens is truly 1.2 mg/L, then the measurement method is biased. This could result from the high concentrations of 2.0 and 1.95 mg/L measured by two laboratories. How do we learn whether this is reasonable? Send out more standard specimens and ask them to try again.

On the other hand, the measurement method might be all right and the true concentration might be higher than 1.2 mg/L. This experiment does not tell us which interpretation is correct. It is not a simple matter to make a standard solution for dissolved oxygen. Dissolved oxygen can be consumed in a variety of reactions. Also, its concentration can change upon exposure to air when the specimen bottle is opened in the laboratory. In contrast, a substance like chloride or zinc will not be lost from the standard specimen, so it is relatively easy to know that the concentration actually delivered to the chemist who makes the measurements is the same concentration contained in the specimen that was shipped. In the case of oxygen at low levels such as 1.2 mg/L, it is not likely that oxygen would be lost from the specimen during handling in the laboratory. If there is a change, the oxygen concentration is more likely to be increased by dissolution of oxygen from the air. We cannot rule out this causing the difference between 1.4 mg/L measured and 1.2 mg/L in the original standard specimens. Nevertheless, the chemists who arranged the test believed they had found a way to prepare stable test specimens, and they were experienced in preparing standards for interlaboratory tests. We have no reason to doubt them. Some more checking of the laboratories seems the best line of action.

REFERENCES

Wilcock, R. J., C. D. Stevenson, and C. A. Roberts (1981). "An Interlaboratory Study of Dissolved Oxygen in Water," *Water Res.*, 15, 321–325.

Chapter 13

Assessing the Average of Differences

Key words: confidence intervals, independent t-test, interlaboratory tests, null hypothesis, t-test, dissolved oxygen, pooled variance

---

A common question is, "Do two different methods of doing A give different results?" For example, two methods for making a chemical analysis are compared to see if the new one is equivalent to the older standard method: algae are grown under different conditions to study a factor that is thought to stimulate growth or two waste treatment processes are tested at different levels of stress caused by a toxic input. In the strict sense, we do not believe that the two analytical methods or the two treatment processes are really identical. What we are really asking then is a question like, "How large might the difference be?" or perhaps "Can we be highly confident that the difference is positive or negative?"

A key idea is that the *design of the experiment* determines the way we compare the two treatments. One experimental design is to make a series of tests using method A and then to independently make a series of tests using method B. Because the data on methods A and B are independent of each other, they are compared by computing the average for each treatment and using an *independent t-test* to assess the difference of the two averages. A second way of designing the experiment is to pair the samples according to time, technician, batch of material, or other factors that might contribute to a difference between the two measurements. Now the test results on methods A and B are produced in pairs that are not independent of each other, so the analysis is done by averaging the differences for each pair of test result. Then a *paired t-test* is used to assess whether the average of these differences is different from zero.

Two samples are said to be paired when each data point in the first sample is matched and related to a unique data point in the second sample. Paired experiments are used when it is difficult to control all factors that might influence the outcome. If these factors cannot be controlled, the experiment is arranged so they are equally likely to influence both of the paired observations.

Paired experiments could be used, for example, to compare two analytical methods. Suppose that the comparison was to be made on specimens collected at a wastewater treatment plant. The influent quality will change from moment to moment. To eliminate variation in influent quality as a factor in the comparative experiment, paired measurements using both analytical methods could be made on the same sample specimen of wastewater. The alternative approach of using method A on wastewater collected on day one and then using method B on wastewater collected at some later time would be inferior because the difference due to analytical method would be confounded, and probably overwhelmed, by day-to-day differences in wastewater quality. The difference between paired same-day tests is not influenced by day-to-day variation. Paired tests are evaluated using the paired t-test, which assesses the average of the paired differences.

To summarize, the test statistic that will be used to compare two treatments is as follows: when assessing *the difference of two averages*, we will use the *independent t-test*; when comparing *the average of paired differences*, we will use the *paired t-test*. Which method is used depends on the design of the experiment. You know which method will be used *before* the data are collected.

Once the appropriate difference has been computed, it is examined to decide whether we can confidently declare the difference to be positive, negative, or whether the difference is so small that we are uncertain about the direction of the difference. The standard procedure for making such comparisons is to construct a *null hypothesis*, which is tested statistically using a t-test. The classical null hypothesis is "the difference between the two methods is zero." We do not expect the two methods to give exactly the same results, so it may seem strange to investigate a hypothesis that is certainly wrong. The philosophy is the same as in law, where the accused is presumed innocent until proven guilty. We cannot prove a person innocent, which is why the verdict is worded "not guilty" when the evidence is insufficient to convict. In a statistical comparison, we cannot prove that the two methods are the same, but we may collect evidence that shows them to be different. The null hypothesis, therefore, is a philosophical device for letting us avoid saying that two things are equal. Instead we conclude, at some stated level of confidence, that "there is a difference" or that "the evidence does not permit me to state with a high degree of confidence that the two methods are different."

An alternate, but equivalent, approach to constructing a null hypothesis is to compute the difference and also the interval in which the difference is expected to fall if the experiment were repeated many, many times. This interval is called the *confidence interval*. For example, we may determine that "A – B = 0.20 and that the true difference falls in the interval 0.12 to 0.28, this statement being made at a 95% level of confidence." This tells us all that is important. We are highly confident that A gives a result that is on average higher than B. And, it tells all this without the sometimes confusing notions of null hypothesis and significance tests.

CASE STUDY — INTERLABORATORY STUDY OF DISSOLVED OXYGEN

An important procedure in certifying the quality of work done in laboratories is the analysis of specimens that contain known amounts of a substance. These specimens are usually introduced into the laboratory routine in a way that keeps the analysts "blind" to the identity of the sample. Often the analyst is blind to the fact that quality assurance samples are included in the assigned work.

Several laboratories were sent a test solution that was prepared to have a low DO concentration (1.2 mg/L) and asked to measure the dissolved oxygen (DO) concentration.[1] Fourteen laboratories made the measurements using the Winkler method (a titration), and 14 others used the electrode method. The question is whether the two methods predict different DO concentrations. Table 13.1 shows a portion of the data. The observations for each method may be assumed random and independent as a result of the way the test was designed. The differences plotted in Figure 13.1 suggest that the Winkler method may give DO measurements that are slightly lower than the electrode method.

The data analysis problem is to compare results obtained with each measurement method against the known concentration and also to compare the two measurement methods that were used. Comparing a mean with a known value was illustrated in Example 2.9 and also in Chapter 12. Here the problem of comparing two different measurement techniques (treatments) is discussed.

Table 13.1 Dissolved Oxygen Data from the Interlaboratory Study

Winkler	1.2	1.4	1.4	1.3	1.2	1.3	1.4	2.0	1.9	1.1	1.8	1.0	1.1	1.4
Electrode	1.6	1.4	1.9	2.3	1.7	1.3	2.2	1.4	1.3	1.7	1.9	1.8	1.8	1.8
Diff. (W – E)	–0.4	0.0	–0.5	–1.0	–0.5	0.0	–0.8	0.6	0.6	–0.6	–0.1	–0.8	–0.7	–0.4

Data from Wilcock et al. (1981).

---

[1] This example is loosely based on a study done by Wilcock et al. (1981).

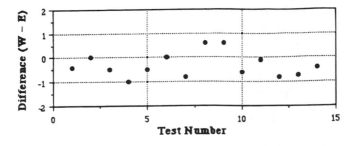

Figure 13.1 The DO data and the differences of the paired values.

THEORY — THE PAIRED t-TEST ANALYSIS

Define δ as the true mean of differences between random variables y_1 and y_2 that were observed within matched pairs of experimental conditions. δ will be zero if the populations from which y_1 and y_2 are drawn are equal. The estimate of δ is $\bar{d}$, the average of differences between paired observations, where

$$\bar{d} = \frac{\Sigma d_i}{n} = \frac{1}{n}\Sigma(y_{1,i} - y_{2,i})$$

Because of measurement error, the value of $\bar{d}$ computed from a particular sample is not likely to be zero, though it will tend toward zero if δ is zero.

The sample variance is

$$s_d^2 = \frac{\Sigma(d_i - \bar{d})^2}{n - 1}$$

and the standard error of the average difference is

$$s_{\bar{d}} = \frac{s_d}{\sqrt{n}}$$

This is used to establish the $1 - \alpha$ confidence limits for δ, which are $\bar{d} \pm s_d\, t_{\alpha/2, n-1}$.

SOLUTION

The differences were calculated by subtracting the Electrode measurements from the Winkler measurements. The average of the paired differences is

$$\bar{d} = \frac{(-0.4) + 0 + (-0.5) + (-1.0) + \ldots + (-0.4)}{14} = -0.326 \text{ mg/L}$$

and the variance of the paired differences is

$$s_d^2 = \frac{(-0.4)^2 + 0^2 + (-0.5)^2 + (-1.0)^2 + \ldots 1(-0.4)^2}{14 - 1} = 0.244 \text{ (mg/L)}^2$$

giving $s_d = 0.494$ mg/L. The standard error of the average of the paired differences is

$$s_{\bar{d}} = \frac{0.494}{\sqrt{14}} = 0.132 \text{ mg/L}$$

The $(1 - \alpha)100\%$ confidence interval is computed using the t distribution with 13 degrees of freedom at the $\alpha/2$ probability point. For a 95% confidence interval, $t_{\nu = 13, \alpha/2 = 0.025} = 2.160$, and the 95% confidence interval of the true difference δ is

$$\bar{d} - t_{13,0.025} \cdot s_{\bar{d}} < \delta < \bar{d} + t_{13,0.025} \cdot s_{\bar{d}}$$

and for the particular values of this example,

$$-0.326 - (2.160)(0.132) < \delta < -0.326 + (2.160)(0.132)$$

$$-0.65 \text{ mg/L} < \delta < -0.04 \text{ mg/L}$$

We are highly confident that the difference between the two methods is not zero because the confidence interval does not include the difference of zero. The methods give different results, and furthermore, the electrode method has given higher readings than the Winkler method.

If the confidence interval had included "zero," the interpretation would be that "we cannot say with a high degree of confidence that the methods are different." We should be reluctant to report that "the methods are the same" or that "the difference between the methods is zero" because what we know about chemical measurements makes it unlikely that these statements are strictly correct. We may decide that the difference is small enough to have no practical importance. Or, the range of the confidence interval might be large enough that the difference, if real, is important, in which case additional tests should be done to resolve the matter.

An alternate method of interpreting the results is to formulate and test the null hypothesis that "the difference between the two averages is zero." The way of stating the conclusion when the 95% confidence interval does not include zero is to say that "the difference was significant at the 95% confidence level." *Significant*, in this context, has a purely statistical meaning. It conveys nothing about how interesting or important the difference will be to an engineer or chemist. Rather than reporting that "the difference was significant" (or not significant), communicate the conclusion more simply and directly by giving the confidence interval. Some reasons for preferring to look at the confidence interval and not do a significance test are given at the end of this chapter.

CASE STUDY TO EMPHASIZE THE BENEFITS OF A PAIRED DESIGN

Paired experiments are used when it is difficult to control all factors that might influence the outcome. A paired experimental design insures that the uncontrolled factors contribute equally to both of the paired observations. The difference between the paired values is unaffected by the uncontrolled disturbances, whereas the differences of unpaired tests would reflect this additional component of experimental error. The following example shows how a large seasonal effect can be blocked out by the paired design.

A once-through cooling system at a power plant is suspected of reducing the population of certain aquatic organisms. The copepod and copepod nauplii population densities (as number of organisms per cubic meter) were measured at the inlet and outlet of the cooling system on 17 different days (Simpson and Dudaitis, 1981). On each sampling day, water specimens were collected with a short time interval. The sampling plan represents a thoughtful effort to block out the effect of day-to-day and month-to-month variations in population counts. It "pairs" the inlet and outlet measurements. Of course, it is impossible to sample the same parcel of water at the inlet and outlet (i.e., the pairing is not exact), but any variation caused by this will be reflected as a component of the random measurement error.

The data are plotted in Figure 13.2. Copepod nauplii definitely are reduced (we don't need a statistical test to known whether we've been run over by a truck or a mouse), but the plot gives the impression that the cooling system may not have reduced the copepod populations. A statistical test might support this impression. Before doing the calculations, consider once more why the paired comparison should be done.

In the sample in Figure 13.2, specimens 1–6 were taken in November 1977, specimens 7–12 in February 1978, and specimens 13–17 in August 1978. A large seasonal variation is apparent. If we were to compute the variance of the inlet and outlet with respect to the average inlet and outlet levels, it would be huge and would consist largely of variations due to seasonal differences. Since we are not trying to evaluate seasonal differences, this would be a poor way to analyze the data. The paired comparison operates on the differences of the daily inlet and outlet counts, and these differences do not reflect the seasonal variation (except, as we shall see in a moment, to the extent that the differences are proportional to the population density).

It is tempting to tell ourselves that, "I wouldn't be foolish enough not to do a paired comparison on data such as these." Of course, we wouldn't when the variation due to the nuisance factor (season, in this case) is both huge and obvious. But, almost every experiment is at risk of being influenced by one or more nuisance factors, and even the most careful experimental technique cannot guarantee that these will not alter the outcome. The paired experimental design will prevent this, and we recommend its use whenever the experiment can be so arranged.

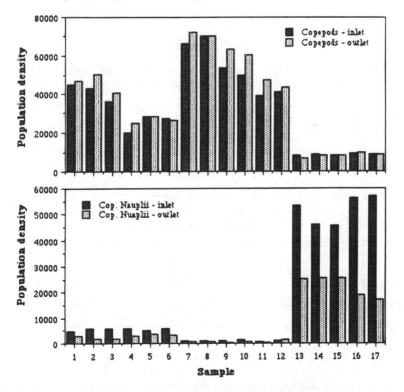

Figure 13.2 Copepod and copepod nauplii population densities (organisms/m³).

Biological counts often need to be transformed to make the variance uniform over the observed range of values. We will keep this in mind as we proceed. Plotting the data is always important because it brings to our attention critical features of the data. The differences of the inlet and outlet counts are plotted in Figure 13.3. Clearly, the differences are larger when the counts are larger, which means that the variance is not constant over the range of population counts observed. Constant variance is one condition of the t-test, since we want each observation to contribute in equal weight to the analysis. Any statistics computed from these data would be dominated by the large differences of the high population counts, and it would be misleading to construct a confidence interval or test a null hypothesis using a mean and variance estimated from the data in their original form. A transformation is needed to make the variance constant over the tenfold range of the counts in the sample. A square root transformation is often used on biological counts (Sokal and Rohlf, 1969), but for these data a log transformation seemed to be better. The bottom section of Figure 13.3 shows that the differences of the log-transformed data are reasonably uniform over the range of the transformed values.

Table 13.2 shows the data, the log-transformed data, and their differences. Base e logarithms were used. The average difference of the log(input) – log(output) is $\bar{d} = \Sigma d_i / 17 = -0.022$. The variance of differences is $s^2 = \Sigma (d_i - \bar{d})^2 / 16 = 0.003$, and the standard error of average difference is $s_{\bar{d}} = s/\sqrt{17} = 0.013$.

The 95% confidence interval is constructed using $t_{\nu=16, \alpha=0.05/2} = 2.12$, which gives

$$\bar{d} \pm t_{16, 0.025} \, s_{\bar{d}} = 0.022 \pm 2.12 \, (0.013)$$

The interpretation is that with 95% confidence, the true difference falls in the region $-0.048 < \delta < 0.005$. This confidence interval includes zero, so we cannot state with a high degree of confidence that outlet levels are different than inlet levels.

COMMENTS

The paired t-test compares the average of the differences of two samples. This is not equivalent to comparing the difference of the average of two samples. Pairing the data

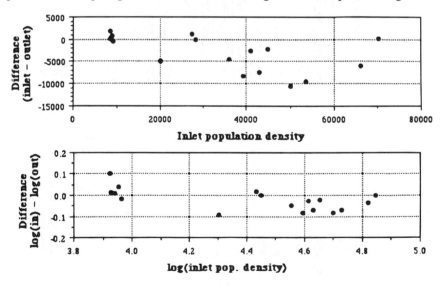

Figure 13.3 The difference in copepod inlet and outlet population density is larger when the population is large, indicating nonconstant variance at different population levels.

Table 13.2 Outline of Computations for a Paired t-Test on the Copepod Data after a Logarithmic Transformation

Sample	Original Counts (no./m^3)			Transformed Data, $z = \log_{10}(y)$		
	$y_{in,i}$	$y_{out,i}$	$d_i = y_{in,i} - y_{out,i}$	$z_{in,i}$	$z_{out,i}$	$d_i = z_{in,i} - z_{out,i}$
1	44909	47069	−2160	10.712	10.759	−0.047
2	42858	50301	−7443	10.666	10.826	−0.160
3	35976	40431	−4455	10.491	10.607	−0.117
4	20048	24887	−4839	9.906	10.122	−0.216
5	28273	28385	−112	10.250	10.254	−0.004
6	27261	26122	1139	10.213	10.171	0.043
7	66149	72039	−5890	11.100	11.185	−0.085
8	70190	70039	151	11.159	11.157	0.002
9	53611	63228	−9617	10.890	11.055	−0.165
10	49978	60585	−10607	10.819	11.012	−0.192
11	39186	47455	−8269	10.576	10.768	−0.191
12	41074	43584	−2510	10.623	10.682	−0.059
13	8424	6640	1784	9.039	8.801	0.238
14	8995	8244	751	9.104	9.017	0.087
15	8436	8204	232	9.040	9.012	0.028
16	9195	9579	−384	9.126	9.167	−0.041
17	8729	8547	182	9.074	9.053	0.021

by running the modified and unmodified treatments side by side creates a similarity of conditions that reduces experimental error. This experimental design should be able to detect a smaller difference than an unpaired design. We do not have free choice of which t-test to use for a particular set of data. The appropriate test is determined by the experimental design.

We almost never really believe the null hypothesis. It is too much to expect that the difference between any two methods is truly zero. Two methods are always different in some decimal place. Tukey (1991) states this bluntly:

> Statisticians classically asked the wrong question — and were willing to answer with a lie . . . They asked "Are the effects of A and B different?" and they were willing to answer "no."
>
> All we know about the world teaches us that A and B are always different — in some decimal place. Thus asking "Are the effects different?" is foolish.
>
> What we should be answering first is "Can we be confident about the direction from method A to method B? Is it up, down, or uncertain?"

If uncertain whether the direction is up or down, it is better to answer "we are uncertain about the direction" than to say "we reject the null hypothesis." If the answer was "direction uncertain," the follow-up question is how big the difference might be. This question is answered by computing confidence intervals.

Most engineers and scientists probably will like Tukey's view of this problem. If you are one of these, instead of accepting or rejecting a null hypothesis, compute and interpret the confidence interval of the difference. We almost always want to know the confidence interval anyway, so this saves work, while relieving us of having to remember exactly what it means to "accept the null hypothesis." And, it lets us avoid using the word "significant."

To further emphasize this, we quote from Hooke (1963) who identifies some of the inadequacies of significance tests.

1. The test is qualitative rather than quantitative. In dealing with quantitative variables, it is often wasteful to point an entire experiment toward determining the existence of an effect when the effect could also be measured at no extra cost. A confidence statement, when it can be made, contains all the information that a significance statement does and more.

2. The word "significance" often creates misunderstandings, owing to the common habit of omitting the modifier "statistical." Statistical significance merely indicates that evidence of an effect is present, but provides no evidence in deciding whether the effect is large enough to be important. *In a given experiment, statistical significance is neither necessary nor sufficient for scientific or practical importance.*

3. Since statistical significance means only that an effect can be seen in spite of the experimental error (or, a signal is heard above the noise), it is clear that the outcome of an experiment depends very strongly on the sample size. Large samples tend to produce significant results, while small samples fail to do so.

Now, having declared that we prefer not to state results as being significant or non-significant, we pass on two tips from Chatfield (1983) that are well worth remembering:

1. A nonsignificant difference is not necessarily the same thing as no difference.
2. A significant difference is not necessarily the same thing as an interesting difference.

REFERENCES

Chatfield, C. (1983). *Statistics for Technology*, 3rd ed., London, Chapman and Hall.

Hooke, R. (1963). *Introduction to Scientific Inference*, San Francisco, Holden-Day.

Simpson, R. D. and A. Dudaitis (1981). "Changes in the Density of Zooplankton Passing Through the Cooling System of a Power-Generating Plant, *Water Res.*, 15,133–138.

Sokal, R. R. and F. J. Rohlf (1969). *Biometry: The Principles and Practice of Statistics in Biological Research*, New York, W. H. Freeman and Co.

Tukey, J. W. (1991). "The Philosophy of Multiple Comparisons," *Stat. Sci.*, 6, 6, 100–116.

Wilcock, R. J., C. D. Stevenson, and C. A. Roberts (1981). "An Interlaboratory Study of Dissolved Oxygen in Water," *Water Res.*, 15, 321–325.

Chapter 14

Assessing the Difference of Two Averages

Key words: confidence interval, independent t-test, mercury

Two methods, treatments, or conditions are to be compared. The last chapter dealt with the experimental design which produces measurements from two treatments that were paired. Sometimes it is not possible to pair the tests, and then the averages of the two treatments must be compared using the *independent t-test*.

CASE STUDY — MERCURY IN DOMESTIC WASTEWATER

Extremely low limits now exist for mercury in wastewater effluents. It is often thought that whenever the concentration of heavy metals is too high the problem can be corrected by forcing industries to stop discharging the offending substance. It is possible, however, for target effluent concentrations to be so low that they might be exceeded by the concentration in domestic sewage. The mercury concentrations listed in Table 14.1 were measured in Madison, WI in 1989. "CWS" indicates samples taken from a residential area that is served by the city water supply, and "PWS" indicates samples taken from a residential area that is served by private wells. For future studies on mercury concentrations in residential areas, it would be convenient to be able to sample in either neighborhood without having to worry about the water supply affecting the outcome. Is there any difference in the mercury content of the two residential areas?

THE t-TEST TO COMPARE THE AVERAGES OF TWO SAMPLES

Two independently distributed random variables y_a and y_b have, respectively, mean values η_a and η_b and variances σ_a^2 and σ_b^2. The usual statement of the problem is in terms of testing the *null hypothesis* that the difference in the means is zero: $\eta_a - \eta_b = 0$, but we prefer viewing the problem in terms of the confidence interval of the difference.

The expected value of the difference between the averages of the two treatments is

$$E(\bar{y}_a - \bar{y}_b) = \eta_a - \eta_b$$

If the data are from random samples, the variance of the averages $\bar{y}_a$ and $\bar{y}_b$ are

$$V(\bar{y}_a) = \frac{\sigma_a^2}{n_a} \quad \text{and} \quad V(\bar{y}_b) = \frac{\sigma_b^2}{n_b}$$

where n_a and n_b are the sample sizes, and the variance of the difference is

$$V(\bar{y}_a - \bar{y}_b) = \frac{\sigma_a^2}{n_a} + \frac{\sigma_b^2}{n_b}$$

Table 14.1 **Mercury Concentrations in Wastewater Originating in an Area Served by the City Water Supply (CWS) and an Area Served by Private Wells (PWS)**

Source	Mercury Concentrations (µg/L)
CWS (n = 13)	0.34 0.18 0.13 0.09 0.16 0.09 0.16 0.10 0.14 0.26 0.06 0.26 0.07
PWS (n = 10)	0.26 0.06 0.16 0.19 0.32 0.16 0.08 0.05 0.10 0.13

Data provided by the Madison Metropolitan Sewerage District in Wisconsin.

Usually the variances, σ_a^2 and σ_b^2, are unknown and must be estimated from the sample data by computing

$$s_a^2 = \frac{\Sigma(y_a - \bar{y}_a)^2}{n_a - 1} \quad \text{and} \quad s_b^2 = \frac{\Sigma(y_b - \bar{y}_b)^2}{n_b - 1}$$

These estimates can be pooled if the population variances are of equal magnitude. Assuming this to be true, the pooled estimate of the variance is

$$s_{pool}^2 = \frac{(n_a - 1)\, s_a^2 + (n_b - 1)\, s_b^2}{n_a + n_b - 2}$$

This is the weighted average of the sample variances, where the weights are proportional to the degrees of freedom of each variance. In general, the number of observations used to compute each average and variance need not be equal. If the variances are very different in magnitude, it is cause for concern that the data may need to be transformed to achieve uniform variance.

The estimated variance of the difference $\bar{y}_a - \bar{y}_b$ is

$$V(\bar{y}_a - \bar{y}_b) = \frac{s_{pool}^2}{n_a} + \frac{s_{pool}^2}{n_b} = s_{pool}^2 \left(\frac{1}{n_a} + \frac{1}{n_b} \right)$$

and the standard error of $\bar{y}_a - \bar{y}_b$ is the square root of this quantity.

$$s_{\bar{y}_a - \bar{y}_b} = s_{pool} \left(\frac{1}{n_a} + \frac{1}{n_b} \right)^{1/2}$$

This procedure is said to be *robust* to moderate non-normality because the central limit effect will tend to make the distributions of the averages and their difference normal even though the parent distributions of y_a and y_b are not normal. Therefore, we can use Student's t distribution to compute the level confidence interval. To construct the $(1 - \alpha)100\%$ confidence interval, use the t statistic for $\alpha/2$ and $\nu = n_a - n_b - 2$ degrees of freedom.

$$(\bar{y}_a - \bar{y}_b) \pm t_\nu = n_a + n_{b-2,\alpha/2} \left[s_{pool} \left(\frac{1}{n_a} + \frac{1}{n_b} \right)^{1/2} \right]$$

SOLUTION — MERCURY DATA

The MWS and PWS measurements will be identified by subscripts m and p. The averages, variances, standard deviations, and standard errors are

MWS: $n_m = 13$ $\bar{y}_m = 0.157$ $s_m^2 = 0.007$ $s_m = 0.084$ $s_{\bar{y}_m} = 0.023$
PWS: $n_p = 10$ $\bar{y}_p = 0.151$ $s_p^2 = 0.008$ $s_p = 0.028$ $s_{\bar{y}_p} = 0.028$

The difference in the averages of the measurements is $\bar{y}_m - \bar{y}_c = 0.157 - 0.151 = 0.006$ µg/L. The variances of the MWS and PWS samples, s_m^2 and s_p^2, are nearly equal, so it can be assumed that they estimate the same population variance and they can be pooled. The pooled variance is the average of the variances weighted in proportion to their degrees of freedom.

$$s_{pool}^2 = \frac{12(0.007) + 9(0.008)}{12 + 9} = 0.00743 \ (\mu g/L)^2$$

The estimated variance of the difference between averages is

$$V(\bar{y}_m - \bar{y}_p) = s_{\bar{y}_n}^2 + s_{\bar{y}_m}^2 = s_{pool}^2\left(\frac{1}{n_m} + \frac{1}{n_p}\right) = 0.00743\left(\frac{1}{13} + \frac{1}{10}\right) = 0.0013 \ (\mu g/L)^2$$

and the standard error of $\bar{y}_m - \bar{y}_p$ is $\sqrt{0.0013} = 0.036$ µg/L.
    The 95% confidence interval of the difference is

$$(\bar{y}_m - \bar{y}_p) \pm t_{\nu=21,\alpha/2=0.025} \cdot s_{\bar{y}m-\bar{y}p}$$

The variance of the difference is estimated with $\nu = 12 + 9 = 21$ degrees of freedom. For the 95% confidence interval, $\alpha = 0.05/2 = 0.025$, and $t_{21,0.025} = 2.080$. The 95% confidence interval is

$$0.006 \pm 2.080 \ (0.036) \quad \text{or} \quad 0.006 \pm 0.075 \ \mu g/L$$

and we can state, with 95% confidence, that the true difference between the two methods falls in the interval of –0.069 and 0.081 µg/L. This confidence interval includes zero, so there is no persuasive evidence in these data that the mercury concentrations are different in the two residential areas. The difference is not statistically significant. Future sampling can be done in either area without worry that the water supply will affect the outcome.

COMMENTS

The mercury data in the case study example showed that the difference between the average concentration in the two residential neighborhoods was so small that one cannot be highly confident there is any difference. In planning future sampling, therefore, one might proceed as though the neighborhoods are identical, though we understand that this cannot be strictly true. Sometimes a difference is statistically significant, but it is small enough that, in practical terms, we do not care. It is statistically significant, but unimportant. Suppose in the mercury case study that the concentrations in both areas were not a fraction of a microgram per liter, but instead were about 0.15 and 0.17 mg/L, and the difference of 0.2 mg/L was statistically significant. Now both concentrations are so high that our concern is with the absolute level, and the relatively small difference between them is not a major concern. This reminds us that "significance" in the statistical sense and "importance" in the practical sense are two different concepts.

In this chapter, the test statistic used to compare two treatments was the *difference of two averages*, and the comparison was made using the *independent t-test*. Independent, in this context, means that all sources of uncontrollable random variation will equally affect each treatment. For example, specimens tested on different days will reflect variation due to any daily difference in materials or procedures in addition to the random variations that always exist in the measurement process. In contrast, in a *paired t-test* (Chapter 13) is arranged to block out some possible sources of variation, for example, by testing both treatments on the same batch of material on the same day.

Chapter 15

Assessing the Difference of Proportions

Key words: bioassay, binomial distribution, censored data, effluent testing, normal distribution, normal approximation, percentages, proportions, ratios, toxicity, t-test

---

Ratios and proportions arise frequently in biological, epidemiological, and public health studies. We may want to study the proportion of people infected at a given dose of virus, the proportion of rats showing tumors after exposure to a suspected carcinogen, or the incidence rate of leukemia near a contaminated well. Engineers would usually study such problems with help from specialists, but they still need to understand the issues and some of the statistical methods that are used to study them.

A case where engineers will use ratios and proportions is when data samples have been censored by a limit of detection. A data set on an upgradient groundwater monitoring well has 90% of all observations censored and a downgradient well has only 75% censored. Does this difference indicate that contamination has occurred in the groundwater flowing between the two wells? Another example where engineers may be involved directly is in interpreting bioassays to test the toxicity of effluents.

CASE STUDY

Biological assays are a means of determining the toxicity of an effluent. There are many ways such tests might be organized: species of test organism, number of test organisms, how many dilutions of effluent to test, specification of response, physical conditions, and so on. Most of these are biological issues. Here we consider some statistical issues in the context of a simple bioassay test.

Organisms will be subjected to two treatment conditions: (1) they are put into an aquarium containing effluent or (2) they are put into a control aquarium that contains clean water. Equal numbers of organisms are assigned randomly to the control and effluent groups. The experimental response is a binary measure: presence or absence of some characteristic. In an acute bioassay, the characteristic measured is the survival of each organism. In a chronic bioassay, the organisms are exposed to nonlethal conditions, and the measured response might be loss of equilibrium, breathing rate, loss of reproductive capacity, rate of weight gain, formation of neoplasms, and so on.

In our example, 80 organisms ($n_1 = n_2 = 80$) were exposed to each treatment condition (control and effluent), and toxicity was measured in terms of survival. The data shown in Table 15.1 were observed. Are the survival proportions in the two groups so different that we can state with a high degree of certainty that the two treatments truly differ in toxicity?

THE BINOMIAL MODEL

The data from a binomial process consist of two discrete outcomes (i.e., binary). In a bioassay, the test organism is either dead or alive after exposure for the specified time. An effluent is either in compliance or it is not. In a given year, a river floods or it does not flood. The binomial probability distribution gives the probability of observing an event x in a set of n experiments. The experiments are commonly called trials. If x is observed, the trial is said to be successful. "Success" in this statistical sense does not

115

Table 15.1 **Data from a Bioassay on Wastewater Treatment Plant Effluent**

	Number			Percent	
Group	**Surviving**	**Not Surviving**	**Totals**	**Surviving**	**Not Surviving**
Control	72	8	80	90	10
Effluent	64	16	80	80	20
Totals	136	24	160	Ave. = 85	15

mean that the outcome is desirable, for a success may be the death of an organism, failure of a machine, or violation of a regulation. It means that we have had success in observing the behavior of interest. The true probability of the event of interest occurring in a given trial is p, and $1 - p$ is the probability of the event not occurring. In the context of most environmental problems, the desired outcome is for x to occur infrequently, which means that we are interested typically in cases where p is small.

The binomial probability function for x as a function of n and p is

$$f(x; n, p) = \binom{n}{x} p^x (1 - p)^{n - x} \qquad x = 0, 1, 2, \ldots, n$$

where

$$\binom{n}{x} = \frac{n!}{x!(n - x)!}$$

indicates the number of ways that x successes can occur in a sample of size n. The probability of a success in a single trial is p (the corresponding probability of failure in a single trial is $1 - p$). The expected number of occurrences in n trials is the mean of x, which is $\mu_x = n\,p$. The variance is $\sigma_x^2 = n\,p\,(1 - p)$. The assumptions made in using the binomial model are that p is constant and that outcomes are independent from trial to trial.

The probability of r or fewer successes in n independent trials, given a probability of success p in a single trial, is given by the cumulative binomial distribution:

$$Pr(x \leq r) = F(x; n, p) = \sum_{0}^{r} \binom{n}{x} p^x (1 - p)^{n - x}$$

Some examples based on the bioassay experiment will demonstrate how these functions are used. The bioassay is modeled as a binomial process. Each test organism is a trial, and the event of interest is its death within the specified test period. It is assumed that (1) theoretical probability of death (and of survival) is assumed equal for all organisms subjected to the same treatment and that (2) the fate of each organism is independent of the fate of other organisms. If n organisms are exposed to a test condition, the probability of observing any specific number of dead organisms, given a true underlying random probability of death for an individual organism, is computed from the binomial distribution.

Table 15.2, reproduced from standard tables of the binomial distribution, gives the *cumulative* probability of x deaths in a group of n = 20 organisms for several values of p. Table 15.3 gives the probability of *exactly* x occurrences for the same conditions as Table 15.2. It can be constructed by differencing the entries in Table 15.2 or by solving the binomial probability equation. For example, if p = 0.05, we expect on average to have one death in a population of 20 test organisms; $\mu_x = 0.05\ (20) = 1$ death. From Table 15.2, the probability of getting one or less death is $Pr(x \leq 1) = 0.74$, and the probability of observing exactly zero deaths is $Pr(x = 0) = 0.36$. These two values are used to compute the probability of exactly one death as follows

$$Pr(x = 1) = Pr(x \leq 1) - Pr(x = 0) = 0.74 - 0.36 = 0.38$$

which is the value given in Table 15.3 for x = 1 and p = 0.05. Likewise, the probability of two or less deaths is $Pr\ (x \leq 2) = 0.92$, and the probability of exactly two deaths is

$$Pr(x = 2) = Pr(x \leq 2) - Pr(x \leq 1) = 0.92 - 0.74 = 0.18.$$

We can look at the problem in another way. Suppose that the toxicity of the effluent corresponds to a underlying probability of death for a single organism in the effluent group of p = 0.2. The expected number of deaths in an experiment with 20 test organisms is $\mu_x = 0.2\ (20) = 4$. The probability of observing exactly four deaths is 0.22. The probability of observing four or fewer deaths is $Pr(x \leq 4) = 0.63$, and the probability of having more than four deaths is $Pr(x > 4) = 1 - 0.63 = 0.37$. The chance of observing exactly one death, which is the expected number for the control group with p = 0.05, is also 0.06 or 6%.

Table 15.3 shows that the range of possible outcomes in a binomial process can be quite large and that the variability about the expected value increases as p increases. This characteristic of the variance needs to be considered in designing bioassay experiments.

Table 15.2 Cumulative Binomial Probability for x Deaths in a Group of 20 Organisms, Given a True Underlying Random Probability of Death for an Individual Organism of p

x	p = 0.05	0.10	0.15	0.20	0.25	0.50
0	0.36	0.12	0.04	0.01	0.00	0.00
1	0.74	0.39	0.18	0.07	0.02	0.00
2	0.92	0.68	0.40	0.21	0.09	0.00
3	0.98	0.87	0.65	0.41	0.23	0.00
4	1.00	0.96	0.83	0.63	0.41	0.01
5	1.00	0.99	0.93	0.80	0.62	0.02
6		1.00	0.98	0.91	0.79	0.06
7		1.00	0.99	0.97	0.90	0.13
8			1.00	0.99	0.96	0.25
9			1.00	1.00	0.99	0.41
10				1.00	1.00	0.59
11					1.00	0.75
12						0.87
13						0.94
14						0.98
15						0.99
16						1.00

Table 15.3 **Probability of Observing Exactly x Deaths in a Group of 20 Organisms, Given a True Underlying Random Probability of Death of p**

x	p = 0.05	0.10	0.15	0.20	0.25	0.50
0	0.36	0.12	0.04	0.01	—	—
1	0.38	0.27	0.14	0.06	0.02	—
2	0.18	0.29	0.22	0.14	0.07	—
3	0.06	0.19	0.25	0.20	0.14	—
4	0.02	0.09	0.18	0.22	0.18	0.01
5		0.03	0.10	0.17	0.21	0.01
6		0.01	0.05	0.11	0.17	0.04
7			0.01	0.06	0.11	0.07
8			0.01	0.02	0.06	0.12
9				0.01	0.02	0.16
10					0.01	0.18
11						0.16
12						0.12
13						0.07
14						0.04
15						0.01
16						0.01

The binomial distribution becomes cumbersome to use when n is large. Most statistics books only provide binomial tables up to $n = 20$. Fortunately, under certain circumstances, the normal distribution provides a good approximation to the binomial distribution. The question is, "when?" The binomial probability distribution is symmetric when $p = 0.5$. For values of $p < 0.5$, it is skewed to the right; for $p > 0.5$, it is skewed left. For a large n, however, the skewness is not great unless p is near to 0 or 1. The distribution approaches symmetry as n becomes larger, the approach being more rapid when p is close to 0.5. Also, as n increases, the binomial distribution can be approximated by a normal distribution with the same mean and variance [$\mu_x = np$ and $\sigma_x^2 = np(1 - p)$]. This approximation gives reasonable results if $np > 5$ *and* $n(1 - p) > 5$. Table 15.3 shows this. The distribution is exactly symmetric for $p = 0.5$ and is already approaching symmetry for $n = 20$ and $p = 0.2$ [where $n(1 - p)$ is only 3.2].

ASSESSING THE DIFFERENCE BETWEEN TWO PROPORTIONS

The binomial distribution expresses the number of occurrences of an event x in n trials, where p is the probability of occurrence in a single trial. Usually the value of p in a binomial process is unknown, so it is often more useful to examine the proportion of occurrences rather than their absolute number, x. The proportion of occurrences, typically called the sample proportion, is defined as x/n and has mean $\mu_{x/n} = p$ and variance $\sigma_{x/n}^2 = p(1 - p)/n$. The sample proportion, x/n, is an unbiased estimator of the underlying population probability, p, in a binomial process.

Note that contrary to our guidelines on notation given in Chapter 2 (which follows those in most statistics books), the population parameter p is not denoted with a Greek letter symbol. We still must distinguish between the population parameter p and the sample proportion, which will be called $\hat{p} = x/n$. The hat (^) reminds us that this is a statistic computed from the data and that it estimates the parameter of interest.

The objective is to determine whether the values of p for two binomial processes (for example, a control and a treatment) are the same or different. We use the difference of the respective sample proportions, $\hat{p}_1 - \hat{p}_2$, to assess this.

Two independent test groups of size n_1 and n_2 are to be compared. Suppose that group 1 represents a control and group 2 is the treatment group (i.e., exposed to effluent). The number of surviving organisms is x_1 in the control and x_2 in the treatment, giving observed sample proportions of $\hat{p}_1 = x_1/n_1$ and $\hat{p}_2 = x_2/n_2$. If we further assume (as in a null hypothesis) that the control and treatment populations have the same true underlying population probability p (i.e., $p = p_1 = p_2$), then $\hat{p}_1$ and $\hat{p}_2$ will be normally distributed with mean p and variances $p(1-p)/n_1$ and $p(1-p)/n_2$, respectively.

The difference between the two sample proportions, $\hat{p}_1 - \hat{p}_2$, will be normally distributed with mean zero and variance:

$$\frac{p(1-p)}{n_1} + \frac{p(1-p)}{n_1} = p(1-p)\left(\frac{1}{n_1} + \frac{1}{n_2}\right)$$

The standardized difference also is normally distributed:

$$z = \frac{(\hat{p}_1 - \hat{p}_2) - 0.0}{\sqrt{p(1-p)\left(\dfrac{1}{n_1} + \dfrac{1}{n_2}\right)}}$$

The difficulty with using the standardized difference is that p is unknown and the denominator cannot be computed until some estimate is found. The best estimate of p is the weighted average of the sample proportions (Rosner, 1990)

$$p = \frac{n_1\hat{p}_1 + n_2\hat{p}_2}{n_1 + n_2}$$

We can compute z as given above under the null hypothesis that the two population probabilities are equal ($p_1 = p_2$). This value is compared with the tabulated standard normal variate, z_α, for a specified significance level α. If $z > z_\alpha$, the difference is too large to accept the hypothesis that the populations are the same. Usually only the case where $p_1 > p_2$ (or $p_2 > p_1$) is of interest, so a one-sided test is used.

The $(1 - \alpha)100\%$ confidence limits for $p_1 - p_2$ for large samples under the normal approximation is

$$\hat{p}_1 - \hat{p}_2 \pm z_{\alpha/2}\sqrt{\frac{\hat{p}_1(1-\hat{p}_1)}{n_1} + \frac{\hat{p}_2(1-\hat{p}_2)}{n_2}}$$

To set an upper bound with a $(1 - \alpha)100\%$ confidence level, use the above equation with z_α.

For the special (but common) case where the two test populations are the same size, $n = n_1 = n_2$, and the standardized difference simplifies to

$$z = \frac{(\hat{p}_1 - \hat{p}_2) - 0.0}{\sqrt{\dfrac{2(p)(1-p)}{n}}}$$

where

$$p = (\hat{p}_1 + \hat{p}_2)/2$$

Fleiss (1981) suggests that when the n is small (say n < 20) the computation of z should be modified to

$$z = \frac{(p_1 - p_2) - \dfrac{1}{n}}{\sqrt{\dfrac{2(p)(1-p)}{n}}}$$

The term (−1/n) in the numerator is the so-called Yates continuity correction, which takes into account the fact that a continuous distribution (the normal) is being used to represent the discrete binomial distribution of sample proportions. For reasonably small n, say n = 20, the correction is −1/n = −0.05, which can be substantial relative to the differences usually observed (e.g., $p_1 - p_2 = 0.15$). Not everyone uses this correction (Rosner, 1990), but Fleiss (1981) says that it is appropriate. If it is omitted, there is a reduced probability of detecting a real difference in rates. As n becomes large, this correction factor becomes negligible because the normal distribution very closely approximates the binomial distribution.

SOLUTION

Eighty organisms ($n_1 = n_2 = 80$) were exposed to each treatment condition (control and effluent) and the sample survival proportions were observed to be

$$\hat{p}_1 = 72/80 = 0.90 \text{ and } \hat{p}_2 = 64/80 = 0.8$$

We would like to use the normal approximation to the binomial distribution. Checking to see whether this is appropriate gives

$$n\hat{p}_1 = 80(0.90) = 72 \text{ and } n(1-p_1) = 80(0.10) = 8$$

and

$$n\hat{p}_2 = 80(0.8) = 64 \text{ and } n(1-p_2) = 80(0.20) = 16$$

Since these are greater than five, the normal approximation can be used. The sample size is large enough that the Yate's continuity correction is not needed.

Might the true difference of the two proportions be zero? Could we observe a difference as large as 0.1 just as a result of random variation? This can be checked by examining the lower limit of the confidence interval for $p_1 - p_2$. We will compute the lower 95% confidence limit. From the observed proportions, we estimate p = (0.9 + 0.8)/2 = 0.85 and 1 − p = 0.15. For a one-sided test at the 95% level, $z_{\alpha=0.05} = 1.64$, and the lower 95% confidence bound is

$$(0.90 - 0.80) - 1.645\sqrt{\frac{2(0.15)(0.85)}{80}} = 0.1 - (1.645)0.056 = 0.008$$

The lower limit is greater than zero, so it is concluded that the observed difference is larger than is reasonably expected to occur by chance and, furthermore, that $p_2 = 0.8$ is less than $p_1 = 0.9$.

Alternately, we could have compared the computed z statistic,

$$z = \frac{0.90 - 0.80}{\sqrt{\dfrac{2(0.15)(0.85)}{80}}} = 1.77$$

with the tabulated value for $\alpha = 0.05$ which is 1.645. Based on finding $z = 1.77 > 1.645$, it is concluded that the difference between the proportions is large enough to say that the two treatments are different.

COMMENTS

The example problem found a small difference in proportions to be statistically significant because the number of test organisms was large. Many bioassays use about 20 test organisms, in which case the observed difference in proportions will have to be quite large before the statistical test will let us conclude with a high degree of confidence that the difference did not arise merely from chance.

Testing proportions is straightforward if the sample size and proportions are large enough for the normal approximation to be used. However, when the proportions approach one or zero, the sample size that satisfies this condition becomes quite large. When the approach illustrated here is not suitable, because the small sample size invalidates the normal approximation, consult Fleiss (1981) or Rosner (1990).

REFERENCES

Bartlett, M. S. (1947). "The Use of Transformations," *Biometrics*, 3, 39–52.

Cohen, J. (1960). *Statistical Power Analysis for the Behavioral Sciences*, New York, New York Academy Press.

Fleiss, J. L. (1981). *Statistical Methods for Rates and Proportions*, New York, John Wiley & Sons.

Mowery, F. D., J. A. Fava, and L. W. Clatlin (1985). "A Statistical Test Procedure for Effluent Toxicity Screening," *Aquatic Toxicology and Hazard Assessment, 7th Symp.*, Philadelphia, ASTM STP, 854.

Rosner, B. (1990). *Fundamentals of Biostatistics*, 3rd ed., Boston, PWS-Kent Publishing Co.

Chapter 16

Multiple Paired Comparisons of k Averages

Key words: data snooping, Dunnett's procedure, multiple comparisons, sliding reference distribution, t-tests, Tukey's procedure

---

The problem of comparing several averages arises in many contexts. In bioassays, we want to compare several treatments against a control. Or, several new electrodes are on the market, and we want to compare their performance. Or, eight combinations of media are to be compared for odor removal from ventilation air.

Knowing how to do a t-test may tempt us to compare several combinations of treatments using a series of paired t-tests. If there are k treatments, the number of pairwise comparisons that could be made is $k(k - 1)/2$. For $k = 4$, there are 6 possible combinations, for $k = 5$ there are 10, for $k = 10$ there are 45, and for $k = 15$ there are 105! Checking 5, 10, 45, or even 105 combinations is manageable, but not recommended. Statisticians call this *data snooping* (Sokal and Rohlf, 1969) or *data dredging* (Tukey, 1991). First, we need to understand why data snooping is dangerous.

Suppose, to take a not too extreme example, that we have 15 different treatments. The number of possible pairwise comparisons that could be made is $15(15 - 1)/2 = 105$. If, before the results are known, we select one comparison to make and do this with a t-test that has a 5% error rate, there is a 5% chance of reaching the wrong decision each time we repeat the data collection experiment for those two treatments. If, however, several pairs of treatments are tested for possible differences using this procedure, the error rate will be quite wrong. Imagine that a two-sample t-test is used to compare the largest of the 15 average values against the smallest average. This difference, the largest of all the 105 possible pairwise differences, is likely to be so different from the smallest that the null hypothesis (that the two treatment means are the same) will be rejected almost every time the experiment was repeated, instead of just at the 5% rate that would apply to making one pairwise comparison selected at random from among the 105 possible comparisons.

The number of comparisons does not have to be large for problems to arise. If there are just three treatment methods and three averages, A is larger than B and C is slightly larger than A [i.e., $\bar{y}_C > \bar{y}_A > \bar{y}_B$]. It is possible for the three possible t-tests to indicate that A gives higher results than B ($\eta_A > \eta_B$), A is not different from C ($\eta_A = \eta_C$), and B is not different from C ($\eta_B = \eta_C$). This apparent contradiction can happen because different variances are used to make the different comparisons. Analysis of variance (Chapter 17) eliminates this problem by using a common variance to make a single test of significance (using the F statistic).

There are two multiple paired comparison problems. One is to compare all possible pairs of k treatments. The other case arises when one of the treatments is a control and the task is to compare each of the other $k - 1$ treatments with the control. In each an allowance is made in the error rate to compensate for the *data snooping* aspect of the analysis.

The "allowance" has the goal of keeping the collective error rate at a stated level. This collective rate can be defined in two ways. Returning to the example of 15 treatments and 105 possible pairwise comparisons, the probability of getting the wrong conclusion for a single randomly selected comparison is the *individual error rate*. The *family error rate* (also called the Bonferroni error rate) is the chance of getting 1 or more of the 105 comparisons wrong in each repetition of data collection for all 15 treatments. The *family error rate* counts an error for each wrong comparison in each repetition of data collection for all 15 treatments. Thus, to make valid statistical comparisons, the individual per comparison error rate must be shrunk to keep the simultaneous *family-wise* or *family* error rate at the desired level.

Here we will illustrate Tukey's multiple t-test and also Dunnett's method of multiple comparisons with a control, by example, with minimal explanation of the statistical theory.

CASE STUDY — MEASUREMENTS OF LEAD BY FIVE LABORATORIES

Five laboratories each made measurements of lead on ten replicate wastewater specimens. The data are given in Table 16.1 along with the means and variance for each laboratory. The ten possible comparisons of mean lead concentrations are given in Table 16.2. Laboratory 3 has the highest mean (4.47 µg/L) and Laboratory 4 has the lowest (3.13 µg/L). Are the differences consistent with what one might expect from random sampling and measurement error or can the differences be attributed to real differences in the performance of the laboratories? We cannot answer this question by making a t-test only on the largest differences because this would give the wrong family error rate. Instead we will make the comparison using Tukey's paired comparison method with a family error rate of 5% (i.e., 95% confidence level).

Table 16.1 **Ten Measurements of Lead Concentration (µg/L) Measured on Identical Wastewater Specimens by Five Laboratories**

	Lab 1	Lab 2	Lab 3	Lab 4	Lab 5
	3.4	4.5	5.3	3.2	3.3
	3.0	3.7	4.7	3.4	2.4
	3.4	3.8	3.6	3.1	2.7
	5.0	3.9	5.0	3.0	3.2
	5.1	4.3	3.6	3.9	3.3
	5.5	3.9	4.5	2.0	2.9
	5.4	4.1	4.6	1.9	4.4
	4.2	4.0	5.3	2.7	3.4
	3.8	3.0	3.9	3.8	4.8
	4.2	4.5	4.1	4.2	3.0
Mean $\bar{y}_i$ =	4.30	3.97	4.47	3.13	3.34
Variance, s_i^2 =	0.82	0.19	0.41	0.58	0.54

Table 16.2 **Ten Possible Differences of Means $(\bar{y}_i - \bar{y}_j)$ Between Five Laboratories**

	Laboratory i (Average = $\bar{y}_i$)				
Laboratory j	1 (4.30)	2 (3.97)	3 (4.46)	4 (3.12)	5 (3.34)
1	—	—	—	—	
2	0.33	—	—	—	
3	−0.16	−0.49	—	—	
4	1.18	0.85	1.34	—	
5	0.96	0.63	1.12	−0.22	—

TUKEY'S PAIRED COMPARISON METHOD

A $(1 - \alpha)100\%$ confidence interval for the true difference between the means of two treatments, say treatments i and j, is

$$(\bar{y}_i - \bar{y}_j) \pm t_{v,\alpha/2} \sqrt{\frac{1}{n_i} + \frac{1}{n_j}}$$

where it is assumed that the two treatments have the same variance which is estimated by pooling the two sample variances

$$s_p^2 = \frac{(n_i - 1)s_i^2 + (n_j - 1)s_j^2}{n_i + n_j - 2}$$

This interval is valid for any *single* comparison. The chance that the interval includes the true value is exactly $1 - \alpha$. But, the chance that all possible $k(k - 1)/2$ intervals will simultaneously contain their true values is less than $1 - \alpha$.

Tukey (1949) has shown that the confidence interval for the difference in two means, η_i and η_j, taking into account that all possible comparisons of k treatments may be made, is given by

$$(\bar{y}_i - \bar{y}_j) \pm \frac{q_{k,v,\alpha/2}}{\sqrt{2}} s_p \sqrt{\frac{1}{n_i} + \frac{1}{n_j}}$$

where $q_{k,v,\alpha/2}$ is the appropriate upper significance level of the *studentized range* for k means and v degrees of freedom in the estimate s_p^2 of the variance σ^2. This formula is exact if the numbers of observations in all the averages are equal and is approximate if the k treatments have different numbers of observations. The value of s_p^2 is obtained by pooling sample variances over all k treatments.

The size of the confidence interval is larger when the range statistic $q_{k,v}$ is used than for when the t statistic is used. This is because the range allows for the possibility that any one of the $k(k - 1)/2$ possible pairwise comparisons might be selected for the test. Critical values of $q_{k,v}$ have been tabulated by Harter (1960) and may be found in the statistical tables of Rohlf and Sokal (1981) and Pearson and Hartley (1966). Table 16.3 gives a few values for computing the two-sided 95% confidence interval.

Table 16.3 **Values of the Studentized Range Statistic $q_{k,v}$ for $k(k - 1)/2$ Two-Sided Comparisons for a Joint 95% Confidence Interval Where There Are a Total of k Treatments (Family Error Rate = 5%; α = 0.05/2 = 0.025)**

	k						
v	2	3	4	5	6	8	10
5	4.47	5.56	6.26	6.78	7.19	7.82	8.29
10	3.73	4.47	4.94	5.29	5.56	5.97	6.29
15	3.52	4.18	4.59	4.89	5.12	5.47	5.74
20	3.43	4.05	4.43	4.70	4.91	5.24	5.48
30	3.34	3.92	4.27	4.52	4.72	5.02	5.24
60	3.25	3.80	4.12	4.36	4.54	4.81	5.01
∞	3.17	3.68	3.98	4.20	4.36	4.61	4.78

Data from Harter (1960).

SOLUTION — TUKEY'S METHOD

For this example, n = 50, k = 5, $s_p^2 = 0.51$, $s_p = 0.71$, $\nu = 50 - 5 = 45$, and $q_{5,45,0.05/2} = 4.49$. This gives the 95% confidence limits of

$$(\bar{y}_i - \bar{y}_j) \pm \frac{q_{k,\nu,\alpha/2}}{\sqrt{2}} s_p \sqrt{\frac{1}{n_i} + \frac{1}{n_2}}$$

$$(\bar{y}_i - \bar{y}_j) \pm \frac{4.49}{\sqrt{2}} (0.71) \sqrt{\frac{1}{10} + \frac{1}{10}}$$

and the difference in the true means is, with 95% confidence, within the interval

$$-1.01 \leq (\bar{y}_i - \bar{y}_j) \leq 1.01$$

We can say, with a high degree of confidence, that any observed difference larger than 1.01 or smaller than −1.01 μg/L is not likely to be zero. We conclude that Laboratories 3 and 1 are higher than 4 and that Laboratory 3 is also different from Laboratory 5. We cannot say which laboratory is correct, or which one is best, without knowing the true concentration of the test specimens.

DUNNETT'S METHOD FOR MULTIPLE COMPARISONS WITH A CONTROL

In many experiments and monitoring programs, one experimental condition (treatment, location, etc.) is a standard or a control treatment. For example, in bioassays, there is always an "unexposed" group of organisms that serve as a control. In river monitoring, one location above a waste discharge point may serve as a control or reference station. The question is whether the mean for organisms exposed to potentially toxic conditions in the bioassay or the mean response at downstream locations in the river is different from the mean of the controls. This question is different than the one answered above, because here there are only k − 1 comparisons to be made and we have a reasonable likelihood that the control will be different from at least one of the other treatments.

The quantities to be tested are the differences $\bar{y}_i - \bar{y}_c$, where $\bar{y}_c$ is the observed average response for the control treatment. The $(1 - \alpha)100\%$ confidence intervals for all k − 1 comparisons with the control are given by

$$(\bar{y}_i - \bar{y}_c) \pm t_{k-1,\nu,\alpha/2} s_p \sqrt{\frac{1}{n_i} + \frac{1}{n_c}}$$

This expression is similar to Tukey's as used in the previous section, except the quantity $q_{k,\nu,\alpha/2}/\sqrt{2}$ is replaced with Dunnett's t. The value of s_p is obtained by pooling over all treatments. A portion of his table for 95% confidence intervals is reproduced in Table 16.4. Dunnett (1964) also gives tables for one-sided tests.

SOLUTION — DUNNET'S METHOD

Rather than create a new example, we reconsider the data in Table 16.1, supposing that Laboratory 2 is a reference (control) laboratory. The estimate of s_p^2 is still the pooled within-laboratory variance obtained by combining sample variances over all five labora-

Table 16.4 **Table of $t_{k-1,v,\,0.05/2}$ for k − 1 Two-Sided Comparisons for a Joint 95% Confidence Level Where There Are a Total of k Treatments, One of Which Is a Control**

	k − 1, Number of Treatments Excluding the Control						
$v - \omega$	2	3	4	5	6	8	10
5	3.03	3.29	3.48	3.62	3.73	3.90	4.03
10	2.57	2.76	2.89	2.99	3.07	3.19	3.29
15	2.44	2.61	2.73	2.82	2.89	3.00	3.08
20	2.38	2.54	2.65	2.73	2.80	2.90	2.98
30	2.32	2.47	2.58	2.66	2.72	2.82	2.89
60	2.27	2.41	2.51	2.58	2.64	2.73	2.80
∞	2.21	2.35	2.44	2.51	2.57	2.65	2.72

Abstracted from Dunnett (1964).

tories. The value obtained is $s_p^2 = 0.51$, which gives $s_p = 0.71$. For k − 1 = 4 treatments to be compared with the control and $v = 45$ degrees of freedom, the value of $t_{4,45,0.025} = 2.55$ is found in Table 16.4. The 95% confidence limits are

$$(\bar{y}_i - \bar{y}_c) \pm t_{k-1,v,\alpha/2}\, s_p \sqrt{\frac{1}{n_c} + \frac{1}{n_i}}$$

$$(\bar{y}_i - \bar{y}_c) \pm 2.55\,(0.71) \sqrt{\frac{1}{10} + \frac{1}{10}}$$

$$-0.81 \le (\bar{y}_i - \bar{y}_c) \le 0.81$$

This we can say with 95% confidence that any observed difference $(\bar{y}_i - \bar{y}_c)$, greater than 0.81 or smaller than −0.81, is unlikely to be zero (due to chance). The four comparisons with Laboratory 2 shown in Table 16.5 indicate that the measurements from Laboratory 4 are smaller than those from the control laboratory.

COMMENTS

Box et al. (1978) describe yet another way of making multiple comparisons. The basic idea is that if k treatment averages had the same mean they would appear to be k observations from the same, nearly normal distribution with standard deviation $\sigma/\sqrt{n}$. The plausibility of this outcome is examined graphically by constructing a normal reference distribution and superimposing upon it a dot diagram of the k average values. The reference distribution is then moved along the horizontal axis to see if there is a

Table 16.5 **Comparing Four Laboratories with a Reference Laboratory's Results**

Laboratory	2	1	3	4	5
Average	3.97	4.30	4.46	3.12	3.34
Difference $(\bar{y}_i - \bar{y}_c)$	—	0.33	0.49	−0.85	−0.63

way to locate it so that all the observed averages appear to be typical random values selected from it. This *sliding reference distribution* is a "... rough method for making what are called multiple comparisons." The Tukey and Dunnett methods are more formal ways of making these comparisons.

Multiple comparisons have special importance in biological assay studies. An interesting article that uses Dunnett's tests and two other multiple comparison procedures to estimate the no observable effects level (NOEL) of chronic toxicity is by Capizzi et al. (1984).

Dunnett (1955) discussed the allocation of observations between the control group and the other $p = k - 1$ treatment groups. For practical purposes, if the experimenter is working with a joint confidence level in the neighborhood of 95% or greater, then the experiment should be designed so that $n_c/n_1 = \sqrt{p}$ approximately, where n_c is the number of observations on the control and n_1 is the number on each of the p noncontrol treatments.

REFERENCES

Box, G. E. P., W. G. Hunter, and J. S. Hunter (1978). *Statistics for Experimenters: An Introduction to Design, Data Analysis, and Model Building,* New York, Wiley Interscience.

Capizzi, T., L. Openheimer, H. Mehta, H. Naimie, and J. L. Fair (1984). "Statistical Considerations in the Evaluation of Chronic Aquatic Toxicity Studies," *Environ. Sci. Technol.,* 19, 35–43.

Dunnett, C. W. (1955). "Multiple Comparison Procedure for Comparing Several Treatments with a Control," *J. Am. Stat. Assoc.,* 50, 1096–1121.

Dunnett, C. W. (1964). "New Tables for Multiple Comparisons with a Control," *Biometrics* 20, 482–491.

Harter, H. L. (1960). "Tables of Range and Studentized Range," *Ann. Math. Stat.,* 31, 1122–1147.

Pearson, E. S. and H. O. Hartley (1966). *Biometrika Tables for Statisticians*, Vol. 1, 3rd ed., Cambridge, England, Cambridge University Press.

Rohlf, F. J. and R. R. Sokal (1981). *Statistical Tables*, 2nd ed., San Francisco, W. H. Freeman and Co.

Sokal, R. R. and F. J. Rohlf (1969). *Biometry,* San Francisco, W. H. Freeman and Co.

Tukey, J. W. (1949). "Comparing Individual Means in the Analysis of Variance," *Biometrics*, 5, 99.

Tukey, J. W. (1991). "The Philosophy of Multiple Comparison," *Stat. Sci.,* 6, 1, 100–116.

Chapter 17

Analysis of Variance to Compare k Averages

Key words: ANOVA, ANOVA table, analysis of variance, average, grand average, F test, F distribution, sum of squares

---

Analysis of variance (ANOVA), is a method for testing two or more treatments to determine whether their sample means could have been obtained from populations with the same true mean. This is done by estimating the amount of variation within treatments and comparing it to the variance between treatments. If the treatments are alike (that is, from populations with the same mean, so far as can be determined by the evidence of the available data), the variation within each treatment will be about the same as the variation between treatments. Whether the "within variance" and the "between variance" are alike is tested using the F statistic, which is a measure of the variability in estimated variances in the same way that the t statistic is a measure of the variability in estimates of means.

Analysis of variance is a rich and widely used field of statistics; ". . . the analysis of variance is more than a technique for statistical analysis. Once understood, analysis of variance provides an insight into the nature of variation of natural events, into Nature in short, which is possibly of even greater value than the knowledge of the method as such. If one can speak of beauty in a statistical method, analysis of variance possesses it more than any other" (Sokal and Rohlf, 1969).

Naturally, full treatment of such a powerful subject has been the subject of entire books and only a brief introduction will be attempted here. We seek to illustrate the key ideas of the method and to show it as an alternative to the multiple t-tests that were discussed in Chapter 16.

CASE STUDY — COMPARISON OF FIVE LABORATORIES

The data shown in Table 17.1 (and in Chapter 16, Table 16.1) were obtained by dividing a large quantity of prepared material into 50 identical aliquots and having five different laboratories each analyze ten randomly selected specimens. By design of the experiment, there is no real difference in specimen concentrations, but still the laboratories have produced results having different mean values and different variances. In Chapter 16, these data were analyzed using a multiple t-test. The t-test focused on comparing the mean levels. Here we will use a one-way ANOVA, which focuses on comparing the variation within laboratories with the variation between laboratories. The analysis is "one-way" because there is one factor — laboratories — to be assessed.

ONE-WAY ANALYSIS OF VARIANCE

Consider an experiment that has k treatments (techniques, methods, etc.) to be compared with n_t replicate observations made under each treatment, giving a total of $N = \Sigma n_t$ observations, where $t = 1, 2, \ldots, k$. There always will be some intrinsic variability within each treatment, but this should be due only to random measurement error. If this is true, this "within treatment" variance will provide a good estimate of the pure random experimental error. To see whether the k techniques are different, find out whether the variation between the k treatments is greater than might be expected in light of the variation which occurs within the treatment.

Table 17.1 **Ten Measurements of Lead Concentration (μg/L) on Identical Specimens from Five Laboratories**

Lab 1	Lab 2	Lab 3	Lab 4	Lab 5
3.4	4.5	5.3	3.2	3.3
3.0	3.7	4.7	3.4	2.4
3.4	3.8	3.6	3.1	2.7
5.0	3.9	5.0	3.0	3.2
5.1	4.3	3.6	3.9	3.3
5.5	3.9	4.5	2.0	2.9
5.4	4.1	4.6	1.9	4.4
4.2	4.0	5.3	2.7	3.4
3.8	3.0	3.9	3.8	4.8
4.2	4.5	4.1	4.2	3.0
$\bar{y}_i = 4.30$	3.97	4.47	3.13	3.34
$s_t^2 = 0.82$	0.19	0.41	0.58	0.54

The within-treatment sum of squares is calculated from the residuals of the observations within a treatment and the average for that treatment. The variance within each treatment is

$$s_t^2 = \sum_{i=1}^{n_t} \frac{(y_{ti} - \bar{y}_i)^2}{n_t - 1}$$

where y_{ti} are the n_t observations under each treatment.

Assuming that all treatments have the same population variance, we can pool the k sample variances to estimate the within-treatment variance, s_w^2,

$$s_w^2 = \frac{\sum_{t=1}^{k} (n_t - 1)\, s_t^2}{\sum_{i=1}^{k} (n_t - 1)}$$

The between-treatment variance, s_b^2, is calculated using the treatment averages and the grand average, $\bar{y}$,

$$s_b^2 = \frac{\sum_{t=1}^{k} n_t\, (\bar{y}_t - \bar{y})^2}{k - 1}$$

If there are an equal number of observations in each treatment, the equations for s_w^2 and s_b^2 simplify to

$$s_w^2 = \frac{(n - 1) \sum_{t=1}^{k} s_t^2}{N - k}$$

and

$$s_b^2 = \frac{n \sum_{t=1}^{k} (\bar{y}_t - \bar{y})^2}{k - 1}$$

1. The estimated pooled variance within treatments, s_w^2, is based on $N - k$ degrees of freedom. It will be unaffected by real differences in means between the treatment groups. Assuming no hidden factors are affecting the results, s_w^2 estimates the pure measurement error variance σ^2.

2. If in truth there are no real differences between the treatment averages other than what would be expected to occur by chance, the variance between treatments, s_b^2, also reflects only random measurement error. As such, it would be nearly the same magnitude as s_w^2 and would give a second estimate of σ^2.

3. If the true means do vary from treatment to treatment, s_b^2 will tend to be inflated and will be larger than s_w^2.

4. The null hypothesis implies that no difference exists between the k means. It is tested by checking to see whether the two estimates of σ^2, that is s_b^2 and s_w^2, are the same magnitude. If they are, there is no strong evidence to support a conclusion that the means are different. On the other hand, an indication that s_b^2 is inflated supports concluding that there is a difference. The two estimates of σ^2 are compared using the analysis of variance table and the F test.

The sample variances have a χ^2 distribution. The ratio of sample variances are distributed according to the F distribution. These are skewed distributions whose exact shape depends on the degrees of freedom involved. The χ^2 and F distributions are related to each other in much the same way that the normal and t distributions are related in the t-test.

The computations for the one-way ANOVA are much more simple than the above equations may suggest. Suppose that an experiment yields the data shown below.

	Treatment		
	A	B	C
	12	13	18
	10	17	16
	13	20	21
	9	14	17
Treatment average	$\bar{y}_A = 11.0$	$\bar{y}_B = 16.0$	$\bar{y}_C = 18.0$
Treatment variance	$s_A^2 = 3.33$	$s_B^2 = 10.0$	$s_C^2 = 4.67$
Grand average	$\bar{y} = 15$		

The order of the experimental runs has been randomized within and between treatments. The *grand average* of all 12 observed values is $\bar{y} = 15$. The averages for each treatment, $\bar{y}_t$, are $\bar{y}_A = 11$, $\bar{y}_B = 16$, and $\bar{y}_C = 18$. The within-treatment variances are

$$s_A^2 = \frac{(12 - 11)^2 + (10 - 11)^2 + (13 - 11)^2 + (9 - 11)^2}{4 - 1} = 10/3 = 3.33$$

$$s_B^2 = \frac{(13 - 16)^2 + (17 - 16)^2 + (20 - 16)^2 + (14 - 16)^2}{4 - 1} = 30/3 = 10.0$$

$$s_{\hat{c}}^2 = \frac{(18 - 18)^2 + (16 - 18)^2 + (21 - 18)^2 + (17 - 18)^2}{4 - 1} = 14/3 = 4.67$$

The pooled within-treatment variance is

$$s_w^2 = (4 - 1)(3.33 + 10 + 4.67)/(12 - 3) = 54/9 = 6.06$$

The variance between treatments is computed from the mean for each treatment and the grand mean as follows:

$$s_b^2 = \frac{4(11 - 15)^2 + 4(16 - 15)^2 + 4(18 - 15)^2}{3 - 1} = \frac{108}{2} = 52$$

Is this between-treatment variance larger than the within-treatment variance? This is judged by comparing the ratio of the two variances (between:within). The ratio of sample variances are distributed according to the F distribution. The tabulation of F values is arranged according to the degrees of freedom in the variances used to compute the ratio. The numerator is the mean square of the between-treatment variance, which has v_1 degrees of freedom. The denominator is always the estimate of the pure random error variance, in this case the within-treatment variance, which has v_2 degrees of freedom. An F value with these degrees of freedom is denoted by $F_{v1,v2,\alpha}$, where α is the upper percentage point at which the test is being made. Usually, α is 0.05 (5%) or 0.01 (1%). Geometrically, α is the area under the $F_{v1,v2}$ distribution that lies beyond the value $F_{v1,v2,\alpha}$.

In this example, we will make the test at the 5% level with degrees of freedom $v_1 = k - 1 = 3 - 1 = 2$ and $v_2 = N - k = 12 - 3 = 9$. The relevant value of F is $F_{2,9,0.05} = 4.74$. The ratio computed for our experiment is F = 52/6 = 8.67 > $F_{2,9,\alpha=0.05}$ = 4.74, so we conclude that $\sigma_b^2 > \sigma_w^2$, which provides sufficient evidence to conclude that the means of the three treatments are not equal. We are entitled only to conclude that $\eta_A \neq \eta_B \neq \eta_C$. This analysis does not tell us whether one treatment is different from the other two (say, $\eta_A \neq \eta_B$, but $\eta_B = \eta_C$) or whether all three are different. To determine which are different requires additional analysis (see Chapter 16).

When ANOVA is done by a commercial computer program, which is the way we usually do it, the results are presented in a special ANOVA table that needs some explanation. For the example problem just presented, this table would be as given in Table 17.2.

The "sum of squares" in Table 17.2 is the sum of the squared deviations in the numerator of each variance estimate. The "mean square" in Table 17.2 are a *sum of squares* divided by the degrees of freedom of that sum of squares; they estimate the within-treatment variance, s_w^2, and the between-treatment variance, s_b^2. Note that the mean square for variation between treatments is 52, which is the same as the between-treatments variance computed above. Also, note that the within-treatment mean square of 6 is equal

Table 17.2 Analysis of Variance Table for the Example Problem

Source of Variation	Sum of Squares	Degrees of Freedom	Mean Square	F Ratio
Between treatments	104	2	52	8.7
Within treatments	54	9	6	
Total	158	11		

to the within-treatment variance computed above. The F ratio is the ratio of these two mean squares and is the same as the F ratio of the two variances computed above. Thus, the presentation in Table 17.2 is equivalent to computing the variances and F ratio as illustrated above.

CASE STUDY SOLUTION

Figure 17.1 is a dot diagram showing the location and spread of the data from each laboratory. It appears that the variability in the results is about the same in each lab, but Laboratories 4 and 5 may be giving low readings. The data are replotted in Figure 17.2 as deviations about their respective means. An analysis of variance will tell us if the means of these laboratories are statistically different.

The variance within a single laboratory should be due to random errors arising in handling and analyzing the specimens. The variation between laboratories might be due to a real difference in performance or it might also be due to random variation. If the variation between laboratories is random, the five laboratory observed means will vary randomly about the grand mean of all 50 measured concentrations ($\bar{y} = 3.84$ μg/L), and

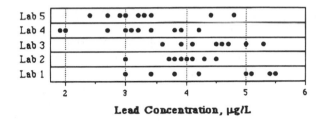

Figure 17.1   Dot plots comparing the results from five laboratories.

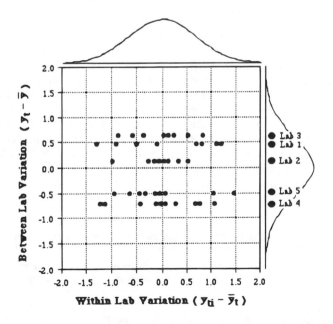

Figure 17.2   Plot of the data from the five laboratories and the distributions of within-laboratory and between-laboratory variation.

furthermore, the variance of the five laboratory's means with respect to the grand mean will be the same as the variance within laboratories.

The ANOVA table for the laboratory data is given in Table 17.3. The F ratio is compared with the critical value $F_{4,45,0.05} = 2.59$. The value found for this experiment, F ratio = 6.809, is much larger than this, so we conclude that the variation between laboratories has been inflated by real differences in the mean level of performance of the labs.

Knowing this result of the analysis of variance, a plausible conclusion would be that Laboratory 4, having the lowest average, is different from the others. But, Laboratories 4 and 5 may both be different from the other three. Or Laboratory 3 may also be different, but on the high side. Unfortunately, ANOVA does not tell us how many or which laboratories are different. We only know that they seem not to be giving the same results.

COMMENTS

When the ANOVA indicates differences between laboratories, additional questions arise.

1. Which labs are different and which are the same? Making multiple pairwise comparisons to answer this was discussed in Chapter 16.
2. Which labs, if any, are giving correct results? Without knowing the true concentration of the samples analyzed, there is no answer to this question. We remind ourselves, however, that the performer who is different may be the champion!
3. Is a special program needed to bring all laboratories into conformance? Maybe. But, maybe the only step needed is to show the participating labs the results of the study. Simply, knowing that there is a *possible* problem usually will stimulate careful checking and improvement. Quality improvement depends more on collecting data and communicating the results than on fixing blame or identifying poor performers. This means that we do not always need to find out which labs are different. (Recall point 2.)

A common question is, "Are the differences between labs large enough to have important consequences in practice?" *Importance* and *statistical significance* are two quite different concepts. Importance depends on the actual use to which the measurements are put. Statistically significant means only that the errors in the measurements are larger than the differences we are trying to detect. These differences are not always important. We sometimes can change significance to nonsignificance by changing the probability level of the test (or by using a different statistical procedure altogether). This obviously would not change the practical importance of a real difference in performance. Furthermore, the importance of a difference will exist whether we have detected it or not.

ANOVA can be applied to problems having many factors. One such example, a four-way ANOVA, is discussed in Chapter 19. Chapter 18 discusses the use of ANOVA to discover the relative magnitude of several sources of variability in a sampling and

Table 17.3 **ANOVA Table for the Comparison of Five Laboratories**

Source of Variation	Sum of Squares	Degrees of Freedom	Mean Square	F Ratio
Between laboratories	13.885	4	3.471	6.809
Within laboratories	22.939	45	0.51	
Total	36.824	49		

measurement procedure. Box et al. (1978) provide an interesting geometric interpretation of the analysis of variance.

REFERENCES

Box, G. E. P., W. G. Hunter, and J. S. Hunter (1978). *Statistics for Experimenters: An Introduction to Design, Data Analysis, and Model Building*, New York, Wiley Interscience.

Johnson, R. A. and D. W. Wichern (1992). *Applied Multivariate Statistical Analysis*, Englewood Cliffs, NJ, Prentice Hall.

Sokal, R. R. and F. J. Rohlf (1969). *Biometry*, San Francisco, W. H. Freeman and Co.

Chapter 18

Estimating Variance Components in Experimental Measurements

Key words: analysis of variance, ANOVA, components of variance, foundry waste, nested design

---

A common problem arises when extreme variation is noted in routine measurements of a material. What are the important sources of variability? How much of the observed variability is caused by real differences in the material, how much is related to sampling, and how much is caused by measurement error? A well-planned experiment and statistical analysis can answer these questions by breaking down the total observed variance into components that are attributed to specific sources.

Such problems are abundant. Hahn (1977) gives one example. A city water supply is monitored to assess the daily concentration of a particular agent. Random specimens are taken daily from a large composite volume of water, and repeat readings are taken on each specimen. Preparing a specimen costs $20 and taking a reading costs $5. A budget of $100/day is available for making these determinations. This budget allows the following possible combinations of specimens and readings: (1) one specimen per day with 16 readings, (2) two specimens per day with 6 readings on each, (3) three specimens per day with 2–3 readings on each, and (4) four specimens per day with 1 reading on each.

The objective is to choose the specimen/reading combination that provides the most precise estimate of the daily concentration when the results of n_r readings on each of n_s specimens are averaged. The variance of the estimated average concentration is a function of the variance contributed from the differences between specimens and the variance contributed by the differences between readings. If the variance between specimens is small compared with the testing variance, more work should be invested in replicate readings rather than in replicate sampling.

CASE STUDY — FOUNDRY WASTES

Foundries produce castings by pouring molten metal into molds made of molding sands and cores sands. Molding sand and core sand can be recycled several times, but eventually are discarded as a mixture called "system sand." Molding sand is a mixture of silica sand, clay, carbon, and water. Core sand is composed of silica sand with a small amount of chemical binder, which may be natural substances (e.g., vegetable or petroleum oils, sodium silicate, ground corn flour and oil, ground hardwood cellulose) and synthetic binders (e.g., phenol formaldehyde, phenol isocyanate, alkyl isocyanate). The typical solid waste of a foundry is two thirds or more system sand, 2–20% core sand and core butts, and up to 11% dust collected in the baghouse that is used for air pollution control.

These wastes are generally put into landfills. Most of the waste is inert (sand), but certain components have the potential of being leached from the landfill and entering subsurface soils and groundwater. Studies of leachate from foundry landfills have shown large variations in chemical composition. The variation may arise from the nonuniformity of the waste materials deposited in the landfill, but it also may have other causes. This raises questions about how large a sample should be collected in the field and whether this sample should be a composite of many small portions. There are also questions about how to partition a large field specimen into smaller portions that are suitable for laboratory analysis. Finally, the laboratory work itself cannot be overlooked as a possible source of variability.

These considerations point to the need for replicate specimens and replicate measurements. An efficient protocol for sampling and analytical replication cannot be designed until the sources of variation are identified and quantified. The major source of variation might be between field specimens, from subspecimens prepared in the laboratory, or from the analytical procedure itself.

Krueger (1985) studied three kinds of solid waste (system sand, core butts, or baghouse dust) from a foundry. One of his objectives was to assess the magnitude of the variance components due to (1) the batch of material sampled (system sand, core butts, or baghouse dust), (2) the specimen preparation in the laboratory (which includes a leaching extraction performed on that specimen), and (3) the analytical test for a specific substance. One of the substances measured was copper. Table 18.1 is an abridgment of Krueger's data on copper concentrations observed in a leaching extract from the three batches of solid waste.

Table 18.1 **Copper Concentrations in the Leachate Extract of Foundry Solid Wastes**

Batch of Solid Waste	Sample Number	Copper Concentrations (mg/L)		
		Sample	Sample Ave.	Batch Ave.
Baghouse dust	1	0.082		
		0.084	0.0830	
	2	0.108		
		0.109	0.1085	
	3	0.074		
		0.070	0.0720	
	4	0.074		
		0.071	0.0725	0.0840
Core butts	1	0.054		
		0.051	0.0525	
	2	0.050		
		0.050	0.0500	
	3	0.047		
		0.050	0.0485	
	4	0.092		
		0.091	0.0915	0.0606
System sand	1	0.052		
		0.050	0.0510	
	2	0.084		
		0.080	0.0820	
	3	0.044		
		0.041	0.0425	
	4	0.050		
		0.044	0.0470	0.0556

Data from Krueger (1985).

VARIANCE COMPONENTS ANALYSIS

Variance components analysis is a method for learning what fraction of the total variance in a measurement process is caused by different factors (components) that contribute random variation into the sampling and testing process. If we have n measurements, denoted by y, the sample variance for the entire data set is s_y^2.

$$s_y^2 = \frac{\Sigma(y_i - \bar{y})^2}{n - 1}$$

In order to learn the relative importance of different factors that are expected to contribute to this total variance, we must carry out an experiment designed to allow the variance of each of these factors to be estimated independently of the other factors. One design that is efficient for this purpose is the nested (or hierarchical) design shown in Figure 18.1.

The analysis can be generalized for k components, but it is more convenient to explain it specifically for a three-factor analysis (note that the case study involves three factors). The three components are the batch of specimen, subspecimens from the batch, and chemical tests on the subspecimens. In general, there are n_b batches, n_s subspecimens, and n_t chemical tests, for a total of $n = n_b n_s n_t$ observations. The nested experimental design shown in Figure 18.1 consists of three batches, with two chemical tests on each of the four subspecimens from each batch, giving a total of $n = (3)(4)(2) = 24$ observations.

The overall error of any particular measurement y will be $e_y = y - \eta$, where η is the true mean of the population of specimens. In practice, we estimate the mean by computing the average of all the measurements in the variance components experiment. A measurement on any 1 of the 24 test specimens produced by the design shown in Figure 18.1 will reflect variability contributed by each component, so

$$e_y = e_b + e_s + e_t$$

where e_b, e_s, and e_t are the error contributions from the batch, sample, and test, respectively. Assuming these errors are random and independent, their variances will add to give the total population variance σ_y^2

$$\sigma_y^2 = \sigma_b^2 + \sigma_s^2 + \sigma_t^2$$

where the subscripts b, s, and t identify the variance components of the batches, speci-

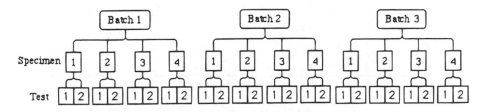

Figure 18.1 Nested design to estimate components of variance.

mens, and chemical tests, respectively. The aggregation of the error and variance components is diagramed in Figure 18.2.

The variation among replicate chemical tests on each sample provides an estimate of σ_t^2. The variation among specimen averages reflects both test and specimen variance and provides an estimate of the quantity $n_t \sigma_s^2 + \sigma_t^2$. The variance among batches embodies all three sources of variance and provides an estimate of the quantity $n_s n_t \sigma_b^2 + n_t \sigma_s^2 + \sigma_t^2$. The case study provides an opportunity to demonstrate the calculation of the variances. A similar example with additional explanation is given by Box et al. (1978).

These calculations can be organized as an analysis of variance table. Table 18.2 shows the algebra of the analysis of variance for the three-factor nested design being discussed here. Some additional nomenclature is also given. In practice, the calculations usually will be done by computer, and many software packages have this capability.

CASE STUDY — SAMPLING WASTE FOUNDRY SAND

The data in Table 18.1 showed that three batches of foundry solid waste were collected (these batches were baghouse dust, core butts, and system sand). Each batch was subdivided in the laboratory into four specimens, and each specimen was analyzed in duplicate. This gives $n_b = 3$, $n_s = 4$, and $n_t = 2$, for a total of $3 \times 4 \times 2 = 24$ observations.

The average of the 24 observations is $\bar{y} = 0.06673$. Table 18.1 also gives the averages for the duplicate measurements on each specimen, $\bar{y}_{bs}$, and the average of the eight measurements made on each batch of waste, $\bar{y}_b$.

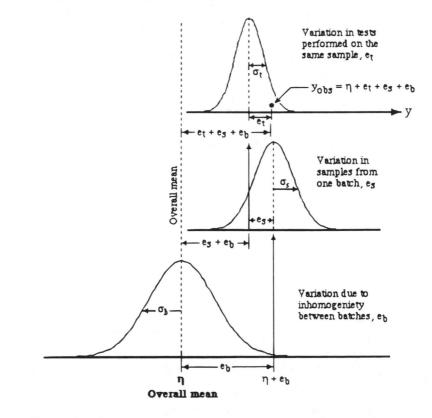

Figure 18.2 Error components contributing to variance in the measured quantity.

Table 18.2 General Analysis of Variance Table for Estimating the Variance Components from a Three-Factor Nested Experimental Design

Source of Variation	Sum of Squares	Degrees of Freedom	Mean Square	Mean Square Estimates
Average	$SS_a = n_b\, n_s\, n_t\, \bar{y}^2$	1		
Batches	$SS_b = n_s\, n_t \sum\limits_{b=1}^{n_b} (\bar{y}_b - \bar{y})^2$	$n_b - 1$	$MS_b = \dfrac{SS_b}{n_b - 1}$	$n_s n_t \sigma_b^2 + n_t \sigma_s^2 + \sigma_t^2$
Specimens	$SS_s = n_t \sum\limits_{b=1}^{n_b} \sum\limits_{s=1}^{n_s} (\bar{y}_{bs} - \bar{y}_s)^2$	$n_b(n_s - 1)$	$MS_s = \dfrac{SS_s}{n_b(n_b - 1)}$	$n_t \sigma_s^2 + \sigma_t^2$
Tests	$SS_t = \sum\limits_{b=1}^{n_b} \sum\limits_{s=1}^{n_s} \sum\limits_{t=1}^{n_t} (y_{bst} - \bar{y}_{bs})^2$	$n_s n_b(n_t - 1)$	$MS_t = \dfrac{SS_t}{n_s\, n_b\, (n_t - 1)}$	σ_t^2
Total	$SS_T = \sum\limits_{b=1}^{n_b} \sum\limits_{s=1}^{n_s} \sum\limits_{t=1}^{n_t} y_{bst}^2$	$n_b n_s n_t$		

The sum of squares for each variance component is

$$\text{Total } SS_T = \sum_{b=1}^{n_b} \sum_{s=1}^{n_s} \sum_{t=1}^{n_t} y_{bst}^2$$

$$= 0.082^2 + 0.084^2 + \ldots + 0.044^2 = 0.11690$$

$$\text{Average } SS_{ave} = n_b \, n_s \, n_t \, \bar{y}^2$$

$$= 3\,(4)(2)(0.06675)^2 = 0.106934$$

$$\text{Batch } SS_b = n_s \, n_t \sum_{b=1}^{n_b} (\bar{y}_b - \bar{y})^2$$

$$= (4)(2)[(0.0840 - 0.6675)^2 + (0.0606 - 0.06675)^2 + (0.0556$$
$$- 0.06675)^2] = 0.003671$$

$$\text{Specimen } SS_s = n_t \sum_{b=1}^{n_b} \sum_{s=1}^{n_s} (\bar{y}_{bs} - \bar{y}_s)^2$$

$$= 2[(0.083 - 0.0840)^2 + (0.1085 - 0.0840)^2 + \ldots + (0.047 - 0.0556)^2]$$
$$= 0.00624$$

$$\text{Test } SS_t = \sum_{b=1}^{n_b} \sum_{s=1}^{n_s} \sum_{t=1}^{n_t} (y_{bst} - \bar{y}_{bs})^2$$

$$= (0.082 - 0.083)^2 + (0.084 - 0.083)^2 + \ldots + (0.044 - 0.0470)^2$$
$$= 0.000057$$

Table 18.3 gives the full analysis of variance table, with sums of squares, degrees of freedom, and mean square values. The mean squares (the sums of squares divided by the respective degrees of freedom) are $MS_b = 0.001836$, $MS_s = 0.000693$, and $MS_t = 0.000005$. The mean squares are used to estimate the variances. Table 18.2 shows that MS_t estimates σ_t^2, MS_s estimates $n_t\sigma_s^2 + \sigma_t^2$, and MS_b estimates $n_s n_t \sigma_b^2 + n_t\sigma_s^2 + \sigma_t^2$. Using these relations with the computed mean square values gives the following estimates:

$$\hat{\sigma}_t^2 = MS_t = 0.000005$$

$$\hat{\sigma}_s^2 = \frac{MS_2 - MS_t}{n_t} = \frac{0.000693 - 0.000005}{2} = 0.000344$$

$$\hat{\sigma}_b^2 = \frac{MS_b - MS_s}{n_s n_t} = \frac{0.001836 - 0.000693}{4 \times 2} = 0.000143$$

The analysis shows that the variance between subspecimens is about 2 times the variance of between batches and about 60 times the variance of the chemical analysis of copper. Variation in the actual copper measurements is small relative to other variance

Table 18.3 **Analysis of Variance Table for the Foundry Waste Example**

Source of Variation	Sum of Squares	df	MS	F
Average	0.10693	1		
Batches	0.00367	2	0.001836	367.2
Samples	0.00624	9	0.000693	138.6
Tests	0.00006	12	0.000005	
Total	0.11690	24		

components. This suggests that extensive replication of copper measurements scarcely helps in reducing the total variance of the solid waste characterization program. Make only enough replicates to maintain quality control, and put the major effort into reducing other variance components.

Knowing the source of the three batches of foundry solid waste, it is reasonable that they should be different; the variation between batches should be large. One goal of sampling is to assess the *difference* between batches of waste material taken from the foundry and perhaps samples taken from the landfill as well. This assessment will be complicated by the difficulty in dividing a large field specimen into representative subspecimens for laboratory analysis. Work should focus on the variability between batches and how they can be subdivided. One approach would be to prepare more subspecimens and enhance the statistical discrimination by averaging.

Table 18.4　**Calculation of the Individual Errors e_t, e_s, and e_b that Are Squared and Added to Compute the Sums of Squares**

Experimental Design			Copper Conc. (mg/L)	Deviations of Tests		Deviations of Samples		Deviations of Batches
Batch	Sample	Test	y (1)	$\bar{y}_s$ (2)	e_t (3)	$\bar{y}_b$ (4)	e_s (5)	e_b (6)
Baghouse	1	1	0.082	0.0830	−0.0010	0.0840	−0.0010	0.0173
Baghouse	1	2	0.084	0.0830	0.0010	0.0840	−0.0010	0.0173
Baghouse	2	1	0.108	0.1085	−0.0005	0.0840	0.0245	0.0173
Baghouse	2	2	0.109	0.1085	0.0005	0.0840	0.0245	0.0173
Baghouse	3	1	0.074	0.0720	0.0020	0.0840	−0.0120	0.0173
Baghouse	3	2	0.070	0.0720	−0.0020	0.0840	−0.0120	0.0173
Baghouse	4	1	0.074	0.0725	0.0015	0.0840	−0.0115	0.0173
Baghouse	4	2	0.071	0.0725	−0.0015	0.0840	−0.0115	0.0173
Core butts	1	1	0.054	0.0525	0.0015	0.0606	−0.0081	−0.0061
Core butts	1	2	0.051	0.0525	−0.0015	0.0606	−0.0081	−0.0061
Core butts	2	1	0.050	0.0500	0.0000	0.0606	−0.0106	−0.0061
Core butts	2	2	0.050	0.0500	0.0000	0.0606	−0.0106	−0.0061
Core butts	3	1	0.047	0.0485	−0.0015	0.0606	−0.0121	−0.0061
Core butts	3	2	0.050	0.0485	0.0015	0.0606	−0.0121	−0.0061
Core butts	4	1	0.092	0.0915	0.0005	0.0606	0.0309	−0.0061
Core butts	4	2	0.091	0.0915	−0.0005	0.0606	0.0309	−0.0061
System sand	1	1	0.052	0.0510	0.0010	0.0556	−0.0046	−0.0111
System sand	1	2	0.050	0.0510	−0.0010	0.0556	−0.0046	−0.0111
System sand	2	1	0.084	0.0820	0.0020	0.0556	0.0264	−0.0111
System sand	2	2	0.080	0.0820	−0.0020	0.0556	0.0264	−0.0111
System sand	3	1	0.044	0.0425	0.0015	0.0556	−0.0131	−0.0111
System sand	3	2	0.041	0.0425	−0.0015	0.0556	−0.0131	−0.0111
System sand	4	1	0.050	0.0470	0.0030	0.0556	−0.0086	−0.0111
System sand	4	2	0.044	0.0470	−0.0030	0.0556	−0.0086	−0.0111

Sums of squares:　From column (3)　　$SS_t = 0.00006$
　　　　　　　　　From column (5)　　$SS_s = 0.00624$
　　　　　　　　　From column (6)　　$SS_b = 0.00367$

COMMENTS

The variance components analysis is an effective method for quantifying the sources of variation in an experimental program. Obviously, this kind of study will be most useful if it is done *at the beginning* of a monitoring program. The information gained from the analysis can be used to plan a cost-effective program for collecting and analyzing samples. Cost effective means that both the cost of the measurement program and the variance of the measurements produced are minimized.

It can happen in practice that the procedure outlined above can estimate variance components that are negative. Such a nonsense result often can be interpreted as a sign of the variance component being zero or having some insignificant positive value. It may result from the lack of normality in the residual errors (Leone and Nelson, 1966). In any case, it should indicate that a review of the data structure and the basic assumptions of the components of the variance test should be made.

REFERENCES

Box, G. E., W. G. Hunter, and J. S. Hunter (1978). *Statistics for Experimenters: An Introduction to Design, Data Analysis, and Model Building*, New York, Wiley Interscience.

Davies, O. L. and P. L. Goldsmith (1972). *Statistical Methods in Research and Production*, 4th ed., rev., New York, Hafner Publishing Co.

Hahn, G. J. (1977). "Random Samplings: Estimating Sources of Variability," *Chemtech*, Sept., 580–582.

Krueger, R. C. (1985). "Characterization of Three Types of Foundry Waste and Estimation of Variance Components," Research Report, Department of Civil and Environmental Engr., University of Wisconsin-Madison.

Leone, F. C. and L. S. Nelson (1966). "Sampling Distributions of Variance Components I. Empirical Studies of Balanced Nested Designs," *Technometrics*, 8, 457–468.

Chapter 19

Multiple Factor Analysis of Variance

Key words: air pollution, dioxin, furan, incineration, samplers, ANOVA (analysis of variance), factorial experiment, sampling error

---

Environmental monitoring is expensive and complicated. Many factors may contribute variation to measured values. An obvious source of variation is the sampling method. An important question is, "Do two samplers give the same result?" This question may arise because a new sampler has come on the market or because a monitoring program needs to be expanded and there are not enough samplers of one kind available.

It might seem natural to compare the two (or more) available sampling methods under a fixed set of conditions. This kind of experiment would estimate random error under only that specific combination of conditions. The samplers, however, will be used under other conditions. A sampler that is effective under one condition may be weak under others. The error of one or both samplers might depend on plant operation, weather, concentration level being measured, or other factors. The variance due to laboratory measurements may be a significant part of the total variance. Interactions between sampling methods and other possible sources of variation should be checked. The experimental design should take into account all these factors.

Comparing two samplers under fixed conditions pursues the wrong goal. A better plan would be to assess performance under a variety of conditions. It is important to learn whether variation between samplers is large or small in comparison with variation due to laboratory analysis, operating conditions, etc. A good experiment would check these possibilities by providing an analysis of variance of all factors that might be important in planning a sampling program.

It is incorrect to imagine that one data point provides one piece of information, and therefore, the information content of a data set is determined entirely by the number of measurements. The amount of information available from a fixed number of measurements increases dramatically if each observation contributes to estimating more than one parameter (mean, variance, etc.). An exciting application of statistical experimental design is to make each observation do double duty or even triple or heavier duty. However, any valid statistical analysis can only extract the information existing in the data at hand. This content is largely determined by the experimental design and cannot be altered by the statistical analysis.

This chapter discusses an experimental design that was used to efficiently evaluate four factors that were expected to be important in an air quality monitoring program. The experiment is based on a *factorial design* (but not the two-level design discussed in Chapter 20). The method of computing the results is not discussed, since this can be done by commercial computer programs. Instead, discussion focuses on how the *four-factor analysis of variance* is interpreted. References are given for the reader who wishes to know how such experiments are designed and how the calculations are done (e.g., Scheffe, 1959).

CASE STUDY — SAMPLING DIOXIN AND FURAN EMISSIONS FROM AN INCINERATOR

Emission of dioxins and furans from waste incinerators is under investigation in many countries. It is important to learn whether different samplers (perhaps used at different incinerators or in different cities or countries) affect the amount of dioxin or furan measured. It is also important to assess whether differences, if any, are independent of other factors (such as incinerator loading rate and different feed materials, which change from one sampling period to another).

The data in Table 19.1 were collected at a municipal incinerator by the Danish Environmental Agency (Pallesen 1987). Two different kinds of samplers were used to take simultaneous samples during four 3.5-hr sampling periods spread over a 3-day period. Operating load, temperature, pressure, etc. were variable. Each sample was analyzed for five dioxin groups (TCDD, PeCDD, HxCDD, HpCDD, and OCDD) and five furan groups (TCDF, PeCDF, HxCDF, HpCDF, and OCDF). The species within each group are chlorinated to different degrees (4, 5, 6, 7, and 8 chlorine atoms per molecule). All analyses were done in one laboratory.

There are four factors being evaluated in this experiment: two kinds of samplers (S), four sampling periods (P), two dioxin and furan groups (DF), and five levels of chlorination within each group (CL). This gives a total of n = 2 × 4 × 2 × 5 = 80 measurements. The data set is completely balanced — all conditions were measured once with no repeats. If there are any missing values in an experiment of this kind or if some conditions are measured more often than others, the analysis becomes more difficult (Milliken and Johnson, 1992).

When the experiment was designed, it was expected that the two samplers would perform similarly but that variation over sampling periods would be large. It was also expected that the levels of dioxins and furans and the amounts of each chlorinated species would be different. There was no prior expectation regarding interactions. A four-factor analysis of variance (ANOVA) was done to assess the importance of each factor and their interactions.

METHOD — ANALYSIS OF VARIANCE

Analysis of variance addresses the problem of identifying which factors contribute significant amounts of variance to measurements. The general idea is to partition the

Table 19.1   Dioxin and Furan Data from a Designed Factorial Experiment

Sample Period Sampler	1		2		3		4	
	A	B	A	B	A	B	A	B
Dioxins								
Sum TCDD	0.4	1.9	0.5	1.7	0.3	0.7	1.0	2.0
Sum PeCDD	1.8	28	3.0	7.3	2.7	5.5	7.0	11
Sum HxCDD	2.5	24	2.6	7.3	3.8	5.1	4.7	6.0
Sum HpCDD	17	155	16	62	29	45	30	40
OCDD	7.4	55	7.3	28	14	21	12	17
Furans								
Sum TCDF	4.9	26	7.8	18	5.8	9.0	13	13
Sum PeCDF	4.2	31	11	22	7.0	12	17	24
Sum HxCDF	3.5	31	11	28	8.0	14	18	19
Sum HpCDF	9.1	103	32	80	32	41	47	62
OCDF	3.8	19	6.4	18	6.6	7.0	6.7	6.7

Note: Values shown are concentrations in ng/m^3 normal dry gas at actual CO_2 percentage.

total variation in the data and assign portions to each of the four factors studied in the experiment and to their interactions.

Total variance is measured by the total residual sum of squares

$$\text{Total SS} = \sum_{\text{all obs.}}^{N} (y_{\text{obs.}} - \bar{y})^2$$

where the residuals are the deviations of each observation from the grand mean

$$\bar{y} = \frac{1}{n} \sum_{\text{all obs.}}^{N} y_{\text{obs}}$$

of the n = 80 observations. This is also called the total adjusted sum of squares (corrected for the mean). Each of the n observations provides 1 degree of freedom. One of them is consumed in computing the grand average, leaving n − 1 degrees of freedom available to assign to each of the factors that contribute variability. The total SS and its n − 1 degrees of freedom are separated into contributions from the factors controlled in the experimental design. For the dioxin/furan emissions experiment, these sums of squares are

$$\text{Total SS} = \text{Periods SS} + \text{Samplers SS} + \text{Dioxin/Furan SS} + \text{Chlorination SS} + \text{Interaction(s) SS} + \text{Error SS}$$

Another approach is to specify a general model to describe the data. It might be simple, such as

$$y_{ijkl} = \bar{y} + \alpha_i + \beta_j + \gamma_k + \lambda_l + (\text{interaction terms}) + e_i$$

where the Greek letters indicate the true response due to the four factors and e_i is the random residual error of the ith observation. The residual errors are assumed to be independent and normally distributed with mean zero and constant variance σ^2 (see, for example, Rao, 1965; Box et al., 1978).

The assumptions of independence, normality, and constant variance are not equally important to the ANOVA. Scheffe (1959) states, "In practice, the statistical inferences based on the above model are not seriously invalidated by violation of the normality assumption, nor, . . . by violation of the assumption of equality of cell variances. However there is no such comforting consideration concerning violation of the assumption of statistical independence, except for experiments in which randomiztion has been incorporated into the experimental procedure."

If measurements had been replicated, it would be possible to make a direct estimate of the error sum of squares, σ^2. In the absence of replication, the usual practice is to use the higher-order interactions as estimates of σ^2. This is justified by assuming, for example, that the fourth-order interaction has no meaningful physical interpretation. It is also common that third-order interactions have no physical significance. If sums of squares of third-order interactions are of the same magnitude as the fourth-order interaction, they can be pooled to obtain an estimate of σ^2 that has more degrees of freedom.

Because no one is likely to do the computations manually for a four-factor analysis of variance, we assume that results are available from some commercial statistical software package. The analysis that follows emphasizes variance decomposition and interpretation rather than model specification.

The first requirement for using available statistical software is recognizing whether the problem to be solved is a one-way ANOVA, two-way ANOVA, etc. This is determined by the number of factors that are considered. In the example problem, there are four factors: S, P, DF, and CL. It is, therefore, a four-way ANOVA.

In practice, such a complex experiment would be designed in consultation with a statistician, in which case the method of analysis will be determined by the experiment plan. The investigator will have no need to guess which method of analysis or which computer program will suit the data. As a corollary, we also recommend that happenstance data (data from unplanned experiments) should not be subjected to analysis of variance. In such data sets, randomization will almost certainly have not been incorporated, and the results are likely to be affected by lurking variables.

DIOXIN CASE STUDY RESULTS

The ANOVA calculations were done on the natural logarithm of the concentrations because this transformation tended to strengthen the assumption of constant variance.

The results shown in Table 19.2 are the complete variance decomposition, specifying all sum of squares (SS) and degrees of freedom (df) for the main effects of the four factors and all interactions between the four factors. These are produced by any computer program capable of handling a four-way ANOVA (e.g., SAS, 1982). The main effects and interactions are listed in descending order with respect to the mean sums of squares (MS = SS/df).

The individual terms in the sums of squares column measure the variability due to each factor plus some random measurement error. The expected contribution of variance due to random error is the random error variance, σ^2, multiplied by the degrees of freedom of the individual factor. If the true effect of the factor is small, its variance will be of the same magnitude as the random error variance. Whether this is the case is determined by comparing the individual variance contributions with σ^2, which is estimated below.

Table 19.2 **Variance Decomposition of the Dioxin/Furan Incinerator Emission Data**

Source of Variation	SS	df	MS	F*
S	18.3423	1	18.3423	573
CL	54.5564	4	13.6391	426
DF	11.1309	1	11.1305	348
DF × CL	22.7618	4	5.6905	178
S × P	9.7071	3	3.2357	101
P	1.9847	3	0.6616	21
DF × P	1.1749	3	0.3916	12.2
DF × S	0.2408	1	0.2408	7.5
P × CL	1.4142	12	0.1179	3.7
DF × P × CL	0.8545	12	0.0712	2.2
S × P × CL	0.6229	12	0.0519	*
S × CL	0.0895	4	0.0224	0.7
DF × S × CL	0.0826	4	0.0206	0.6
DF × S × P × CL	0.2305	12	0.0192	*
DF × S × P	0.0112	3	0.0037	*

* F calculated using $\sigma^2 = 0.032$, which is estimated with 27 degrees of freedom.

There was no replication in the experiment, so no independent estimate of σ^2 can be computed. Assuming that the high-order interactions reflect only random measurement error, we can take the fourth-order interaction, DF $\times$ S $\times$ P $\times$ CL, as an estimate of the error sum of squares, giving $\hat{\sigma}^2 = 0.2305/12 = 0.0192$. We note that several other interactions have mean squares of about the same magnitude as the DF $\times$ S $\times$ P $\times$ CL interaction, and it is tempting to pool these. There are, however, no hard and fast rules about which terms may be pooled. It depends on the data analyst's concept of a model for the data. Pooling more and more degrees of freedom into the random error term will tend to make $\hat{\sigma}^2$ smaller. This carries risks of distoring the decision regarding significance, and we will simply use the value of Pallesen (1987) who pooled only the third-order interactions S*P*CL and S*P*DF to obtain $\hat{\sigma}^2 = (0.2305 + 0.6229 + 0.0112)/(12 + 12 + 3) = 0.8646/27 = 0.032$.

The estimated error variance, $\hat{\sigma}^2 = 0.032 = 0.18^2$, on the logarithmic scale can be interpreted as a measurement error with a standard deviation of about 18% in terms of the original concentration scale.

The main effects of all four factors are all significant at the 0.05% level. The largest source of variation is due to differences between the two samplers. Clearly, it is not acceptable to consider the samplers as equivalent. Presumably, Sampler B gives higher concentrations (see Table 19.1), implying a greater efficiency of contaminant recovery. The differences between samplers is much greater than difference between sampling periods, even though "periods" represents a variety of operating conditions.

The interaction of the sampler with dioxin/furan groups (S $\times$ DF) was small, but statistically significant. The interpretation is that the difference between the samplers changes depending on whether the contaminant level is dioxin or furan. The S $\times$ P interaction is also significant, indicating that the difference between samplers was not constant over the four sampling periods.

The *a priori* expectation was that the dioxin and furan groups (DF) would have different levels and that the amounts of the various chlorinated species (CL) with chemical groups would not be equal. The large mean squares for DF and CL supports this.

COMMENTS

When the experiment was planned, variation between sampling periods was expected to be large and differences between samplers was expected to be small. The data showed both expectations to be wrong. The major source of variation was between the two samplers. Variation between periods was small, though statistically significant.

Several interactions were statistically significant. These, however, have no particular practical importance until the matter of which sampler to use is settled. Presumably, after further research, one of the samplers will be accepted and the other rejected, or one will be modified. If one of the samplers were modified to make it perform more like the other, this analysis of variance would not represent the performance of the modified equipment.

Analysis of variance is a useful tool for breaking down the total variability of designed experiments into interpretable components. For well-designed (complete and fully balanced) experiments, this partitioning is unique and allows clear conclusions to be drawn from the data.

REFERENCES

Box, G. E. P., W. G. Hunter, and J. S. Hunter (1978). *Statistics for Experimenters: An Introduction to Design, Data Analysis, and Model Building*, New York, Wiley Interscience.

Milliken, G. A. and D. E. Johnson (1992). *Analysis of Messy Data — Volume I: Design Experiments*, New York, Van Nostrand Reinhold.

Pallesen, L. (1987). "Statistical Assessment of PCDD and PCDF Emission Data," *Waste Manage. Res.*, 5 367–379.

Rao, C. R. (1965). *Linear Statistical Inference and Its Applications*, New York, John Wiley & Sons.

SAS Institute Inc. (1982). *SAS User's Guide: Statistics*, Cary, NC.

Scheffe, H. (1959). *The Analysis of Variance*, New York, John Wiley & Sons.

Chapter 20

Factorial Experimental Designs

Key words: fly ash, factorial designs, permeability, design matrix, density, main effects, interactions, cube plot

---

Experiments are done to (1) screen a set of factors (independent variables) and learn which produce an effect, (2) estimate the magnitude of effects produced by experimental factors, (3) develop an empirical model, and (4) develop a mechanistic model. Factorial experimental designs are efficient tools for meeting the first two objectives. Many times they are also excellent for objective three, and at times, they can provide a useful strategy for building mechanistic models.

Factorial designs allow a large number of variables to be investigated in few experimental runs. They have the additional advantage that no complicated calculations are needed to analyze the data produced. In fact, the experimental design is so efficient that the important effects are sometimes apparent even before any calculations are done. The efficiency stems from using settings of the independent variables that are completely uncorrelated with each other. In mathematical terms, the experimental designs are called *orthogonal*. The consequence of the orthogonal design is that the main effect of each experimental factor, and also the possible interactions between factors, can be estimated independent of the other effects.

CASE STUDY — COMPACTION OF FLY ASH

There is some interest in using pozzolanic fly ash from a large coal-fired electric generating plant to build impermeable liners for storage lagoons and landfills. Pozzolanic fly ash reacts with water and sets into a rock-like material. With proper compaction, this material can be made very impermeable. A typical criterion is that the liner must have a permeability of no more than 10^{-7} cm/sec. This is easily achieved using small quantities of fly ash in the laboratory, but in the field there are difficulties because the pozzolanic chemical reaction is so fast that the fly ash can start to set before it is properly placed and compacted. If the reaction reaches a certain point before the material is compacted, the permeability will probably exceed the target of 10^{-7} cm/sec.

As a first step, it was decided to study the importance of water content (%), compaction effort (psi), and reaction time (min) before compaction. These three factors were each investigated at two levels. An obvious name for this is a *two-level, three-factor experimental design*. Three factors at two levels gives a total of eight experimental conditions.

The eight conditions are given in Table 20.1, where W denotes water content (4 or 10%), C denotes compaction effort (60 or 260 psi), and T denotes reaction time (5 or 20 min). Also given are the measured densities in pounds per cubic feet. The permeability of each test specimen was also measured, but the data will not be presented here, except to say that permeability was inversely proportional to density. Test details are given by Edil et al. (1987).

The results of the experiment are presented as a cube plot in Figure 20.1. Each corner of the cube represents one experimental condition. The plus (+) and minus (−) signs indicate the levels of the factors. The top of the cube represents the four tests at high compression, whereas the bottom represents the four tests at low pressure. The front of the cube shows the four tests at low reaction time, while the back shows long reaction time.

Without any additional data analysis, it is apparent that each of the three factors has some effect on density. Of the investigated conditions, the best is high water content,

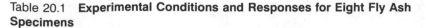

Table 20.1 **Experimental Conditions and Responses for Eight Fly Ash Specimens**

	Factor			Response
Run	W (%)	C (psi)	T (min)	Density (lb/ft³)
1	4	60	5	107.9
2	10	60	5	120.8
3	4	260	5	118.6
4	10	260	5	126.5
5	4	60	20	99.8
6	10	60	20	117.5
7	4	260	20	107.6
8	10	260	20	118.9

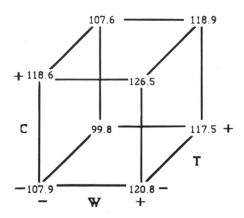

Figure 20.1 Cube plot for the eight experimental conditions of the 2^3 factorial design and showing the measured densities.

high compaction effort, and short reaction time. Densities are higher at the top of the cube than on the bottom, showing that higher pressure increases density. Density is lower at the back of the cube than at the front, showing that long reaction time reduces density. Higher water content increases the density. These differences between the response at high and low levels are called *main effects*. They can be quantified and tested for statistical significance.

It is possible that density is affected by how the factors act in combination. For example, is the effect of water content at 20 min reaction time the same as at 5 min? If not, there is said to be a *two-factor interaction* between water content and reaction time. Water content and compaction might interact, as might compaction and time.

METHOD — A FULL 2^k FACTORIAL DESIGN

The k independent variables whose possible influence on a response variable is to be assessed are referred to as *factors*. An experiment with k factors, each set at two levels, is called a *two-level factorial design*. A *full factorial design* involves making runs at 2^k different experimental conditions, which represent all combinations of the k factors at both high and low levels. The high and low levels are conveniently denoted by + and − or by +1 and −1. The factors can be continuous (pressure, temperature, concentration, etc.) or discrete (additive present, source of raw material, stirring used, etc.).

There are designs for investigating k factors, each set at two levels, that use less than 2^k runs. These designs, which are not "full," are called *fractional factorial designs*; they are discussed in Chapter 21. An experiment in which each factor is set at three levels would be a three-level factorial design (Box and Draper, 1987; Davies, 1960). Only two-level designs will be considered here.

Experimental Design

The *design matrix* lists the setting of each factor in a standard order. Table 20.2 contains the design matrix for a full factorial design with k = 3 factors at two levels and a k = 4 factor design. The three-factor design uses $2^3 = 8$ experimental runs to investigate three factors. The 2^4 design uses 16 runs to investigate four factors. Note the efficiency — only 8 runs to investigate three factors or 16 runs to investigate four factors.

The design matrix provides the information needed to set up each experimental test condition. Run number 5 in the 2^3 design, for example, is to be conducted with factor 1 at its low (–) level, factor 2 at its low (–) setting, and factor 3 at its high (+) setting. The runs should be carried out in randomized order to avoid the possibility that unknown or uncontrolled changes in experimental conditions might bias the factor influence. For example, a gradual increase in response over time might wrongly be attributed to factor 3 if runs were carried out in the standard order sequence. The lower responses would occur in the early runs where factor 3 is at the low setting, while the higher responses would tend to coincide with the + settings of factor 3.

Data Analysis

When the experiment has been completed and the response measurements are in, the statistical analysis consists of estimating the effects of the factors and assessing their significance. For a 2^3 experiment, we can use the *cube plots* in Figure 20.2 to illustrate the nature of the estimates of the three main effects. Later we show how the effects are estimated by linear algebra, which is more convenient in larger designs.

Table 20.2 **Design Matrices for 2^3 and 2^4 Full Factorial Designs**

Run Number	Factor			Run Number	Factor			
	1	2	3		1	2	3	4
1	–	–	–	1	–	–	–	–
2	+	–	–	2	+	–	–	–
3	–	+	–	3	–	+	–	–
4	+	+	–	4	+	+	–	–
5	–	–	+	5	–	–	+	–
6	+	–	+	6	+	–	+	–
7	–	+	+	7	–	+	+	–
8	+	+	+	8	+	+	+	–
				9	–	–	–	+
				10	+	–	–	+
				11	–	+	–	+
				12	+	+	–	+
				13	–	–	+	+
				14	+	–	+	+
				15	–	+	+	+
				16	+	+	+	+

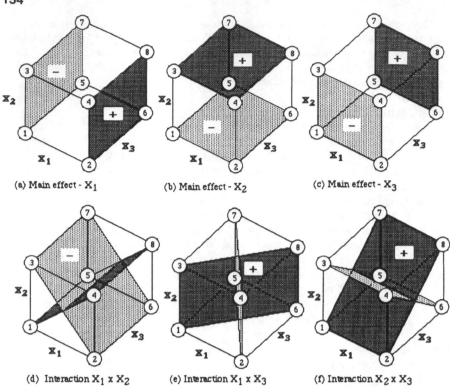

(a) Main effect - X_1

(b) Main effect - X_2

(c) Main effect - X_3

(d) Interaction $X_1 \times X_2$

(e) Interaction $X_1 \times X_3$

(f) Interaction $X_2 \times X_3$

Figure 20.2 Cube plots showing the main effects and two-factor interactions of a 2^3 factorial experimental design.

The main effect of a factor measures the average change in the response caused by changing that factor from its low to its high setting. This experimental design gives four separate estimates of each effect. Referring to Table 20.2, it is seen that the only difference between runs 1 and 2 are the level of factor 1. Therefore, the difference in the response measured in these two runs is an estimate of the effect of factor 1. Likewise, the effect of factor 1 is estimated by comparing runs 3 and 4, runs 5 and 6, and runs 7 and 8. These four estimates of the effect are averaged to estimate the *main effect of factor 1*.

This can also be shown graphically. The main effect of factor 1, shown in Figure 20.2a, is the average of the responses measured where factor 1 is at its *high* (+) setting minus the average of the *low* (−) setting responses. Similarly, the main effects of factor 2 (Figure 20.2b) and factor 3 (Figure 20.2c) are the differences between the average at the *high* and *low* settings for factors 2 and 3. Note that the effects are the change resulting from changing a factor from the *low* to the *high* level. It is not, as we are accustomed to seeing in regression models, the change associated with a one unit change in the level of the factor.

The interactions measure the nonadditivity of the effects of factors 1 and 2. A significant *two-factor interaction* indicates antagonism or synergism between two factors; their combined effect is not the sum of their separate contributions. The interaction between factors 1 and 2 (Figure 20.2d) is the average difference between the effect of factor 1 at the high setting of factor 2 and the effect of factor 1 at the low setting of factor 2. Equivalently, it is the effect of factor 2 at the high setting of factor 1 minus the effect of factor 2 at the low setting of factor 1. This interpretation holds for the two-factor interactions between factors 1 and 3 (Figure 20.2e) and factors 2 and 3 (Figure 20.2f). (There is also a *three-factor interaction*. Ordinarily, this is expected to be small compared to the two-factor interactions and the main effects. This is not diagramed in Figure 20.2.)

The effects are estimated using the *model matrix* shown in Table 20.3. The structure of the matrix is determined by the *model* being fitted to the data. The model to be considered here is linear, and it consists of the average plus three main effects (one for each factor) plus three two-factor interactions plus a three-factor interaction. The model matrix gives the signs that are used to calculate the effects.

This model consists of model column vectors, usually equal in number to the number of experimental runs, and a column vector of the response values. The elements of the column vectors, X_i, can always be coded to be +1 or –1, and the signs are determined from the design matrix, Table 20.3. X_0 is always a vector of +1. X_1 has the signs associated with factor 1 in the design matrix, X_2 has those associated with factor 2, and X_3 has those of factor 3 (and so on for higher-order full factorial designs). These vectors are used to estimate the main effects.

Interactions are presented in the model matrix by cross-products. The elements in X_{12} are the products of X_1 and X_2 (for example, $(-1)(-1) = -1$, $(1)(-1) = -1$, $(-1)(1) = -1$, $(1)(1) = 1$, and so on). Similarly, X_{13} is found from $X_1 \times X_3$ and X_{23} is $X_2 \times X_3$. Likewise, X_{123} is found by multiplying the elements of X_1, X_2, and X_3 (or the equivalent, $X_{12} \times X_3$, or $X_{13} \times X_2$). The order of the X vectors in the model matrix is not important, but the order shown (a column of +1, the factors, the two-column products, followed by higher-order products) is a standard and convenient form.

From the eight response measurements y_1, y_2, . . . , y_8, we can now form eight statistically independent quantities by multiplying the y vector by each of the X vectors. The reason these eight quantities are statistically independent derives from the fact that the X vectors are orthogonal.[1] The independence of the estimated effects is a consequence of the orthogonal arrangement of the experimental design.

This multiplication is done by applying the signs of the X vector to the responses in the y vector and then adding the signed y's. For example, y multiplied by X_0 gives the sum of the responses: $X_0 \cdot y = y_1 + y_2 + \ldots + y_8$. Dividing the quantity $X_0 \cdot y$ by eight gives the *average* response of the whole experiment. Multiplying the y vector by an X vector yields the sum of the four differences between the four y's at the +1 levels

Table 20.3 Model Matrix for a 2^3 Full Factorial Design

Run	X_0	X_1	X_2	X_3	X_{12}	X_{13}	X_{23}	X_{123}	y
1	+1	–1	–1	–1	+1	+1	+1	–1	y_1
2	+1	+1	–1	–1	–1	–1	+1	+1	y_2
3	+1	–1	+1	–1	–1	+1	–1	+1	y_3
4	+1	+1	+1	–1	+1	–1	–1	–1	y_4
5	+1	–1	–1	+1	+1	–1	–1	+1	y_5
6	+1	+1	–1	+1	–1	+1	–1	–1	y_6
7	+1	–1	+1	+1	–1	–1	+1	–1	y_7
8	+1	+1	+1	+1	+1	+1	+1	+1	y_8

---

[1] That is, the product of any two column vectors is zero. For example, $X_3 \cdot X_{123} = (-1)(-1) + \ldots$ $+ (+1)(+1) = 1 - 1 - 1 + 1 + 1 - 1 - 1 + 1 = 0$.

and the four y's at the −1 levels. The average *effect* is the average of the four differences; that is, the effect of factor $\mathbf{X}$ is $\mathbf{y}*\mathbf{X}/4$.

The eight effects and interactions that can be calculated from a full eight-run factorial design are

Average

$$\mathbf{y} \cdot \mathbf{X_0} = \bar{y} = \frac{y_1 + y_2 + y_3 + y_4 + y_5 + y_6 + y_7 + y_8}{8}$$

Main effect of factor 1

$$\mathbf{y} \cdot \mathbf{X_1} = \frac{-y_1 + y_2 - y_3 + y_4 - y_5 + y_6 - y_7 + y_8}{4}$$

$$= \frac{y_2 + y_4 + y_6 + y_8}{4} - \frac{y_1 + y_3 + y_5 + y_7}{4}$$

Main effect of factor 2

$$\mathbf{y} \cdot \mathbf{X_2} = \frac{y_3 + y_4 + y_7 + y_8}{4} - \frac{y_1 + y_2 + y_5 + y_6}{4}$$

Main effect of factor 3

$$\mathbf{y} \cdot \mathbf{X_3} = \frac{y_5 + y_6 + y_7 + y_8}{4} - \frac{y_1 + y_2 + y_3 + y_4}{4}$$

Interaction of factors 1 and 2

$$\mathbf{y} \cdot \mathbf{X_{12}} = \frac{y_1 + y_4 + y_5 + y_8}{4} - \frac{y_2 + y_3 + y_6 + y_7}{4}$$

Interaction of factors 1 and 3

$$\mathbf{y} \cdot \mathbf{X_{13}} = \frac{y_1 + y_3 + y_6 + y_8}{4} - \frac{y_2 + y_4 + y_5 + y_7}{4}$$

Interaction of factors 2 and 3

$$\mathbf{y} \cdot \mathbf{X_{23}} = \frac{y_1 + y_2 + y_7 + y_8}{4} - \frac{y_3 + y_4 + y_5 + y_6}{4}$$

Interaction of factors 1, 2, and 3

$$\mathbf{y} \cdot \mathbf{X_{123}} = \frac{y_2 + y_3 + y_5 + y_8}{4} - \frac{y_1 + y_4 + y_6 + y_7}{4}$$

If the variance of the individual measurements is σ^2, the variance of the mean is

$$\mathrm{Var}(\overline{y}) = \left(\frac{1}{8}\right)^2 (\mathrm{Var}(y_1) + \mathrm{Var}(y_2) + \ldots + \mathrm{Var}(y_8)) = \left(\frac{1}{8}\right)^2 8\sigma^2 = \frac{\sigma^2}{8}$$

and the variance of each main effect and interaction is

$$\mathrm{Var(effect)} = \left(\frac{1}{4}\right)^2 (\mathrm{Var}(y_1) + \mathrm{Var}(y_2) + \ldots + \mathrm{Var}(y_8)) = \left(\frac{1}{4}\right)^2 8\sigma^2 = \frac{\sigma^2}{2}$$

The experimental designs just described do not produce an estimate of σ^2 because there was no replication at any experimental condition. In this case, the significance of effects and interactions is determined from a normal plot of the effects (Box et al., 1987). This plot is illustrated later.

CASE STUDY SOLUTION

The responses at each setting and the calculation of the main effects are shown on the cube plots, Figure 20.3. The average density is

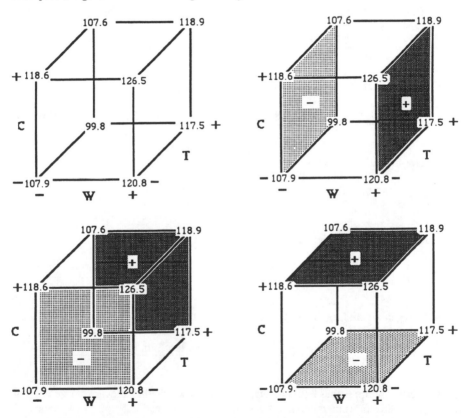

Figure 20.3 Cube plots of the 2^3 factorial experimental design. The values at the corners of the cube are the measured densities at the eight experimental conditions. The shaded faces indicate how the main effects are computed by subtracting the average of the four values at the low setting (– sign; light shading) from the average of the four values at the high setting (+ sign; dark shading).

Average $(\mathbf{y} \cdot \mathbf{X_0})$

$$114.7 = \frac{107.9 + 120.8 + 118.6 + 126.5 + 99.8 + 117.5 + 107.6 + 118.9}{8}$$

The estimates of the three main effects, the three two-factor interactions, and the one three-factor interaction are

Main effect of water $(\mathbf{y} \cdot \mathbf{X_1})$

$$12.44 = \frac{120.8 + 126.5 + 117.5 + 118.9}{4} - \frac{107.9 + 118.6 + 99.8 + 107.6}{4}$$

Main effect of compaction $(\mathbf{y} \cdot \mathbf{X_2})$

$$6.40 = \frac{118.6 + 126.5 + 107.6 + 118.9}{4} - \frac{107.9 + 120.8 + 99.8 + 117.5}{4}$$

Main effect of time $(\mathbf{y} \cdot \mathbf{X_3})$

$$-7.50 = \frac{99.8 + 117.5 + 107.6 + 118.9}{4} - \frac{107.9 + 120.8 + 118.6 + 126.5}{4}$$

Two-factor interaction of water $\times$ compaction $(\mathbf{y} \cdot \mathbf{X_{12}})$

$$-2.85 = \frac{107.9 + 126.5 + 99.8 + 118.9}{4} - \frac{120.8 + 118.6 + 117.5 + 107.6}{4}$$

Two-factor interaction of water $\times$ time $(\mathbf{y} \cdot \mathbf{X_{13}})$

$$2.05 = \frac{107.9 + 118.6 + 117.5 + 118.9}{4} - \frac{120.8 + 126.5 + 99.8 + 107.6}{4}$$

Two-factor interaction of compaction $\times$ time $(\mathbf{y} \cdot \mathbf{X_{23}})$

$$-1.80 = \frac{107.9 + 120.8 + 107.6 + 118.9}{4} - \frac{118.6 + 126.5 + 99.8 + 117.5}{4}$$

Three-factor interaction of water $\times$ compaction $\times$ time $(\mathbf{y} \cdot \mathbf{X_{123}})$

$$-0.35 = \frac{120.8 + 118.6 + 99.8 + 118.9}{4} - \frac{107.9 + 126.5 + 117.5. + 107.6}{4}$$

Before we interpret these effects, we want to know whether they are large enough not to have arisen from random effects. If we had an estimate of the variance of measurement error, the variance of each effect could be estimated and hypothesis tests or confidence intervals could be used to make this assessment. In this experiment, there are no replicated measurements, so it is not possible to compute an estimate of the variance. Lacking a variance estimate, another approach is used to judge the significance of the effects.

If the effects are random, that is arising from random measurement errors, they might be expected to be normally distributed, just as other random variables are expected to be normally distributed. Random effects would will plot as a straight line on a normal probability plot. The normal plot is constructed by ordering the effects (excluding the average), computing the probability plotting points (as shown in Chapter 5), and making a plot on normal probability paper. Since probability paper is not always handy and many computer graphics programs do not make probability plots, it is handy to plot the effects against the normal order scores (or rankits). Table 20.4 shows both the probability plotting positions and the normal order scores for the effects.

Figure 20.4 is a plot of the estimated effects estimated against the normal order scores. Random effects will fall along a straight line on this plot. Significant (nonrandom) effects will fall off the line. These will be the effects that are largest (in absolute value). In this case, a straight line covers the two- and three- factor interactions. None of the interactions are significant. The significant effects are the main effects of water content, compaction effort, and reaction time. (We note that it is possible to draw a straight line that covers the main effects and leaves the interactions off the line. Such an interpretation — significant interactions and insignificant main effects — is not physically plausible. Furthermore, effects of magnitude near zero cannot be significant when effects with larger absolute values are not.)

The final interpretation of the results are

1. The density at the center point of the experiment (W = 7%, P = 160 psi, and T = 12.5 min) is the average $\bar{y}$ = 114.7.
2. Increasing water content from 4 to 10% increases the density by an average of 12.44 lb/ft^3.
3. Increasing compaction effort from 60 to 260 psi increases density by an average of 6.4 lb/ft^3.
4. Increasing reaction time from 5 to 20 min decreases density by an average of 7.5 lb/ft^3.
5. These main effects are additive (because the interactions are zero). Therefore, increasing both water content and compaction effort from their low to high values increases density by 12.44 + 6.4 = 18.84 lb/ft^3.

Table 20.4   Effects, Their Plotting Positions, and Normal Order Scores

Order number i	1	2	3	4	5	6	7
Effect	−7.5	−2.85	−1.80	−0.35	2.05	6.40	12.44
Identity of effect	3	12	23	123	13	2	1
P = 100(i − 0.5)/7	0.07	0.21	0.36	0.50	0.64	0.79	0.93
Normal order scores	−1.352	−0.757	−0.353	0	0.353	0.757	1.352

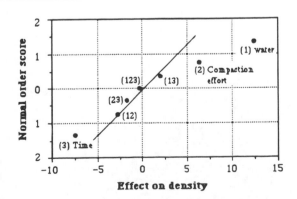

Figure 20.4 Normal probability plot of the estimated main effects and interactions.

COMMENTS

Two-level factorial experiments are a way of discovering the effect of a large number of variables with a minimum of experiments. In general, a *k variable two-level factorial experiment* will require 2^k experimental runs. A 2^2 experiment evaluates 2 variables in 4 runs, a 2^3 experiment evaluates 3 variables in 8 runs, a 2^4 design evaluates 4 variables in 16 runs, and so on. The designs are said to be *full* or *saturated*. From this small number of runs, it is possible to estimate the average level of the response, k main effects, all the possible two-factor interactions, and all higher-order interactions. Not only can all these effects be estimated — they are estimated independently of each other. Each main effect independently estimates the effect of one factor, and only one factor.

The statistical significance of the estimated effects can be evaluated by making the normal plot. If the effects represent only random variation, they will plot as a straight line. If a factor has caused an effect to be larger than expected due to random error alone, the effect will not fall on a straight line. Effects of this kind are interpreted as being significant. Another way to evaluate significance is to compute a confidence interval or a reference distribution. This is shown in Chapter 21.

Factorial designs should be the backbone of an experimenter's design strategy. There is much that is worth learning about this fascinating subject. A few of the more important features are discussed in this book, but only in enough detail to demonstrate their simplicity and power. Chapter 21 shows how four factors can be evaluated with only eight runs. Experimental designs of this kind are called *fractional factorials*. Chapter 22 extends this idea. Chapter 31 explains how factorial designs can be used sequentially to explore a process and optimize its performance.

REFERENCES

Box, G. E. P. and N. R. Draper (1987). *Empirical Model Building and Response Surfaces*, New York, John Wiley & Sons.

Box, G. E. P., W. G. Hunter, and J. S. Hunter (1978). *Statistics for Experimenters: An Introduction to Design, Data Analysis, and Model Building,* New York, Wiley Interscience.

Davies, O. L. (1960). *Design and Analysis of Industrial Experiments*, New York, Hafner Publishing Co.

Edil, T. B., P. M. Berthouex, and K. Vesperman (1987). "Fly Ash as a Potential Waste Liner," Proc. Conf. Geotechnical Practice in Waste Disposal, pp. 447–461, Geotechnical Special Pub. No. 13, ASCE.

Chapter 21

Fractional Factorial Experimental Designs

Key words: dissolved oxygen, fractional factorial design, replication, ruggedness testing, alias structure, confounding, interactions

---

Ruggedness testing is a means of determining which of many steps in an analytical procedure must be carefully controlled and which can be treated with less care. Each aspect or step of the technique needs checking. These problems usually involve a large number of variables and an efficient experimental approach is needed. Fractional factorial designs provide such an approach.

CASE STUDY — SAMPLING HIGH DISSOLVED OXYGEN CONCENTRATIONS

It was necessary, as part of a pilot plant study, to measure the oxygen concentration in the influent to the reactor. The influent was under 20 psig pressure and was aerated with pure oxygen. The dissolved oxygen (DO) concentration was expected to be about 40 mg/L. Sampling methods that are satisfactory at low DO levels (e.g., below saturation) will not work in this situation. Also, conventional methods for measuring DO are not designed to measure such high concentrations above about 20 mg/L. The sampling method that was developed involved withdrawing the highly oxygenated stream into a volume of deoxygenated water, thereby diluting the DO so it could be measured using conventional methods. The DO of the original (undiluted) specimen was estimated by multiplying the measured DO by the dilution factor. There was a possibility that as the pressure dropped from 20 psig in the reactor to atmospheric pressure in the dilution bottle, small bubbles could form and oxygen would be lost. It was essential to mix the pressurized solution with the dilution water in a way that would eliminate, or at least minimize, this loss.

One possible technique would be to try to capture the oxygen before bubbles formed or escaped by introducing the sample at a high rate into a stirred bottle containing a large amount of dilution water. On the other hand, the technique would be more convenient if stirring could be eliminated, if a low sample flow rate could be used, and if only a small amount of dilution water was needed. Perhaps one or all of these simplifications could be made. An experiment was needed that would indicate which, if any, of these variables were important in a particular context. The outcome of this experiment should indicate how the sampling technique could be simplified without loss of accuracy.

Four variables in the sampling procedure seemed critical: (1) stirring rate S, (2) dilution ratio D, (3) specimen input location L, and (4) sample flow rate F. A two-level, four-variable fractional factorial design (2^{4-1}) was used to evaluate the importance of the four variables. This design required measurements at eight combinations of the independent variables. The high and low settings of the independent variables are shown in Table 21.1. The experiment was conducted according to the design matrix in Table 21.2, where the factors (variables) S, D, L, and F are identified as **1**, **2**, **3**, and **4**, respectively. The run order was randomized. DO measurements were duplicated on each specimen. The average and difference between duplicates for each run are shown in Table 21.2.

Table 21.1 **Experimental Settings for the Independent Variables**

Setting	Stirring S	Dilution Ratio D	Sample Input Location L	Sample Flow Rate F
Low level (−)	Off	2:1	Surface	2.6 mL/sec
High level (+)	On	4:1	Bottom	8.2 mL/sec

Table 21.2 **Experimental Design and Measured Dissolved Oxygen Concentrations**

Run	S (1)	D (2)	L (3)	F (4)	Duplicates (mg/L) y_{1i}	Duplicates (mg/L) y_{2i}	Ave. DO (mg/L) $\bar{y}_i$	Diff. (mg/L) d
1	−	−	−	−	38.9	41.5	40.2	−2.6
2	+	−	−	+	45.7	45.4	45.45	0.3
3	−	+	−	+	47.8	48.8	48.3	−1.0
4	+	+	−	−	45.8	43.8	44.8	2.0
5	−	−	+	+	45.2	47.6	46.4	−2.4
6	+	−	+	−	46.9	48.3	47.6	−1.4
7	−	+	+	−	41.0	45.8	43.4	−4.8
8	+	+	+	+	53.5	52.4	52.95	1.1

METHOD — FRACTIONAL FACTORIAL DESIGNS

A fractional factorial design is an experimental layout where a full factorial design is augmented with one or more factors (independent variables) to be analyzed without increasing the number of experimental runs. These designs are labeled 2^{k-p}, where k is the number of factors that could be evaluated as a full factorial design of size 2^k and p is the number of additional factors to be included. When a fourth factor is to be incorporated in a 2^3 design of eight runs, the resulting design is a 2^{4-1} fractional factorial, which also has $2^3 = 8$ runs. The full 2^4 factorial would have 16 runs. The 2^{4-1}, having only eight runs, is a half fraction of the full four-factor design.

Since this design is a half fraction of the full four-factor design, we must determine which half of the 16 experiments are to be done. To preserve the balance of the design, there must be four experiments at the high setting of X_4 and four experiments at the low setting. Note that any combination of four high and four low that we choose for factor 4 will correspond exactly to one of the column combinations for factors X_1, X_2, and X_3 already used in the matrix of the 2^3 factorial design (see Chapter 20, Table 20.2). Which combination should we select? Standard procedure is to choose the three-factor interaction $X_1X_2X_3$ for setting the levels of X_4. Having the levels of X_4 the same as the levels of $X_1X_2X_3$ means that the separate effects of X_4 and $X_1X_2X_3$ cannot be estimated. We can only estimate their combined effect. Their individual effects are *confounded*. Confounded means confused with or tangled up with in a way that we cannot separate without doing more experiments.

The design matrix for a 2^{4-1} design is shown in Table 21.3. The signs of the factor 4 column vector of levels are determined by the product of column vectors for the

Table 21.3   **Design Matrix for a 2^{4-1} Fractional Factorial Design**

	Factor (Independent Variable)			
Run	**1**	**2**	**3**	**4**
1	−	−	−	−
2	+	−	−	+
3	−	+	−	+
4	+	+	−	−
5	−	−	+	+
6	+	−	+	−
7	−	+	+	−
8	+	+	+	+

column **1**, **2**, and **3** factors. (Also, it is the same as the three-factor interaction column in the full 2^3 design.) For example, the signs for run 4 (row 4) are (+) (+) (−) = (−).

The model matrix is given in Table 21.4. The eight experimental runs allow estimation of eight effects, which are computed as the product of a column vector X_i and the **y** vector just as was explained for the full factorial experiment discussed in Chapter 20. The other effects also are computed as for the full factorial experiment, but they have a different interpretation, which will be explained now.

In order to evaluate four factors with only eight runs, we give up the ability to estimate independent main effects. The main effect of X_4 is confounded with the three-factor interaction $X_1X_2X_3$, because we intentionally designed the experiment to create this confounding. This confounding will be denoted by **4 + 123**. Now notice in the design matrix that column vector **1** is identical to the product of column vectors **2**, **3**, and **4**. The effect that is computed as $X_1 \cdot y$ is the main effect of factor 1 plus the three-way interaction of factors 2, 3, and 4. We denote this as **1 + 234**. The four effects that are computed by multiplying the column vectors X_1, X_2, X_3, and X_4 are not clean, independent estimates of the main effects. Each is the sum of a main effect and a three-factor interaction as follows:

Table 21.4   **Model Matrix for the 2^{4-1} Fractional Factorial Design**

Run	Ave.	$\dfrac{1}{S}$	$\dfrac{2}{D}$	$\dfrac{3}{L}$	4 = 123 F (SDL)	12 = 34 SD (LF)	13 = 24 SD (DF)	23 = 14 DL (SF)
1	+	−	−	−	−	+	+	+
2	+	+	−	−	+	−	−	+
3	+	−	+	−	+	−	+	−
4	+	+	+	−	−	+	−	−
5	+	−	−	+	+	−	−	−
6	+	+	−	+	−	+	+	−
7	+	−	+	+	−	−	−	+
8	+	+	+	+	+	+	+	+

Note: Defining relation: **I = 1234** (= S × D × L × F).

$$1 + 234 \qquad 2 + 134 \qquad 3 + 124 \qquad 4 + 123$$

The main effects are confounded with three-factor interactions. The second term (i.e., the **234** in **1 + 234**) is induced by forcing the additional factor into the design without increasing the number of runs. This term indicates what we give up by using a fractional design.

There is further confounding in this design. It can be defined using the relation between X_4 and $X_1X_2X_3$ in the design matrix. We express this as a *generating function* as **4 = 123**. If we multiply both sides of this relation by **4**, we get the defining relation for the confounding in this design: **I = 1234**, where **I** is a column vectors of +'s.

The defining relation reveals that the two-factor interactions are confounded with each other:

$$12 + 34 \qquad 13 + 24 \qquad 23 + 14$$

meaning that the two-way interaction of factors 1 and 2 is confounded with the two-way interaction of factors 3 and 4, and so on.

The consequence of this intentional confounding is that we cannot assume our estimated main effects are unbiased, unless we assume that the three-factor interactions are negligible. Fortunately, three-way interactions often are small and can be ignored. There is no safe basis for ignoring any of the two-factor interactions, so the effects calculated as two-factor interactions must be interpreted with caution.

CASE STUDY SOLUTION

The average response at each experimental setting is shown in Figure 21.1. The small boxes identify the four tests that were conducted at the high flow rate; the low flow rate tests are the four unboxed values. Calculation of the effects was explained in Chapter 20 and will not be repeated here.

This experiment has replication at each experimental condition, so we turn attention to estimating the variance of the measurement error and of the estimated effects. The differences between duplicates, d_i, can be used to estimate the variance of the average response for each run. For a single pair of duplicate observations, y_{1i} and y_{2i}, the sample variance is

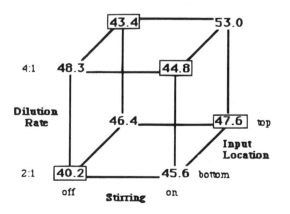

Figure 21.1 A 2^{4-1} fractional factorial design and the average of duplicated measurements at each of the eight design settings.

$$s_i^2 = \frac{1}{2}d_i^2$$

where $d_i = y_{1i} - y_{2i}$ is the difference between the duplicate observations. The average of the duplicate observations is

$$\bar{y} = \frac{y_{1i} + y_{2i}}{2}$$

and the variance of the average of the duplicates is

$$s_{\bar{y}} = \frac{s_i^2}{2} = \frac{d_i^2}{4}$$

For n pairs of duplicate observations, we can combine each individual estimate to get a pooled estimate of the variance of the average

$$s_{\bar{y}}^2 = \frac{1}{n}\sum_{i=1}^{n}\frac{d_i^2}{4} = \frac{1}{4n}\sum_{i=1}^{n}d_i^2$$

For this experiment, n = 8, giving

$$s_{\bar{y}}^2 = \frac{1}{4(8)}(2.6^2 + 0.3^2 + \ldots + 1.1^2) = 1.369$$

and $s_{\bar{y}} = 1.17$.

The main and interaction effects are estimated using the model matrix given in Table 21.4. The average is

$$\bar{y} = \frac{1}{8}\sum_{i=1}^{8}\bar{y}_i$$

and the estimate of each effect is

$$\text{Effect}(j) = \frac{1}{4}\sum_{i=1}^{8}X_{ij}\bar{y}_i$$

where X_{ij} is the ith element of the vector in column j and has values of -1 or $+1$.

The variance of the average $\bar{y}$ is

$$(1/8)^2\, 8\, \sigma_{\bar{y}}^2 = \frac{\sigma_{\bar{y}}^2}{8}$$

and the variance of the other effects is

$$(1/4)^2\, 8\, \sigma_{\bar{y}}^2 = \frac{\sigma_{\bar{y}}^2}{2}$$

Using $s_{\bar{y}}^2$ as our estimator $\sigma_{\bar{y}}^2$ gives the standard errors of the average and the estimated effects

$$SE(\overline{y}) \sqrt{\frac{s_{\overline{y}}^2}{8}} = 0.41$$

and

$$SE(Effect(j)) = \sqrt{\frac{s_{\overline{y}}^2}{2}} = 0.83$$

The estimated effects and their standard errors are given in Table 21.5.

Assuming that the three-factor interactions are negligible, the effects of the four main factors S, D, L, and F may be interpreted as being free of confounding with interactions. This assumption is reasonable since significant three-factor interactions rarely exist. By this, we mean that it is likely that three interacting factors will have a tendency to offset each other and produce a combined effect that is comparable to experimental error. Thus, when the assumption of negligible three-factor interaction is valid, we can achieve in 8 runs what theoretically requires 16 runs (the full 2^4 factorial), and the savings in experimental effort is gratifying.

The same is not true for the two-factor interactions, which are confounded pairwise. For example, the effect we have called DL is not the interaction of factors D and L. It is the interaction of D and L plus the interaction of S and F. If the effect DL (or the effect of SF) is statistically significant, some care must be taken with its interpretation. This same problem exists for all three of the two-factor interaction effects. This is the price of running a fractional factorial 2^{4-1} experiment using only 8 runs instead of the full $2^4 = 16$ run design which would estimate all effects without confounding.

The 95% confidence interval for the true value of the effects is bounded by

$$Effect(j) \pm t_{v=8,\, \alpha/2=0.025} \, SE(Effect(j))$$
$$Effect(j) \pm 2.036\,(0.83) = Effect(j) \pm 1.91$$

from which we can state, with 95% confidence, that effects larger than 1.91 or smaller than −1.91, represent real effects. All four main effects and the two-factor interaction FS-DL are significant.

Alternately, the estimated effects may be viewed in relation to their relevant reference distribution shown in Figure 21.2. This distribution was constructed by scaling a t distribution with 8 degrees of freedom according to the estimated standard error of the

Table 21.5 **Estimated Effects and Their Standard Errors**

Effect	Contributing Factors and Interactions	Estimated Effect	Estimated Standard Error
Average + **1234**	Average(I) + S × D × L × F	46.2	0.41
1 + 234	S + D × L × F	3.2	0.83
2 + 134	D + S × L × F	2.4	0.83
3 + 124	L + S × D × F	2.9	0.83
4 + 123	F + S × D × L	4.3	0.83
12 + 34	S × D + L × F	−0.1	0.83
13 + 24	S × L + D × F	2.2	0.83
23 + 14	D × L + S × F	−1.2	0.83

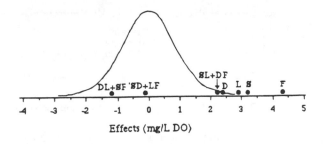

Effects (mg/L DO)

Figure 21.2 Reference distribution that would describe the effects and interactions if they were all random. Effects or interactions falling on the tails of the reference distribution are judged to be real.

effects, which means, in this case, using a scaling factor of 0.83. The calculations are shown in Table 21.6.

It is clear from the reference distribution that the main effects of all four variables are far out on the tails of the distribution, indicating that they are statistically significant. The bounds of the confidence interval, ± 1.91, could be plotted on this reference distribution, but since the results are so clear this is not necessary. Stirring (S) on average elevates the response by 3.2 mg/L. Changing the dilution rate (D) from 2:1 to 4:1 causes an increase of 2.4 mg/L. Setting the sample input location (L) at the bottom yields a response 2.9 mg/L higher than a surface input location. And, increasing the sample flow rate (F) from 2.6 to 8.2 mL/sec causes an increase of about 4.3 mg/L.

Whenever a two-factor interaction appears significant, interpretation of the main effects must be reserved until the interactions have been examined. The reason for this is that a significant two-factor interaction means that the main effect of one interacting factor varies as a function of the level of the other factor. For example, factor X_1 might have a large positive effect at a low value of X_2, but a large negative effect at a high value of X_2. The estimated main effect of X_1 could be near zero (because of averaging

Table 21.6 Constructing the Reference Distribution for Scale Factor = 0.83

t Distribution		Scaled Reference Distribution[a]	
Value of t	**t Ordinate**	**t × 0.83**	**Ordinate/0.83**
0	0.387	0	0.466
0.25	0.373	0.21	0.449
0.5	0.337	0.42	0.406
0.75	0.285	0.62	0.343
1.0	0.228	0.83	0.275
1.25	0.173	1.04	0.208
1.50	0.127	1.24	0.153
1.75	0.090	1.45	0.108
2.00	0.062	1.66	0.075
2.25	0.043	1.87	0.052
2.50	0.029	2.08	0.035
2.75	0.019	2.29	0.023
3.0	0.013	2.49	0.016

[a] Scaling both ordinate and abscissa makes the area under the reference distribution equal to 1.00.

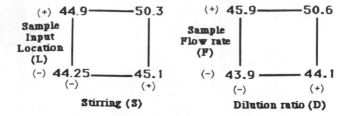

Figure 21.3 Two factor interactions of S and L and of D and F.

the positive and negative effects), and looking at it alone might lead us to wrongly conclude that the factor is inactive.

In this case study, one interaction effect SL+DF appears large enough to be real. This is done by viewing the experiment as one conducted in the interacting factors only. Interpreting the SL interaction is done by examining Figure 21.3, where the responses averaged over all runs with the same signs for L and S. The same examination must be made of the DF interaction, which is also shown in Figure 21.3.

We notice that at the low dilution ratio, the sample flow rate is not very important, and at the low sample flow rate, the dilution ratio is not very important. But, when the sample flow rate is high *and* the dilution ratio is large, the response increases dramatically. A similar interpretation can be made for the L × S interaction. Stirring in conjunction with injecting the sample at the bottom gives higher oxygen measurements, while the other three combinations of L and S show much the same DO levels.

In short, the significant interactions mean that the effect of any factor depends on the level of some other factor. It may be, however, that only one of the interactions LS or FD is real and the other is an artifact caused by the confounding of the two interactions. Further experiments would be needed to untangle this indefinite situation.

However, for the problem at hand, a resolution is not needed because all four factors S, D, L, and F do influence the oxygen measurements and all four should be set at their + levels. That is, the best *of the measurement techniques investigated* would be to inject the sample at the bottom of the bottle, use stirring, use the 4:1 ratio of dilution water to sample, and use the high sample flow rate (8.2 mL/sec). Even better measurement conditions may exist. More tests, using even higher dilution ratios, faster stirring, and a higher sample flow rate, might yield even better results.

COMMENTS

This example has shown how four variables can be evaluated with just 8 runs. The 8 runs were a half fraction of the 16 runs that would be used to evaluate four factors in a full factorial design. This design provided estimates of the main effects of the four factors just as would have been obtained from the full 2^4 design, if we are willing to make the rather mild assumption that three-factor interactions are negligible.

There was a price paid for forcing the extra factor into the eight-run design. Only a total of eight effects and interactions can be estimated from the eight runs. These include the average and the four main effects. The estimates of the two-factor interactions are confounded pairwise with each other, and their interpretation is not clear as it would have been from the full factorial design.

Often we are interested primarily in the main effects, at least in the early iterations of the learning cycle. If the main effects are significant and some of the two-factor interactions hold interest, additional runs could be performed to untangle the interpretation. For example, the other half of the 2^{4-1} fraction could be run. The two half fractions would combine to give the full factorial design, from which all main effects and two-factor interactions could be estimated without confusion. (Actually, since the 16 runs would enable 16 effects and interactions to be evaluated, the three-factor and four-factor interactions could also be estimated.)

In this experiment, running a half fraction made sense. It was experimentally efficient to learn something about the four factors before making a commitment to making 16 runs. Fractional factorial designs are often used in an iterative experimental strategy. Chapter 22 illustrates a 2^{5-1} design for evaluating five factors in 16 runs. Box, Hunter, and Hunter (1978) give two examples of 2^{7-4} designs for evaluating 8 variables in $2^3 = 8$ runs. They also show other fractional factorial designs, their confounding pattern, and give a detailed explanation of how the confounding pattern is discovered.

REFERENCES

Box, G. E.P. and J. S. Hunter (1961). "The 2^{k-p} Fractional Factorial Designs, Part I," *Technometrics*, 3, 3, 311–351.

Box, G. E. P. and J. S. Hunter (1961). "The 2^{k-p} Fractional Factorial Designs, Part II," *Technometrics*, 3, 4, 449–458.

Box, G. E. P., W. G. Hunter, and J. S. Hunter (1978). *Statistics for Experimenters: An Introduction to Design, Data Analysis, and Model Building*, New York, Wiley Interscience.

Chapter 22

Screening of Important Variables

Key words: fly ash, permeability, factorial designs, fractional factorial design, log transformation, normal plot

---

Often several independent variables are potentially important in determining the performance of a process or the properties of a material. The goal is to screen these variables efficiently in order to discover which, if any, do alter performance. *Fractional factorial* experiments are efficient for this purpose. The designs and case study presented here are an extension of the factorial experiment designs discussed in Chapters 20 and 21.

CASE STUDY — USING FLY ASH TO MAKE AN IMPERMEABLE BARRIER

Fly ash from certain kinds of coal is pozzolanic, which means that it will set into a rocklike material when mixed with proper amounts of water. Preliminary tests have shown that pozzolanic fly ash can be mixed with sand or soil to form a material that has a permeability of 10^{-7} cm/sec or lower. Such mixtures might be useful for lining storage lagoons and landfills, if the material is resistant to destruction by being alternately frozen and thawed or wetted and dried. The effect of these conditions on the permeability of fly ash and sand/fly ash mixtures needed to be tested. Two types of fly ash were being considered for this use. An additional factor that was to be tested was the addition of bentonite, a clay that has been frequently used to build impermeable barriers.

A two-level experiment was planned to evaluate the five variables at the levels listed below. The goal was to formulate a durable mixture that had a low permeability.

1. type of fly ash: A or B
2. percentage of fly ash in the mixture: 100% ash or 50% ash and 50% sand
3. bentonite addition (percent of total mixture weight): none or 10%
4. wet/dry cycle: yes or no
5. freeze/thaw cycle: yes or no

A full two-level, five-factor experiment would require testing $2^5 = 32$ different conditions. Each permeability test would take one to three weeks to complete, and only a few permeability test columns were available, so doing 32 runs was not attractive. Reducing the number of variables to be investigated or the number of levels tested for each variable tested was not an acceptable way to reduce the total experimental effort. Some way was needed to investigate the five factors without doing 32 runs. A fractional factorial design provided the solution. Table 22.1 shows the experimental settings for a 16-run 2^{5-1} fractional factorial experimental design.

METHOD — DESIGNS FOR SCREENING IMPORTANT VARIABLES

A full factorial experiment using two levels to investigate k factors requires 2^k experimental runs. The data produced are sufficient to independently estimate 2^k parameters (in this case, the average, k main effects, and $2^k - k - 1$ interactions). The number of main effects and interactions for a few full designs are tabulated in Table 22.2.

There are good reasons other than the amount of work involved to not run large designs like 2^5, 2^6, and so on. First, three-factor and higher level interactions are almost

Table 22.1 2^{5-1} Fractional Factorial Design and the Measured Permeabilities of the Fly Ash Mixtures

Run No.	Type of Fly Ash (X_1)	% Fly Ash (X_2)	Wet/Dry Cycle (X_3)	Freeze/Thaw Cycle (X_4)	Bentonite Addition (X_5)	Permeability cm/sec $\times 10^{10}$ (y)
1	A	50	No	No	10%	1000
2	B	50	No	No	None	160
3	A	100	No	No	None	140
4	B	100	No	No	10%	77
5	A	50	Yes	No	None	1400
6	B	50	Yes	No	10%	550
7	A	100	Yes	No	10%	320
8	B	100	Yes	No	None	22
9	A	50	No	Yes	None	1400
10	B	50	No	Yes	10%	390
11	A	100	No	Yes	10%	580
12	B	100	No	Yes	None	7.5
13	A	50	Yes	Yes	10%	2800
14	B	50	Yes	Yes	None	160
15	A	100	Yes	Yes	None	710
16	B	100	Yes	Yes	10%	19

Table 22.2 Number of Main Effects and Interactions Estimated from a Full 2^k Factorial Design

				Interactions					
k	Number of Runs	Average	Main Effects	2-Factor	3-Factor	4-Factor	5-Factor	6-Factor	7-Factor
3	8	1	3	3	1				
4	16	1	4	6	4	1			
5	32	1	5	10	10	5	1		
7	128	1	7	21	35	35	21	7	1

never significant so we have no interest in getting data just to estimate them. Second, most two-factor interactions are not significant either, and some of the main effects will not be significant. Fractional factorial designs provide an efficient strategy to reduce the work by focusing on the relatively few effects that are realistically expected to be important.

Suppose that out of 32 parameters that could be estimated we expect that 5 or 6 might be important. If we could intelligently select the right subset of experimental runs, we would be able to nicely estimate these few effects. The problem, then, is how to select the subset of experiments.

We could do half the full design, which gives what is called a *half fraction*. If there are five factors, the full design requires 32 runs, but the half fraction requires $(1/2)32 = 2^{-1}2^5 = 2^4$ or $2^{5-1} = 16$. From it, 16 parameters can be estimated. Halving the design again would give $16/2 = 8$ runs $= 2^{5-2}$. Taking a more extreme example, seven factors could be investigated using 128 runs with a 2^7 design, with 64 runs using a 2^{7-1} design, with 32 runs using a 2^{7-2} design, or with 16 runs using a 2^{7-3} design. Each of these is a *fractional factorial design*, more specifically a two-level fractional factorial design.

Common sense tells us that a fractional factorial experiment cannot be equivalent to running the full factorial design. We cannot gain something for nothing. What do we give up as payment for doing fewer runs? A complete answer requires a good deal of explanation. The interested reader will find this explanation in Chapters 10–13 of Box et al. (1978) and in Box and Hunter (1961). Here we give only a brief explanation.

First, we give up being able to make independent estimates of the higher-order interactions. At some level of fractioning, we must also give up the independent estimates of the two-factor interactions. If our primary interest is in knowing the main effects, this price is more than acceptable. It is a terrific bargain. In an experiment that is designed to identify the most important variables, as opposed to an experiment that is intended to produce a model involving variables that are already known to be important, we are satisfied to know only the main effects. If, later, we wish to know more about the interactions, we can always run the missing half fraction of the full design.

We now show how this works out for the 2^5 design of the case study problem. The full design is shown in the left-hand part of Table 22.3. All 32 combinations of the five factors set at two levels are included. The right-hand section of Table 22.3 shows one of the two equivalent half fraction 2^{5-1} designs that can be selected from the full design. The runs selected from the full design are marked with asterisks in the left column, and they are identified by run number in the fractional design. (The equivalent fractional design would be constructed by taking the runs from the full design that are not marked with asterisks.) Note that in each column of the full design, half the values are 1 and half are −1. This balance must be preserved (and it has been) when we select the half fraction.

The 16 runs of the fractional design are constructed as follows:

1. A full 2^4 design was written for the four variables **1**, **2**, **3**, and **4**.
2. The column of signs for the **1234** interaction was written, and these were used to define the levels of variable **5**. Thus, we made **5 = 1234**.

The consequence of making **5 = 1234** is that the value we compute as an estimate of the main effect of factor 5 includes the interaction effect of 1234. That is, the quantity we calculate as the main effect of factor 5 is the main effect plus (or minus) the fourth-factor interaction of factors 2, 3, 4, and 5. We say that the main effect of 5 is *confounded* with the 1234 interaction. Or, we can say that **5** and **1234** are *aliases* of each other. There are other interactions that are confounded, and we need to understand what they are.

The confounding pattern between the columns in the model matrix is determined by the defining relation **I = 12345**. That means that multiplying columns **1234** gives column **5**, that is **5 = 1234**. It is also true that **1 = 2345**, **2 = 1345**, **3 = 1245**, and **4 = 1235**. Each of the main effects is confounded with a four-factor interaction. Normally, such higher-order effects are expected to be small enough to be neglected, in which case the design produces independent estimates of the main effects. Thus, no price is paid for the additional efficiency of the fractional design in terms of estimating the main effects.

The two-factor effects are also confounded in the design. For example, multiplying columns **123** gives a column identical to **45**. As a consequence, the quantity that estimates the 45 interaction also includes the three-factor interaction of 123 (the 123 and 45 interactions are confounded with each other). Also, **12 = 345**, **13 = 245**, **14 = 235**, **15 = 234**, **23 = 145**, **24 = 125**, **25 = 134**, **34 = 125**, and **35 = 124**. Each two-factor interaction is confounded with a three-factor interaction. If a two-factor interaction appeared to be significant, its interpretation would have to take into account the possibility that the effect may be due in part to the higher-order effect.

Table 22.3 **Comparison of the Full 2^5 Factorial Design and the Fractional 2^{5-1} Design**

| | | Full 2^5 Factorial Design | | | | | | Fractional 2^{5-1} Design | | | |
| | | Factor | | | | | | Factor | | | |
Run	1	2	3	4	5	Run	1	2	3	4	5
1	−1	−1	−1	−1	−1	17	−1	−1	−1	−1	1
* 2	1	−1	−1	−1	−1	2	1	−1	−1	−1	−1
* 3	−1	1	−1	−1	−1	3	−1	1	−1	−1	−1
4	1	1	−1	−1	−1	20	1	1	−1	−1	1
* 5	−1	−1	1	−1	−1	5	−1	−1	1	−1	−1
6	1	−1	1	−1	−1	22	1	−1	1	−1	1
7	−1	1	1	−1	−1	23	−1	1	1	−1	1
* 8	1	1	1	−1	−1	8	1	1	1	−1	−1
* 9	−1	−1	−1	1	−1	9	−1	−1	−1	1	−1
10	1	−1	−1	1	−1	26	1	−1	−1	1	1
11	−1	1	−1	1	−1	27	−1	1	−1	1	1
* 12	1	1	−1	1	−1	12	1	1	−1	1	−1
13	−1	−1	1	1	−1	29	−1	−1	1	1	1
* 14	1	−1	1	1	−1	14	1	−1	1	1	−1
* 15	−1	1	1	1	−1	15	−1	1	1	1	−1
16	1	1	1	1	−1	32	1	1	1	1	1
* 17	−1	−1	−1	−1	1						
18	1	−1	−1	−1	1						
19	−1	1	−1	−1	1						
* 20	1	1	−1	−1	1						
21	−1	−1	1	−1	1						
* 22	1	−1	1	−1	1						
* 23	−1	1	1	−1	1						
24	1	1	1	−1	1						
25	−1	−1	−1	1	1						
* 26	1	−1	−1	1	1						
* 27	−1	1	−1	1	1						
28	1	1	−1	1	1						
* 29	−1	−1	1	1	1						
30	1	−1	1	1	1						
31	−1	1	1	1	1						
* 32	1	1	1	1	1						

Note: The runs selected from the full design are marked with asterisks. They are identified by run number in the fractional design.

Using the 2^{5-1} design instead of the full 2^5 saves us 16 runs, but at the cost of having the main effects confounded with four-factor interactions and having the two-factor interactions confounded with three-factor interactions. If our objective is principally to learn about the main effects and if the four-factor interactions are small, the design is highly efficient for identifying the most important among several experimental factors. Furthermore, since each estimated main effect is the average of eight virtually independent comparisons, the precision of the estimates can be excellent.

CASE STUDY SOLUTION

The measured permeabilities are plotted in Figure 22.1. Since the permeabilities vary over several orders of magnitude, the data are best displayed on a logarithmic plot. Clearly, fly ash A has higher permeabilities than fly ash B. Some duplicate tests that are not reported here indicated that the variance tended to be proportional to the permeability. Because of this, a log transformation was used to stabilize the variance.

The main effects of each variable were of primary interest. Two-factor interactions were of minor interest. Three-factor and higher-order interactions were expected to be negligible. A half fraction, or 2^{5-1} fractional factorial design, consisting of 16 runs was used. There are 16 data points, so it is possible to estimate 16 parameters. The "parameters" in this case are the mean, five *main effects*, and ten *two-factor interactions*. Table 22.4 gives the model matrix in terms of the coded variables. The products X_1X_2, X_1X_3, etc. indicate two-factor interactions. Also listed are the permeability, y, and ln(y).

The computation of the average and the main effects was explained in Chapters 20 and 21. The average permeability is 5.50 on the log scale. The estimated and main effects (confounded with four-factor interactions) and two-factor interactions (confounded with three-factor interactions), also on the log scale, are given in Table 22.5.

In the absence of an estimate of σ^2, the significance of these effects is judged by making a normal plot, as recommended by Box et al. (1978). This plot is shown in Figure 22.2. If the effects are random values, arising only from random measurement error, they will fall along a straight line. Significant (i.e., nonrandom) effects fall off the line. These effects are interpreted to indicate that the magnitude of the effect is large enough to be considered a real result of the factor and not the result of random measurement error.

Two factors are seen to be significant — type of fly ash (factor 1) and percentage of fly ash in the mixture (factor 2). The permeability of fly ash A is higher than fly ash B, as can be clearly seen in Figure 22.1. Using 100% fly ash gives a lower permeability than the 50% fly ash mixture. Freeze/thaw cycle, wet/dry cycle, and the addition of bentonite did not affect the permeability of the compacted specimens. The **12 + 345** interaction also appears to be significant. Since the main effects of factors 3, 4, and 5 were insignificant, this interaction effect is probably just the interaction of fly ash type and percent used.

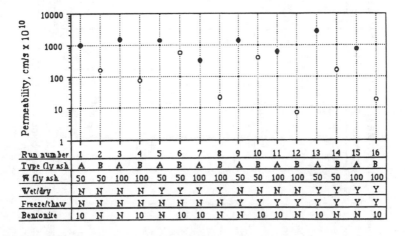

Figure 22.1 Logarithmic plot of the permeability data from the 2^{5-1} factorial experiment. (Solid circles are fly ash A; open circles are fly ash B.)

Table 22.4 Model Matrix for the 2^{5-1} Fractional Factorial Experiment on Permeability of Fly Ash Mixtures with the Measured and Log-Transformed Permeabilities

X_0	X_1 1	X_2 2	X_3 3	X_4 4	X_5 5	X_1X_2 12	X_1X_3 13	X_1X_4 1	X_1X_5 15	X_2X_3 23	X_2X_4 24	X_2X_5 25	X_3X_4 34	X_3X_5 35	X_4X_5 45	y	ln(y)
1	-1	-1	-1	-1	1	1	1	1	-1	1	1	-1	1	-1	-1	1000	6.908
1	1	-1	-1	-1	-1	-1	-1	-1	-1	1	1	1	1	1	1	160	5.075
1	-1	1	-1	-1	-1	-1	1	1	1	-1	-1	-1	1	1	1	140	4.942
1	1	1	-1	-1	1	1	-1	-1	1	-1	-1	1	1	-1	-1	77	4.344
1	-1	-1	1	-1	-1	1	-1	1	1	-1	1	1	-1	-1	1	1400	7.244
1	1	-1	1	-1	1	-1	1	-1	1	-1	1	-1	-1	1	-1	550	6.31
1	-1	1	1	-1	1	-1	-1	1	-1	1	-1	1	-1	1	-1	320	5.768
1	1	1	1	-1	-1	1	1	-1	-1	1	-1	-1	-1	-1	1	23	3.091
1	-1	-1	-1	1	-1	1	1	-1	1	1	-1	1	-1	1	-1	1400	7.244
1	1	-1	-1	1	1	-1	-1	1	1	1	-1	-1	-1	-1	1	390	5.966
1	-1	1	-1	1	1	-1	1	-1	-1	-1	1	1	-1	-1	1	580	6.363
1	1	1	-1	1	-1	1	-1	1	-1	-1	1	-1	-1	1	-1	7.5	2.015
1	-1	-1	1	1	1	1	-1	-1	-1	-1	-1	-1	1	1	1	2800	7.937
1	1	-1	1	1	-1	-1	1	1	-1	-1	-1	1	1	-1	-1	160	5.075
1	-1	1	1	1	-1	-1	-1	-1	1	1	1	-1	1	-1	-1	710	6.565
1	1	1	1	1	1	1	1	1	1	1	1	1	1	1	1	19	2.944

Note: The defining relation is $I = 12345$.

Table 22.5 **Main and Two-Factor Interaction Effects for the Fly Ash Permeability Study**

Effect of Factor	Confounded Factors	Estimated Effect (on Log Scale)
Average		5.5
Main effects (+four-factor interactions)		
Factor 1	1 + 2345	−1.13
Factor 2	2 + 1345	−0.98
Factor 3	3 + 1245	0.14
Factor 4	4 + 1235	0.04
Factor 5	5 + 1234	0.32
Two-factor interactions (+three-factor interactions)		
Factors 1 and 2	12 + 345	−0.28
Factors 1 and 3	13 + 245	−0.13
Factors 1 and 4	14 + 235	−0.38
Factors 1 and 5	15 + 234	0.21
Factors 2 and 3	23 + 145	−0.04
Factors 2 and 4	24 + 135	−0.06
Factors 2 and 5	25 + 134	0.02
Factors 3 and 4	34 + 125	0.00
Factors 3 and 5	35 + 124	−0.22
Factors 4 and 5	45 + 123	−0.56

Note: Data is log transformed.

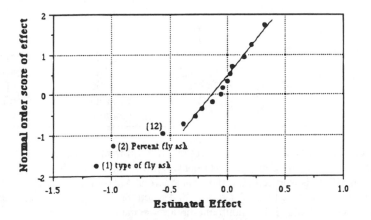

Figure 22.2 Normal plot of the effects computed from the log-transformed permeability values.

COMMENTS

Fractional factorial experiments offer an efficient way of evaluating a large number of variables with a reasonable number of experimental runs. In this example, we have evaluated the importance of five variables in 16 runs. This was a half fraction of a full

178

2^5 factorial experiment, which would use 32 runs. Since there is "no free lunch," something must have been given up in order to study five variables with just 16 runs. What was it? In the design used in the case study, factor five was *confounded* with the factors 1, 2, 3, and 4 in order to construct the 2^{5-1} design. The consequence of this is that main effects of all five factors are confounded with a four-factor interaction and all the two-factor interactions are confounded with the three factor interactions. If the higher-order interactions are small, which is typically expected to be the case, the design produces excellent estimates of the main effects and, therefore, serves its purpose of identifying the most important variables.

The reader should be interested to see other fractional factorial designs. Table 22.6 summarizes some possible designs. It shows five designs that use only eight runs. In eight runs, we can evaluate three or four factors and get independent estimates of the main effects. If we try to evaluate five, six, or seven factors in just eight runs, the main effect will be confounded with second-order interactions. This can often be an efficient design for a screening experiment. The table also shows seven designs that use 16 runs. These can handle four or five factors without confounding the main effects. Lack of confounding is indicated by "OK" in the last two columns of the table, while "confounded" indicates that the mentioned effect is confounded with at least one second-order interaction.

Table 22.6 Some Fractional Factorial Designs

No. of Factors	Design	No. of Runs	Main Effects	Second-Order Interactions
3	2^3	8	OK	OK
4	2^{4-1}	8	OK	Confounded
5	2^{5-1}	8	Confounded	Confounded
6	2^{6-3}	8	Confounded	Confounded
7	2^{7-4}	8	Confounded	Confounded
4	2^4	16	OK	OK
5	2^{5-1}	16	OK	OK
6	2^{5-2}	16	OK	Confounded
7	2^{7-3}	16	OK	Confounded
8	2^{8-4}	16	OK	Confounded
9	2^{9-5}	16	Confounded	Confounded
10	2^{10-6}	16	Confounded	Confounded

This only suggests the main idea of confounding in fractional experimental designs. There is a great deal more that is worth knowing, and the clearly developed discussion in Box et al. (1978) is worth careful study. "A Practical Aid for Experimenters" (Bisgaard, 1987) summarizes more than 40 factorial and fractional designs and gives the effects that can be estimated and their confounding arrangement.

REFERENCES

Bisgaard, S. (1987). *A Practical Aid for Experimenters*, Center for Quality and Productivity Improvement, University of Wisconsin-Madison.

Box, G. E. P. and J. S. Hunter (1961). "The 2^{k-p} Fractional Factorial Designs: Part I," *Technometrics*, 3, 311–351.

Box, G. E. P. and J. S. Hunter (1961). "The 2^{k-p} Fractional Factorial Designs: Part II," *Technometrics*, 3, 449–358.

Box, G. E. P., W. G. Hunter, and J. S. Hunter (1978). *Statistics for Experimenters: An Introduction to Design, Data Analysis, and Model Building*, New York, Wiley Interscience.

Draper, N. R. and H. Smith (1981). *Applied Regression Analysis*, 2nd ed., New York, John Wiley & Sons.

Chapter 23

Correlation Coefficients

Key words: correlation, correlation coefficient, covariance, r, R^2, regression

---

Two variables have been measured, and a plot of the data suggests that there is a linear relationship between them. For example, an increase in one variable is associated with an increase in the other. A statistic that quantifies the strength of the relationship between the variables is the *correlation coefficient.*

Care must be taken lest correlation is confused with causation. Observing that y increases when x increases does not mean that a change in x causes the increase in y. Both x and y may change as a result of change in a third variable z. This reminds us that correlation may, but does not necessarily, indicate causation.

CASE STUDY — COMPARISON OF MEASUREMENTS

Duplicate specimens of wastewater were analyzed for 5-day biochemical oxygen demand (BOD_5), but slightly different dilutions and bacterial inoculum were used on each of the paired measurements. The data are given in Table 23.1. The correlation between the two measurements might be checked to see how closely they are related. There are, of course, other statistical analyses that could be made, including a paired t-test.

COVARIANCE AND CORRELATION

A measure of the linear dependence between two variables x and y, for example, the dissolved solids concentration (x) in a river and river flow (y), is the covariance between x and y. The sample covariance of x and y, denoted by Cov(x,y), is

$$\text{Cov(x,y)} = \frac{\Sigma(x_i - \eta_x)(y_i - \eta_y)}{N}$$

where η_x and η_y are the population means of the variables x and y and N is the size of the population. If x and y were independent, Cov(x,y) would be zero. Note that the converse is not true, however, for we could observe Cov(x,y) = 0 when they are related by, say, a quadratic or exponential function.

The covariance is dependent on the scales chosen. Suppose that x and y are distances measured in inches. If x were converted from inches to feet, the covariance would be divided by 12. If both x and y were converted to feet, the covariance would be divided by $12^2 = 144$. This makes it impossible, in practice, to know whether a value of covariance is large, which would indicate a strong linear relation between two variables, or small, which would indicate a weak association.

A *scaleless covariance*, called the *correlation coefficient* $\rho(x,y)$ or simply ρ, is obtained by dividing the covariance by the two population standard deviations σ_x and σ_y, respectively. The possible values of ρ range from –1 to +1. If x were independent of y, ρ would be zero. Values approaching –1 or +1 indicate a strong correspondence of x with y. A positive correlation ($0 < \rho \leq 1$) indicates that the large values of x are associated with large values of y. In contrast, a negative correlation ($-1 \leq \rho < 0$) indicates that the large values of x are associated with small values of y.

Table 23.1 **Five-Day BOD Measurements on Duplicate Specimens of Wastewater**

BOD-A	200	122	154	176	130	169	166	119	113	110	113	98	181
BOD-B	185	116	158	185	140	179	173	119	119	113	116	122	161

BOD-A	122	157	185	163	178	194	149	119	136	113	130	161
BOD-B	110	176	197	167	179	212	167	116	122	119	119	172

The true values of the population means and standard deviations are estimated from the available data by computing the means $\bar{x}$ and $\bar{y}$ and the standard deviations s_x and s_y. The *sample correlation coefficient* between x and y, for sample size n, is

$$r(x,y) = \frac{\dfrac{\Sigma(x_i - \bar{x})(y_i - \bar{y})}{n - 1}}{s_x \, s_y}$$

CASE STUDY — SOLUTION

Figure 23.1 is a scatterplot of BOD-A vs BOD-B. It shows a good relationship between the measurements on duplicate specimens. Both measurements are affected by random error, but this does not pose any problem in computing the correlation coefficient. We

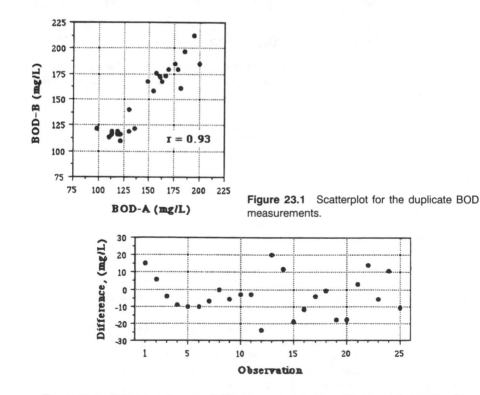

Figure 23.1 Scatterplot for the duplicate BOD measurements.

Figure 23.2 Differences between BOD measurements on duplicate specimens (A – B).

find r = 0.93, indicating a strong positive correlation, which is consistent with the impression gained from the graphical display.

Another interesting issue is whether there is any consistent difference between measurements A and B. Correlation does not answer this question, but it can be investigated by other methods. Figure 23.2 is a plot of the differences (A − B) of measurements on the duplicate specimens. Seventeen of the 25 differences are negative and only 8 are positive. The average difference is 3.0. A paired t-test shows that this difference is not significant, but it would bear watching if additional sampling were planned.

COMMENTS

Correlation coefficients are a familiar way of characterizing the association between two variables. Familiarity sometimes leads to misuse so we remind ourselves that:

1. The correlation coefficient is a valid indicator of association between variables only when that association is linear. If two variables were functionally related according to $y = a + bx + cx^2$, the computed value of the correlation coefficient would not likely approach ± 1, even if the experimental errors were vanishingly small. A scatterplot of the data will reveal whether a low value of r results from large random scatter in the data or from a nonlinear relationship between the variables.
2. Correlation, no matter how strong, does not imply causation. Evidence of causation comes from knowledge of the underlying mechanistic behavior of the system. These mechanisms are best discovered by doing experiments that have a sound statistical design and not from doing correlation (or regression) on data from unplanned experiments.

Ordinary linear regression is similar to correlation in that there are two variables involved and the possibility of a relationship between them is to be investigated. In regression, the two variables of interest are assigned particular roles. One, X, is treated as the independent (or predictor) variable and the other, Y, is the dependent (or predicted) variable. Regression analysis assumes that only Y is affected by measurement error, while X is considered to be controlled or measured without error. Regression of X on Y is not strictly valid when there are errors in both variables (though it is often done). When the errors in X are small (say s_x less than one third s_y), the results are useful. When the errors in X are large relative to those in Y, statements about probabilities of confidence intervals on regression coefficients and standard errors on predictions may be badly wrong.

Above, we said that having errors in both variables invalidates *ordinary* linear regression. There are special regression methods to deal with the *errors-in-variables* problem (Mandel, 1964; Fuller, 1987).

Suppose that ordinary linear regression had been used to investigate the association between the A and B measurements of BOD. The regression equation is $y_B = 3.97 + 0.998\ y_A$ with $R^2 = 0.87$. Or, we could have done the regression with the variables switched to get $y_A = 15.277 + 0.88\ y_B$, also with $R^2 = 0.87$. Notice that $R^2 = 0.87$ happens to be the squares of the correlation coefficient between the two variables ($r^2 = 0.934^2 = 0.87$). In effect, regression has revealed the same information about the strength of the association, even though R^2 and r^2 are different statistics with different interpretations.

Having done regression, we may be tempted to interpret the slope and intercept of the fitted line. For example, we might like to interpret the intercept (3.17) as the bias between measurements A and B and to test its significance to see if it could be considered zero. We cannot do this, because having errors of the same magnitude in both measurements invalidates the confidence intervals on parameters estimated by regression. In

184

Statistical Method	Random Measurement Errors in		
	X	Y	X & Y
Correlation	OK	OK	OK
Regression	$y = f(x)$	$x = f(y)$	Not OK

Figure 23.3 Correlation, regression, and random measurement error.

this situation, therefore, regression does not provide any useful additional information beyond that available from correlation analysis.

Figure 23.3 indicates how the presence of measurement error should influence how we view correlation and regression. Correlation is valid when both variables have random measurement errors. There is no need to think of one variable as X and the other as Y or of one as predictor and the other as predicted. The two variables stand equal, and this helps remind us that correlation and causation are not equivalent concepts.

REFERENCES

Chatfield, C. (1983). *Statistics for Technology*, London, Chapman and Hall.
Folks, J. L. (1981). *Ideas of Statistics*, New York, John Wiley & Sons.
Fuller, W. A. (1987). *Measurement Error Models*, New York, Wiley.
Mandel, J. (1964). *The Statistical Analysis of Experimental Data*, New York, Dover Publications.

Chapter 24

Assessing Serial Correlation

Key words: autocorrelation, correlation, covariance, correlation coefficient, serial correlation, lagged variable, independence

---

When data are collected sequentially, there is often a tendency for those taken close together (in time or space) to be more alike than those taken farther apart. Hourly temperatures, for example, may show great variation over a long period of time, while temperatures 1 hr apart are very similar. Some automated monitoring equipment make measurements so frequently that adjacent values are practically identical. This tendency to be alike is called *serial dependence* or *autocorrelation*. One measure of the serial dependence is the autocorrelation coefficient.

CASE STUDY — SERIAL DEPENDENCE OF BOD DATA

A total of 120 biochemical oxygen demand (BOD) measurements were made at 2-hr intervals as part of a study of treatment plant dynamics. The data are listed in Table 24.1 and plotted in Figure 24.1. The horizontal scale of the graph is observation number and not time. Since the sampling interval is 2 hr, the vertical grid lines mark 24-hr intervals, and the total length of the study is 10 days. There seems to be, as one would expect, a 24-hr cycle; measurements taken 12 sampling intervals (24 hr) apart are similar. The immediate problem is to examine this daily cycle and then assess the strength of the correlation between BOD values separated by 1 up to at least 12 sampling intervals.

CORRELATION AND AUTOCORRELATION COEFFICIENTS

Correlation exists when two variables, say x and y, show a relation with each other. This relation can be seen in a scatterplot of y against x. Correlation is estimated by the *sample correlation coefficient*:

$$r(x,y) = \frac{\Sigma(x - \bar{x})(y - \bar{y})/(n - 1)}{s_x s_y}$$

where $\bar{x}$, $\bar{y}$, s_x, and s_y are the sample means and standard deviations for x and y, and n is the sample size. The correlation coefficient, r, is a dimensionless number that can range from −1 to +1. A value of r = 0 would indicate complete independence between x and y, whereas r = 1 would indicate perfect correspondence.

Autocorrelation is the correlation of a variable with itself. If sufficient data are available, this tendency toward serial dependence can be demonstrated by plotting each observation y_t against the immediately preceding one y_{t-1}. (Note that plotting y_t vs y_{t+1} is equivalent to plotting y_t vs y_{t-1}.) Similar plots can be made for observations two units apart (y_t vs y_{t-2}), three units apart, and so on. If measurements were made daily, a plot of y_t vs y_{t-7} might indicate a tendency for serial dependence in the form of a weekly cycle. If y represented monthly averages, y_t vs y_{t-12} would hold special interest for what it might reveal about an annual cycle. The distance between the observations that are examined for correlation is called the *lag*. This "distance" is measured as number of sampling intervals and not as real time elapsed.

Table 24.1 120 BOD Observations Made at 2 hr Intervals

						Sampling Interval						
Day	1	2	3	4	5	6	7	8	9	10	11	12
1	200	122	153	176	129	168	165	119	113	110	113	98
2	180	122	156	185	16	177	194	149	119	135	113	129
3	160	105	127	162	132	184	169	160	115	105	102	114
4	112	148	217	193	208	196	114	138	118	126	112	117
5	180	160	151	88	118	129	124	115	132	190	198	112
6	132	99	117	164	141	186	137	134	120	144	114	101
7	140	120	182	198	171	170	155	165	131	126	104	86
8	114	83	107	162	140	159	143	129	117	114	123	102
9	144	143	140	179	174	164	188	107	140	132	107	119
10	156	116	179	189	204	171	141	123	117	98	98	108

Note: Time runs left to right.

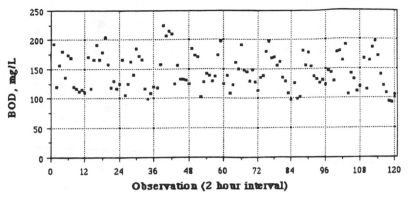

Figure 24.1   A record of influent BOD data sampled at 2-hr intervals.

The correlation coefficients of the lagged observations are called *autocorrelation coefficients*, denoted as γ_k. These are estimated by the lag k *sample autocorrelation coefficient* as

$$r_k = \frac{\Sigma(y_t - \overline{y})(y_{t-k} - \overline{y})}{\Sigma(y_t - y)^2}$$

Again, r_k can be positive or negative, ranging from –1 to +1. Usually, the autocorrelation coefficients are calculated for k = 1, k = 2, and on up to a reasonably high value. Note that $r_0 = 1$ (the correlation of x_t with x_t is perfect).

The set of coefficients, r_k, is called the autocorrelation function. It is common to graph this to show r_k as a function of lag k. The shape of the autocorrelation function is used to identify the form of the time series model that describes the data. This aspect of the analysis will not be discussed. The interested reader should consult Box and Jenkins (1976), Cryer (1986), or Box et al. (1994).

CASE STUDY SOLUTION — EXAMINATION OF THE BOD DATA

Figure 24.2 shows plots of BOD at time t, denoted as BOD(t), against the BOD at 1, 3, 6, and 12 sampling intervals earlier, denoted as BOD(t–1), BOD(t–3), BOD(t–6), and BOD(t–12). The time intervals between these observations are 2, 6, 12, and 24 hr, respectively.

The sample autocorrelation coefficients are given on each plot. At Lag 1 (2-hr interval), a strong autocorrelation is seen and the autocorrelation coefficient is large ($r_1 = 0.49$). There is no relation between observations taken at 6 hr (3 lags); this is clear from the graph and from the autocorrelation coefficient which is essentially zero ($r_3 = -0.03$). At lag 6 (12 hr), the autocorrelation is strong and negative ($r_1 = -0.42$). The negative correlation indicates that observations taken 12 hr apart tend to be opposite in magnitude, one being high and one being low. Samples taken 24 hr apart are positively correlated ($r_1 = 0.25$). The positive correlation shows that when one observation is high, the observation 24 hr ahead (or 24 hr behind) is also high. Conversely, if the observation is low, the observation 24 hr distant is also low.

The sample autocorrelation coefficients are plotted in Figure 24.3 in a form called the *autocorrelation function*. It shows the autocorrelation coefficients of observations that are from 2 (Lag 1) up to 48 hr apart (Lag 24). The correlations for the first 12 lags show a definite diurnal pattern, reflecting the periodic nature of inputs to the wastewater

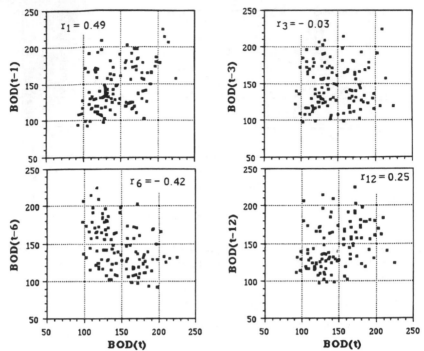

Figure 24.2 Plots of BOD at time t, denoted as BOD(t), against the BOD at 1, 3, 6 and 12 sampling intervals early, denoted as BOD(t–1), BOD(t–3), BOD(t–6) and BOD(t–12). The time intervals between these observations are 2, 6, 12, and 24 hr, respectively.

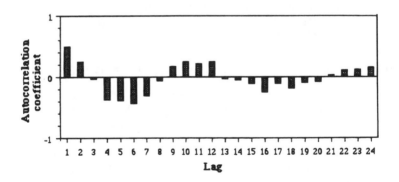

Figure 24.3 The autocorrelation coefficients for Lags 1 through 24 (2 hr–48 hr).

collection system. The pattern of correlations for the first 12 lags repeats for the second 12, but the correlations become less strong as the observations get farther apart (i.e., as the lag increases). Lag 13 is the correlation of observations 26 hr apart. It should be similar to the correlation of samples 2 hr apart, but less strong because of the greater time interval between the samples. The Lag 24 and Lag 12 correlations should be similar, and they are, but the correlation of observations 48 hr apart (Lag 24) is getting weak. This system behavior makes physical sense because many factors (e.g., weather, daily work patterns) change from day to day, thus gradually reducing the system memory.

COMMENTS

It is important to understand that serial correlation may be real even when it is undetected. Unless the number of observations is large (say more than 50), serial correlation is very difficult to detect. Undetected serial correlation is a distinct possibility in small samples, in particular, in environmental systems which tend to have a long memory and to drift slowly in response to changes in season and other factors. Because the identification and confirmation of serial correlation are difficult in small samples (short time series), we can be easily misled into wrongly assuming that autocorrelation is negligible.

When present, autocorrelation can be very upsetting to statistical conclusions, especially to conclusions based on t-tests and F tests. This is why randomization is so important in designed experiments. The t-test, for example, is based on an assumption that the observations are normally distributed, random, and independent. "Independent" refers to an absence of serial correlation. If observations are independent and random, we have no serious problems when using statistical procedures to compare means. Serial correlation, which indicates lack of independence of observations, will bias the estimate of the variance and invalidate the t-test. Box et al. (1978) present a convincing example to demonstrate this pitfall. Chapter 41 on intervention analysis discusses this problem in the context of assessing the shift in the level of a time series related to a manmade intervention in the system.

Linear regression also assumes that the residuals are independent. If serial correlation exists, but we are unaware of it and proceed as though it is absent, all statements about probabilities (hypothesis tests, confidence intervals, etc.) may be wrong. This is illustrated in Chapter 31.

REFERENCES

Box, G. E. P., W. G. Hunter, and J. S. Hunter (1978). *Statistics for Experimenters: An Introduction to Design, Data Analysis, and Model Building*, New York, Wiley Interscience.

Box, G. E. P. and G. M. Jenkins (1976). *Time Series Analysis: Forecasting and Control*, rev. ed., San Francisco, Holden Day.

Box, G. E. P., G. M. Jenkins, and G. C. Reinsel (1994). *Time Series Analysis: Forecasting and Control*. Englewood Cliffs, NJ, Prentice-Hall.

Cryer, J. D. (1986). *Time Series Analysis*, Boston, Duxbury Press.

Chapter 25

Estimating Parameters Using the Method of Least Squares

Key words: joint confidence region, least squares, model, nonlinear least squares, parameter estimation, precision, regression

---

One of the most common problems in statistics is to fit an equation to some data. It is one step, but not the most creative or important step, in the model-building process. The problem might be as simple as fitting a straight-line calibration curve, where the independent variable is the known concentration of a standard solution and the dependent variable is the observed response of an instrument. Or, it might be to fit an unsteady-state nonlinear model, for example, to describe the addition of oxygen to wastewater with a particular kind of aeration device, where the independent variables are water depth, air flow rate, mixing intensity, and temperature.

What do we wish to do with the fitted regression equation? We might (1) desire to predict y in the future from passive observation of $x_1 \ldots x_k$ or (2) to discover how deliberate changes in $x_1 \ldots x_k$ will affect y with the intention of actually modifying the system to get better values for y. Serious errors can result if we mistakenly assume that the causal and correlative system, which operated during the data taking and during the period when predictions are made, has not been interfered with. About this important point, Box said (1966), "To find out what happens to a system when you interfere with it you have to interfere with it (not just passively observe it)."

The equation being fitted to the data may be an *empirical model* (simply descriptive) or a *mechanistic model* (based on a fundamental picture of the process being modeled). A *response variable* or *dependent variable* (y) has been measured at several settings of one or more *independent variables* (x), also called *input variables* or *regressors* or *predictor variables*. *Regression* is a process of fitting an equation to the data. Sometimes regression is called *curve fitting* or *parameter estimation*.

Whatever the form of the model, certain basic ideas apply to fitting both linear and nonlinear models. Nonlinear regression is not conceptually different or more difficult than linear regression. The purpose of this chapter is to explain this. Other chapters will provide specific examples of linear and nonlinear regression.

Many books have been written on regression analysis, and introductory statistics textbooks take care to explain the method. Because this information is rather widely known and readily available, some equations are given in this chapter without much explanation or derivation. The reader who wants more details should refer to the books listed at the end of the chapter.

THE REGRESSION MODEL

The simplest regression problem is that data pairs (y_i, x_i) are to be described by a simple fitted function $y = f(x)$. The models to be fitted can be simple functions of a single independent variable or have many independent variables with higher-order and nonlinear terms, as in the examples given below.

Linear

$$\eta = \beta_0 + \beta_1 x + \beta_2 x^2 \qquad \eta = \beta_0 + \beta_1 x_1 + \beta_2 x_2 + \beta_{12} x_1 x_2$$

Nonlinear

$$\eta = \frac{\theta_1}{1 - \exp(-\theta_2 x)} \qquad \eta = \exp(-\theta_2 x_1)(1 - x_2)^{\theta_1}$$

The terms *linear* and *nonlinear* refer to the parameters in the model and not to the independent variables. In regression, numerical values are known for the dependent and independent variables once the experiment or survey has been completed. It is the parameters — the βs and the θs — that are unknown and must be computed. The model $y = \beta x^2$ is nonlinear in x, but linear in the parameter β. In terms of doing the computations, it does not matter that it is nonlinear with respect to x because the x^2 terms are replaced by numerical values when the parameter estimation is done. It is a *linear* model and it can be fitted by linear regression. In contrast, the model $y = x^\theta$ is nonlinear in θ, and θ must be estimated by nonlinear regression (or we must find a way to transform the model to make it linear).

To help maintain the distinction between linear and nonlinear, we use a different symbol to denote the parameters. In the general linear model $\eta = f(\mathbf{x}, \boldsymbol{\beta})$, $\mathbf{x}$ is a vector of independent variables and the $\boldsymbol{\beta}$ are parameters of the model that will be estimated by carrying out the regression analysis. The estimated values of the parameters β_1, β_2, ..., will be denoted by b_1, b_2, ... Likewise, a general nonlinear model is $\eta = f(\mathbf{x}, \boldsymbol{\theta})$, where $\boldsymbol{\theta}$ is a vector of parameters, the estimates of which are denoted by k_1, k_2, ...

It is usually assumed that a well-conducted experiment produces values of x_i that are essentially without error, while the observations of y_i are affected by *random error*. Under this assumption, for the ith experimental run, the observed value, y_i, is the sum of the true underlying value of the response, η_i, and a residual error, e_i.

$$y_i = \eta_i + e_i \qquad i = 1, 2, \ldots, n$$

If the model is correct, the residual e_i will be nothing more than the random measurement error. If the model is incorrect, e_i will reflect all terms that are needed but missing from the model specification. This means that after we have fitted a model, the residuals contain a large amount of diagnostic information. If they are random and independent, it is persuasive evidence that the proposed model adequately fits the data. If they show some pattern, the pattern will suggest how the model should be modified to improve the fit.

We know, or tentatively propose, a model for η. Suppose that we propose the linear model $\eta = \beta_0 + \beta_1 x$. The observed responses to which the model will be fitted are

$$y_i = \beta_0 + \beta_1 x_i + e_i$$

which has residuals

$$e_i = y_i - \beta_0 + \beta_1 x$$

Similarly, if one proposed the nonlinear model $\eta = \theta_1 \exp(-\theta_2 x)$,

$$y_i = \theta_1 \exp(-\theta_2 x_i) + e_i$$

the residuals will be

$$e_i = y_i - \theta_1 \exp(-\theta_2 x_i)$$

The relation of the residuals to the data and the fitted model are shown in Figure

25.1. The observations are shown as solid circles, and the lines represent the model functions evaluated at particular numerical values of the model's parameters. The vertical distances from the observations and the line representing the model, shown as dotted lines, are the residuals, $e_i = y_i - \eta_i$. The position of the line obviously will depend upon the particular values that are used for β_0 and β_1 in the linear model and for θ_1 and θ_2 in the nonlinear model. The regression problem is to select the values for these parameters that "best fit" the available observations. "Best" is measured in terms of making the residuals small according to a criterion that will be explained in a moment.

If the model is correct, the quantity $e_i = y_i - \eta_i$ is *random experimental error*. If the model is not correct, e_i includes random error plus bias due to model misspecification. In either case, e_i is known as the *residual* or *residual error*. The random errors typically are assumed to be normally and independently distributed with constant variance over the range of values studied. One way to check the adequacy of the model is to check the properties of the residuals of the fitted model by plotting them against the predicted values and against the independent variables.

THE METHOD OF LEAST SQUARES

The "best" estimates of the model parameters are those that minimize the sum of the squared residuals:

$$ S = \sum_{i=1}^{n} (e_i)^2 = \sum_{i=1}^{n} (y_i - \eta_i)^2 $$

Each term in the summation is the difference between the observed value y_i and the value η_i computed from the model at a particular setting of the independent value x_i. If the residuals (e_i) are normally and independently distributed with constant variance, the parameter estimates are unbiased and have minimum variance. This approach to estimating the parameters is known as the *method of least squares*.

The principles outlined thus far about the method of least squares apply equally to linear and nonlinear models. The difference between linear and nonlinear regression lies in how the least squares parameter estimates are calculated. The essential difference is shown by example.

For models that are linear in the parameters, there is a simple algebraic solution for the least squares parameter estimates. Suppose that we wish to estimate β in the model $\eta = \beta x$. The sum of squares function is

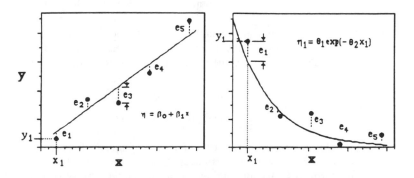

Figure 25.1   A linear model (left) and a nonlinear model (right).

$$S(\beta) = \Sigma(y_i - \beta x_i)^2 = \Sigma(y_i^2 - 2\beta x_i y_i + \beta^2 x_i^2)$$

The parameter value that minimizes S is the *least squares estimate* of the true value of β. This estimate will be denoted by b. We will solve the sum of squares function for this estimate, b, by setting the derivative with respect to β equal to 0 and solving for b

$$\frac{dS(\beta)}{d\beta} = 0 = 2\Sigma(b\, x_i^2 - x_i y_i)$$

This equation is called the *normal equation*. Note that this equation is linear with respect to b, and the reader can confirm the algebraic solution

$$b = \frac{\Sigma x_i y_i}{\Sigma x_i^2}$$

Since the values of x_i and y_i are known once the experiment is complete, this equation provides a generalized method for direct and exact calculation of the least squares parameter estimate.

If the model has two parameters to be estimated, there will be two normal equations, which will be linear with respect to the two parameters to be estimated. Therefore, an algebraic solution is easily derived. (This solution is particularly well known when the two parameters describe a straight line.) As the number of parameters increases, an algebraic solution is still available, but it becomes less interesting to work out and use, so the linear regression calculations are done using matrix algebra.

Unlike linear models, no unique algebraic solution of the normal equations exists for nonlinear models. For example, if $\eta = \exp(-\theta x)$, the method of least squares requires that we find the value of θ that minimizes S.

$$S(\theta) = \Sigma[y_i - \exp(-\theta x_i)]^2 = \Sigma\{y_i^2 - 2\,y_i\exp(-\theta x_i) + [\exp(-\theta x_i)]^2\}$$

The derivative of this nonlinear model with respect to the parameter to be estimated, $dS/d\theta$, does not lead to a neat algebraic solution, and the minimum of S must be found by an iterative numerical search.

EXAMPLES

The similarities and the differences of linear and nonlinear regression will be shown with side-by-side examples. Table 25.1 contains hypothetical data that will be used to illustrate each method.

Assume that there are theoretical reasons why a linear model of the data in Figure 25.2a should go through the origin and why an exponential decay model of Figure 25.2b data should have the value $y = 1$ at $t = 0$. The models and their sum of squares functions are

$$\eta = \beta x + e \qquad \min S(\beta) = \Sigma(y_i - bx_i)^2$$

$$\eta = \exp(-\theta x) + e \qquad \min S(\theta) = \Sigma[y_i - \exp(-kx_i)]^2$$

where again, for convenience, we have written the sum of squares function in terms of the observations, y_i, and the parameter estimates, b and k, respectively. For the linear model, the sum of squares functions written in terms of the observed data and the parameter β is

Table 25.1 **Example Data and the Sum of Squares Calculations for a Linear Model and a Nonlinear Model**

Linear Model

x_i	$y_{calc,i}$	$y_{obs,i}$	e_i	$(e_i)^2$
Trial value: $\beta = 0.115$				
2	0.23	0.150	−0.080	0.00640
4	0.46	0.461	0.001	0.00000
6	0.69	0.559	−0.131	0.01716
10	1.15	1.045	−0.105	0.01102
14	1.61	1.364	−0.246	0.06052
19	2.185	1.919	−0.266	0.07076
			Sum of squares =	0.16586
Trial value: $\beta = b = 0.1$ (optimal)				
2	0.2	0.150	−0.05	0.00250
4	0.4	0.461	0.061	0.00372
6	0.6	0.559	−0.041	0.00168
10	1.0	1.045	0.045	0.00202
14	1.4	1.364	−0.036	0.00130
19	1.9	1.919	0.019	0.00036
			Minimum sum of squares =	0.01158

Nonlinear Model

x_i	$y_{calc,i}$	$y_{obs,i}$	e_i	$(e_i)^2$
Trial value: $\theta = 0.32$				
2	0.5273	0.620	0.0927	0.00859
4	0.2783	0.510	0.2320	0.05381
6	0.1466	0.260	0.1134	0.01286
10	0.0408	0.180	0.1392	0.01939
14	0.0113	0.025	0.0137	0.00019
19	0.0023	0.041	0.0387	0.00150
			Sum of squares =	0.09633
Trial value: $\theta = k = 0.2$ (optimal)				
2	0.6703	0.620	−0.0503	0.00253
4	0.4493	0.510	0.0607	0.00368
6	0.3012	0.260	−0.0410	0.00170
10	0.1353	0.180	0.0447	0.00199
14	0.0608	0.025	−0.0358	0.00128
19	0.0224	0.041	0.0186	0.00035
			Minimum sum of squares =	0.01153

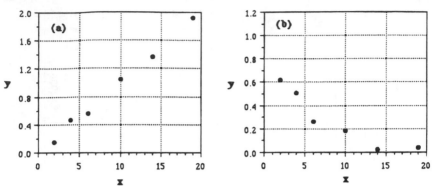

Figure 25.2 Plots of data to be fitted to linear (a) and nonlinear (b) models.

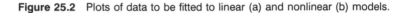

$$S(\beta) = (0.15 - 1\beta)^2 + (0.461 - 2\beta)^2 + (0.559 - 6\beta)^2$$
$$+ (1.045 - 10\beta)^2 + (1.364 - 14\beta)^2 + (1.919 - 19\beta)^2$$

For the nonlinear model, it is

$$S(\theta) = (0.62 - e^{-1\theta})^2 + (0.51 - e^{-4\theta})^2 + (0.269 - e^{-6\theta})^2$$
$$+ (0.180 - e^{-10\theta})^2 + (0.025 - e^{-14\theta})^2 + (0.041 - e^{-16\theta})^2$$

We know that an algebraic solution exists for the linear model, but to show the essential similarity between linear and nonlinear parameter estimation, the least squares parameter estimates of both models will be determined by a straightforward numerical minimization of the sum of squares functions. We simply plot $S(\beta)$ over a range of values of β and do the same for $S(\theta)$ over a range of θ.

Two iterations of this calculation are shown in Table 25.1. The top part of the table shows the trial calculations at $\beta = 0.115$ and $\theta = 0.32$. One clue that these are poor estimates is that the residuals are not random; too many of the linear model regression residuals are negative, and all the nonlinear model residuals are positive. The bottom part of the table is the result obtained for the least squares estimates, b = 0.1 and k = 0.2.

Figure 25.3 shows the smooth curves obtained by following this approach. The minimum sum of squares — the minimum point on the curve — is called the *residual sum of squares*, and the corresponding parameter values are called the *least squares estimates*. The least squares estimates are b = 0.1 and k = 0.2. The *fitted models* (*regression models*) are $\hat{y} = 0.1x$ and $\hat{y} = \exp(-0.2x)$. $\hat{y}$ is the predicted value of the model using the least squares parameter estimates.

The sum of squares function of a univariate linear model always will be a parabola (the curve drawn in Figure 25.3a is a parabola). For nonlinear models, the sum of squares function is not parabolic and it is not, in general, symmetrical. A parabola would not describe the curve in Figure 25.3b very well.

When a model has two parameters, the sum of squares function can be drawn as a surface in three dimensions or as a simple contour map in two dimensions. For a linear model, the surface will be a parabaloid or ellipsoid. For example, the equivalued contours of S of a two-parameter linear model would plot as concentric ellipses. For nonlinear models, the sum of squares surface is not defined by any regular geometric function and it may have very interesting contours.

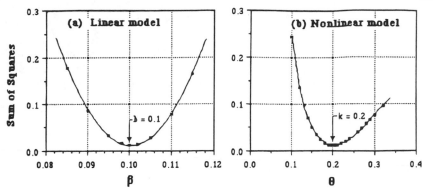

Figure 25.3 The values of the sum of squares plotted as a function of the trial parameter values. The least squares estimates are b = 0.1 and k = 0.2. The sum of squares function is parabolic for the linear model, but not for the nonlinear model.

THE PRECISION OF ESTIMATES OF A LINEAR MODEL

Calculating the "best" values of the parameters is only part of the job. The precision of the parameter estimates needs to be understood. It is not hard to do this, even for a nonlinear model. In fact, Figure 25.3 is the basis for showing the confidence region of the two simple models already considered.

For the one parameter linear model, the variance of b is

$$Var(b) = \frac{\sigma^2}{\Sigma x_i^2}$$

The summation is over all squares of the settings of the independent variable x, and σ^2 is an estimate of the *experimental error variance*.

Ideally, σ^2 would be estimated from independent replicate experiments at some settings of the x variable. Unfortunately, in our example (and in many real experiments), no replicate observations are available, so another approach is used. An estimate of σ^2 can be obtained from the residual sum of squares, if one is willing to assume that the model is correctly specified. If the model is correct, the residuals will be random errors, and the average of these squared residuals is an estimate of the error variance σ^2. Thus, σ^2 may be estimated by dividing the residual sum of squares, S_R, (the minimum sum of squares) by its degrees of freedom, υ. In general, $\upsilon = n - p$, where n is the number of observations and p is the number of estimated parameters.

In our example, $S_R = 0.011584$, p = 1 parameter, n = 6, $\upsilon = 6 - 1 = 5$, and the estimate of the experimental error variance is

$$s^2 = \frac{S_R}{n - p} = \frac{0.011584}{5} = 0.0023168$$

The estimated variance of b is

$$Var(b) = \frac{s^2}{\Sigma x_i^2} = \frac{0.0023168}{713} = 0.00000325$$

and the standard error of b is

$$SE(b) = \sqrt{0.00000325} = 0.0018$$

One statement of the precision of b would be to simply report the interval bounded by b $\pm$ SE(b) = 0.1003 $\pm$ 0.0018. An improvement on this would be to determine a confidence interval on β using Student's t distribution. The $(1 - \alpha)100\%$ confidence limits are given by

$$b \pm t_{\upsilon,\alpha/2} \cdot SE(b)$$

For 95% confidence limits, $\alpha = 0.05$. For our example, $\upsilon = 5$ $t_{5,0.025} = 2.571$, and the 95% confidence limits are $0.1 \pm (2.571)(0.0018)$ or 0.1 ± 0.0046. The confidence limits extend from 0.095 to 0.105.

Figure 25.4a expands the scale of Figure 25.3a in order to more clearly see the confidence interval computed from the t statistic. The sum of squares function and the confidence interval computed using the t statistic are both symmetric about the minimum of the curve. The upper and lower bounds of the confidence interval define two intersections with the sum of squares curve. The sum of squares at these two points happen to be identical because of the symmetry that always exists for a linear model. This level of the sum of squares function is the *critical sum of squares*, S_c. Its significance is that all values of the parameter β that give a value of S that is less than S_c fall within the 95% confidence interval.

THE PRECISION OF ESTIMATES OF A NONLINEAR MODEL

A full discussion of the precision of parameter estimates in a nonlinear model is given in Chapter 26. Here we must be content to point out that the least squares function for the nonlinear model, shown in Figure 25.3b, is not symmetrical about the least squares parameter estimate. The confidence interval for the parameter, θ, will also be asymmetrical. This is one difference between the linear and nonlinear parameter estimation problems. The essential similarity, however, is that we can still define a critical sum of squares, and it will still be true that all parameter values giving S that is less than S_c fall within the confidence interval. Chapter 26 explains how the critical sum of squares level is determined from the minimum sum of squares and an estimate of the experimental error variance.

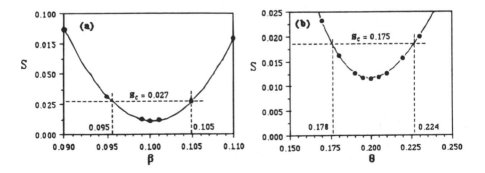

Figure 25.4 Sum of squares functions from Figure 25.3 replotted on a larger scale to show the confidence intervals of β in the linear model (a) and θ in the nonlinear model (b).

COMMENTS

The method of least squares is used in the analysis of data from planned experiments and also in the analysis of data from unplanned happenings. The requirement for the least squares parameter estimates to be the best estimates is that the residual errors, $e = y - \eta$, are random, independent, and with constant variance. Whether the data are planned or unplanned, the residual, e, really describes the effect of a large number of "latent" variables (lurking variables) which we know nothing about. It is the tacit assumption that the requirements for the validity of least squares analysis are satisfied for unplanned data that produces a great deal of trouble (Box, 1966).

There are many conceptual similarities between linear least squares regression and nonlinear regression. In both, the parameters are estimated by minimizing the sum of squares function, which was illustrated in this chapter using two one-parameter models. The basic concepts extend to models with more parameters.

In the case of linear models, the linear algebra used to compute the parameter estimates is so efficient that the work effort is not noticeably different to estimate one or ten parameters. For nonlinear models, the least squares problem in higher dimensions does become more troublesome because the sum of squares surface is not always smooth (it may be bumpy, elongated, or even be unbounded). Conceptually, however, the bigger models are not different than the one-parameter cases examined in this chapter, and efficient algorithms and computer programs are available to tackle the bigger models.

The matter of evaluating the precision of parameter estimates in multiparameter models is discussed in Chapter 26, which again examines both linear and nonlinear cases. If there are two or more parameters, the sum of squares function defines a surface. A *joint confidence region* for the parameters can be constructed by tracing along this surface at the critical sum of squares level. If the model is linear, the joint confidence regions are still based on parabolic geometry. For two parameters, a contour map of the joint confidence region will be described by ellipses. In higher dimensions, it is described by ellipsoids. For linear models, just as there is an exact solution for the parameter estimates, there is an exact solution for the $100(1-\alpha)\%$ confidence interval.

For nonlinear models, the sum of squares surface can have some interesting shapes, but the precision of the estimated parameters is still evaluated by attempting to visualize the sum of squares surface, preferably by making contour maps and tracing approximate joint confidence regions on this surface. For nonlinear models, we can only compute approximate *confidence intervals*, but these approximate intervals nevertheless should be computed because some knowledge of the precision of the estimated values is always useful.

REFERENCES

Box, G. E. P. (1966). "The Use and Abuse of Regression," *Technometrics*, 8, 4, 625–629.

Chatterjee, S. and B. Price (1977). *Regression Analysis by Example*, New York, John Wiley & Sons.

Draper, N. R. and H. Smith (1981). *Applied Regression Analysis*, 2nd ed., New York, John Wiley & Sons.

Meyers, R. H. (1986). *Classical and Modern Regression with Applications*, Boston, Duxbury Press.

Mosteller, F. and J. W. Tukey (1977). *Data Analysis and Regression: A Second Course in Statistics*, Reading, MA, Addison–Wesley.

Neter, J., W. Wasserman, and M. H. Kutner (1983). *Applied Regression Models*, Homewood, IL, Richard D. Irwin Co.

Rawlings, J. O. (1988). *Applied Regression Analysis: A Research Tool*, Pacific Grove, CA, Wadsworth and Brooks/Cole.

Chapter 26

The Precision of Estimated Parameters

Key words: BOD model, Monod model, confidence interval, joint confidence region, least squares, linear regression, nonlinear regression, nonlinear least squares, parameter estimation, parameter correlation, precision, straight line

---

Calculating the "best" values of the parameters is only half the job of fitting and evaluating a model. The precision of these estimates must be known and understood. Unfortunately, this aspect of the job is often overlooked. The precision of estimated parameters, in a linear or nonlinear model, is indicated by the size of their joint confidence region. "Joint" indicates that all the parameters in the model are considered simultaneously. Fortunately, it is not difficult to construct a complete picture of the joint confidence region.

THE CONCEPT OF A JOINT CONFIDENCE REGION

When we use one set of data to estimate the parameters in a model, even simple models like $y = \beta_0 + \beta_1 x + e$ or $y = \theta_1[1-\exp(-\theta_2 x)] + e$, the regression procedure delivers a pair of parameter values. If a different set of data were collected, using the same settings of x, different y values would result, and a different pair of parameter values would be estimated. If this were repeated with many data sets, many pairs of parameter estimates would be produced. If these pairs of parameter estimates were plotted as x and y on Cartesian coordinates, they would cluster about some central point which would be very nearly the true parameter values. Most of the pairs would be near this central value, but some could fall a considerable distance away. This happens because of random variation in the y measurements.

We would also notice on the plot that the location of the plotted pairs of parameter values is not entirely random. The data (if they are useful for model building) will restrict the plausible values to lie within a certain region. The intercept and slope of a straight line, for example, must be within certain limits or the line will not even pass through the data, let alone fit it reasonably well. Furthermore, there will be a relationship between the estimates of the slope and the intercept of a straight line. Specifically, if the slope is decreased somewhat in an effort to better fit the data, inevitably the intercept will increase slightly to preserve a good fit of the line. Thus, low values of slope paired with high values of intercept are plausible, but high slope paired with high intercept is not. This relationship between the parameter values is called *parameter correlation*. It may be strong or weak, depending mainly upon the settings of the x variables at which experimental trials are run.

Figure 26.1 shows some joint confidence regions that might be observed for a two-parameter model. Figures 26.1a and 26.1b show typical elliptical confidence regions of linear models; Figures 26.1c and 26.1d are for nonlinear models which may have confidence regions of irregular shape. A small joint confidence region indicates precise parameter estimates. The orientation and shape of the confidence region also are important. It may show that one parameter is estimated precisely, while another is only known roughly, as in Figure 26.1b where θ_2 is estimated more precisely than θ_1. In general, the size of the confidence region decreases as the number of observations used increases, but it also depends on the actual choice of levels at which measurements are made. This is especially important for nonlinear models. The elongated region in Figure 26.1d could result from placing the experimental runs in locations that are not informative.

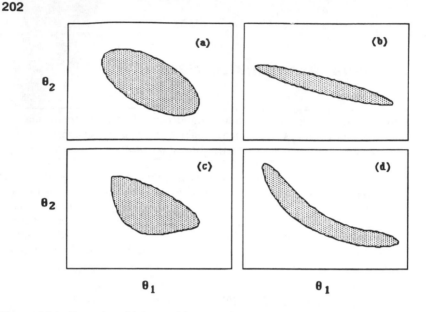

θ_2

θ_2

θ_1 θ_1

Figure 26.1 Examples of joint confidence regions for two parameter models. The elliptical regions (a) and (b) are typical of linear models. The irregular shapes of (c) and (d) might be observed for nonlinear models.

The sum of squares value that bounds the $100(1-\alpha)\%$ joint confidence region is

$$S_c = S_R + S_R\left[\frac{p}{n-p}\,F_{p,n-p,\alpha}\right] = S_R\left[1 + \frac{p}{n-p}\,F_{p,n-p,\alpha}\right]$$

where p is the number of parameters estimated, n is the number of observations, $F_{p,n-p,\alpha}$ is the upper α percent value of the F distribution with p and $n-p$ degrees of freedom and S_R is the residual sum of squares. Here we are using $S_R/(n-p)$ as an estimate of σ^2.

This defines an *exact* $100(1-\alpha)\%$ confidence region for a *linear model*. When the model is nonlinear, some of results that were true for the linear model are no longer correct (Draper and Smith, 1981). When the error e_i of the nonlinear model is normally distributed, $\hat{\sigma}^2 = S_R/(n-p)$ is no longer an unbiased estimate of σ^2. For a nonlinear model, confidence regions can still be defined by this expression, but the confidence level will not be $1-\alpha$. In general, we will not know the exact confidence level, so the region defined is called an *approximate $1-\alpha$ confidence region*.

THEORY — A LINEAR MODEL

Standard statistics texts all give a thorough explanation of linear regression, including a discussion of how the precision of the estimated parameters is determined. Here we only review these ideas in the context of a straight line model: $y = \beta_0 + \beta_1 x + e$. Assuming the errors, e, are normally distributed, with mean zero and constant variance, the best parameter estimates are obtained by the method of least squares. The parameters β_0 and β_1 are estimated by b_0 and b_1

$$b_0 = \bar{y} - b_1\bar{x}$$

and

$$b_1 = \frac{\Sigma(x_i - \bar{x})(y_i - \bar{y})}{\Sigma(x_i - \bar{x})^2}$$

The true value, η, estimated from a measured value of x is $\hat{y} = b_0 + b_1 x$.

The statistics b_0, b_1, and $\hat{y}$ are all normally distributed random variables with means equal to β_0, β_1, and η, respectively, and variances

$$\text{Var}(b_0) = \left(\frac{1}{n} + \frac{\bar{x}^2}{\Sigma(x_i - \bar{x})^2}\right)\sigma^2$$

$$\text{Var}(b_1) = \left(\frac{1}{\Sigma(x_i - \bar{x})^2}\right)\sigma^2$$

and

$$\text{Var}(\hat{y}_k) = \left(\frac{1}{n} + \frac{(x_k - \bar{x})^2}{\Sigma(x_i - \bar{x})^2}\right)\sigma^2$$

The value of σ^2 is typically unknown and must be estimated from the data. If replicate measurements are available, they can be used to get an estimate. More often, σ^2 is estimated by the residual mean square, s^2, which has $\upsilon = n - 2$ degrees of freedom (2 degrees of freedom are lost by estimating two parameters)

$$s^2 = \frac{\Sigma(y_i - \hat{y}_i)^2}{n - 2}$$

The $100(1-\alpha)\%$ confidence intervals for β_0 and β_1 are given by

$$\beta_0: \quad b_0 \pm t_{\upsilon,\alpha/2}\left[\left(\frac{1}{n} + \frac{\bar{x}^2}{\Sigma(x_i - \bar{x})^2}\right)s^2\right]^{1/2}$$

$$\beta_1: \quad b_1 \pm t_{\upsilon,\alpha/2}\left[\left(\frac{1}{\Sigma(x_i - \bar{x})^2}\right)s^2\right]^{1/2}$$

These interval estimates may seem to suggest that the joint confidence region is rectangular, but this is not so. It is elliptical. The exact solution for the $100(1 - a)\%$ joint confidence region for β_0 and β_1 is enclosed by the ellipse given by

$$n(b_0 - \beta_0)^2 + 2(\Sigma x_i)(b_0 - \beta_0)(b_1 - \beta_1) + (\Sigma x_i^2)(b_1 - \beta_1)^2 = 2 s^2 F_{2,n-2,\alpha}$$

where $F_{2,n-2,\alpha}$ is the tabulated value of the F statistic with 2 and $n - 2$ degrees of freedom. This solution is exact.

The confidence interval for prediction of the the mean response, η_0, at a particular value x_0 is

$$\eta_0: \quad (b_0 + b_1 x_0) \pm t_{\upsilon,\alpha/2}\left[\left(\frac{1}{n} + \frac{(x_0 - \bar{x})^2}{\Sigma(x_i - \bar{x})^2}\right)s^2\right]^{1/2}$$

The confidence interval for the prediction of a future single observation, $\hat{y}_f, = b_0 + b_1 x_f$, to be recorded at setting x_f is

$$\hat{y}_f : \quad (b_0 + b_1 x_f) \pm t_{u,a/2}\left[s^2 + \left(\frac{1}{n} + \frac{(x_f - \bar{x})^2}{\Sigma(x_i - \bar{x})^2} \right)s^2 \right]^{1/2}$$

Note that this confidence interval is larger by the factor $t_{u,a/2}s^2$ than the confidence interval for the mean response, η_0, because the prediction error includes the error in estimating the mean response *plus* measurement error in y.

CASE STUDY — A LINEAR MODEL

Data from calibration of a high pressure liquid chromotography (HPLC) instrument and the fitted model are shown in Table 26.1 and in Figure 26.2.

The results of fitting the model $y = \beta_0 + \beta_1 x + \epsilon$ are shown in Table 26.2. The fitted equation for the calibration line is

$$\hat{y} = b_0 + b_1 x = 0.566 + 139.759 \ x$$

and is shown with the data in Figure 26.2. Also shown are the 95% confidence bounds for the mean and future values.

An estimate of the variance of the measured values is needed in order to make any statements about the precision of the estimated parameters or to compute confidence intervals for the line. Since there is no true replication in this experiment, the residual mean square is used as an estimate of the variance. That is, $s^2 = 1.194$, which is estimated

Table 26.1 **Calibration Data**

Dye conc.	0.18	0.35	0.055	0.022	0.29	0.15	0.044	0.028
HPLC peak area	26.666	50.651	9.628	4.634	40.206	21.369	5.948	4.245
Dye conc.	0.044	0.073	0.13	0.088	0.26	0.16	0.10	
HPLC peak area	4.786	11.321	18.456	12.865	35.186	24.245	14.175	

Data from Bailey et al. (1978).

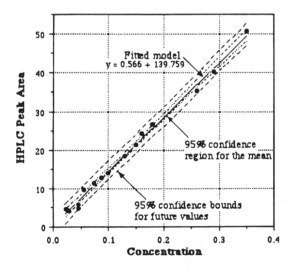

Figure 26.2 Fitted model with 95% confidence bounds for the mean and future values.

Table 26.2 **Results of the Linear Regression Analysis**

Multiple R² = 0.994	Adjusted Multiple R² = 0.994		Standard Error of Estimate = 1.092727			
Variable	**Coefficient**	**Std Error**	**Std Coef**	**Tolerance**	**T**	**P (2 Tail)**
Constant	0.566494	0.473448	0.000000		1.19653	0.25286
X	139.758656	2.889037	0.997234	1.000000	0.048	0.00000

Analysis of Variance

Source	Sum of Squares	Df	Mean Square	F Ratio	P
Regression	2794.309045	1	2794.309045	2340.190092	0.000000
Residual	15.522678	13	1.194052		

with $\upsilon = 15 - 2 = 13$ degrees of freedom. Using this value in the equations given above gives the estimated variances of the parameters:

$$Var(b_0) = 0.2625$$

and

$$Var(b_1) = 2.8898$$

The appropriate value of the t statistic for estimation of the 95% confidence intervals of the parameters is $t_{\upsilon=13,\alpha/2=0.025} = 2.16$, and the confidence intervals are

$$\beta_0 = 0.567 \pm 1.023 \quad \text{or} \quad -0.456 < \beta_0 < 1.587$$
$$\beta_1 = 139.759 \pm 6.242 \quad \text{or} \quad 133.52 < \beta_1 < 146.00$$

The joint confidence interval for the parameter estimates is given by the shaded area in Figure 26.3. Notice that it is not rectangular, as might be suggested by the individual interval estimates. It is elliptical and is bounded by the contour with sum of squares value $S_c = S_R[1 + (2/13)(3.81)] = 24.62$. The equation of this ellipse, based on $n = 15$, $b_0 = 0.567$, $b_1 = 139.758$, $s^2 = 1.194$, $\Sigma x_i = 1.972$, $\Sigma x_i^2 = 0.40249$, and $F_{2,13,0.05} = 3.8056$, is

$$15(0.567 - \beta_0)^2 + 2(1.972)(0.567 - \beta_0)(139.758 - \beta_1)$$
$$+ (0.40429)(139.758 - \beta_1)^2 = 2(1.194)(3.8056)$$

The confidence interval for the mean response η_0 at a single chosen value of x_0 $= 0.2$ is

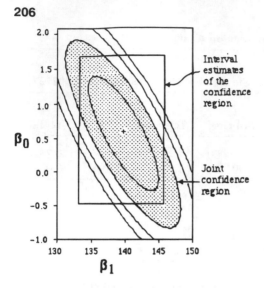

β_0

β_1

Figure 26.3 Contour map of the mean sum of squares surface. The rectangle is bounded by the marginal confidence limits of the parameter considered individually. The shaded area is the 95% joint confidence region for the two parameters of the model and is enclosed by the contour $S_c = 15.523[1 + (2/13)(3.81)] = 24.62$.

$$0.567 + 139.758(0.2) \pm 2.16\left[\left(\frac{1}{15} + \frac{(0.2 - 0.1316)^2}{0.1431}\right)1.194\right]^{1/2} = 28.519 \pm 0.744$$

Thus, the interval 27.775 to 29.263 can be said to contain η, when $x_0 = 0.2$, with 95% confidence.

The confidence interval for a future single observation recorded at a chosen value, say $x_f = 0.2$, is

$$0.567 + 139.758(0.2) \pm 2.16\left[1.194 + \left(\frac{1}{15} + \frac{(0.2 - 0.1316)^2}{0.1431}\right)1.194\right]^{1/2}$$

$$= 28.519 \pm 2.265$$

and we can state, with 95% confidence, that the interval 26.254 to 30.784 will contain the future single observation recorded at $x_0 = 0.2$

CASE STUDY — A BACTERIAL GROWTH MODEL

Here are some data obtained by operating a chemostat at steady-state conditions.

S_i(mg/L COD)	28	55	83	110	138
μ_i(1/hr)	0.053	0.060	0.112	0.105	0.099

The well-known Monod model has been proposed to fit the data

$$\mu = \frac{\mu_{max}\, S}{K_s + S}$$

where

μ = growth rate (per hour) existing at substrate concentration S
S = substrate concentration (mg/L BOD, COD, TOC, etc.)
μ_{max} = maximum growth rate (per hour)
K_s = saturation constant (in units of the substrate concentration)

The sum of squares function that will be minimized to estimate the parameters μ_{max} and K_s is

$$S = \sum_{i=1}^{n} \left[\mu_i - \frac{\mu_{max} S_i}{K_s + S_i} \right]^2$$

The model was fitted using a nonlinear regression program (SYSTAT 1992) to obtain the equation shown in Figure 26.4.

When the model is nonlinear, some of results that were true for the linear model are no longer correct (Draper and Smith, 1981). Confidence regions can still be defined by

$$S_c = S_R \left[1 + \frac{p}{n-p} F_{p,n-px} \right]$$

but the confidence level will not be exactly $1 - \alpha$. Therefore, the region is called an *approximate $100(1 - \alpha)\%$ joint confidence region* for a nonlinear model.

It is rather easy to demonstrate the joint concept confidence region for the two-parameter Monod model shown in Figure 26.4. The least squares parameter estimates are $\hat{\mu}_{max} = 0.153$/hr and $\hat{K}_s = 55.4$ mg/L. The ^ (i.e., the "hats") are reminders that these are the estimates of the true underlying, but unknown, values of μ_{max} and K_s. For our data and experimental design, $S_R = 0.00079$. For $n = 5$ and $p = 2$, $F_{2,3\alpha=0.05} = 9.55$. The bounding sum of squares value is $S_c = 0.00079[1 + (2/3)(9.55)] = 0.00582$.

The left-hand panel of Figure 26.5 shows a contour map of the sum of squares surface for the Monod model. This is constructed by computing sum of squares values over a grid of pairs of values for μ_{max} and K_s and then using a contour plotting program to draw the map. To define the approximate joint confidence region, we just need to trace out the sum of squares contour that has a value of 0.00582. This is shown in the right-hand panel of Figure 26.5. This is a *joint* confidence region because it considers the parameters as a pair. If we collected a very large number of data sets with five observations at the locations used in the example, 95% of the pairs of estimated parameter values would be expected to fall within the joint confidence region.

The size of the region generally indicates how precisely the parameters have been estimated. This confidence region is extremely large — it does not close even when K_s is extended to 500 — and this indicates that the parameter values are not estimated very precisely. We are not entitled to have much confidence in the estimated numerical values.

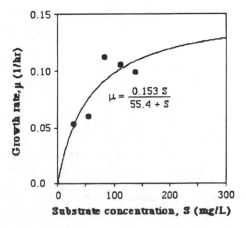

Figure 26.4 The Monod model fitted to the example data.

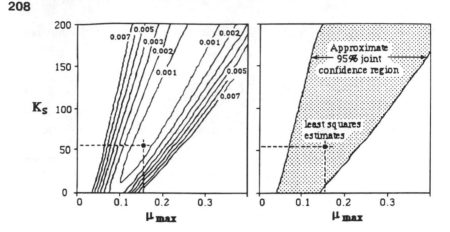

Figure 26.5 Contour map of the sum of squares surface and the approximate 95% joint confidence region for the Monod model shown in Figure 26.4.

WHAT DETERMINES THE SIZE AND SHAPE OF THE CONFIDENCE REGION?

Suppose that after an experiment has been done, and the model has been fitted, we do not like the confidence region because it is too large, it is unbounded, or it has a funny shape. In short, it reveals that the parameter estimates are not estimated with sufficient precision to have adequate predictive value. What can be done?

The size and shape of the confidence region depend on (1) the measurement precision, (2) the number of observations made, and (3) the location of the observations along the scale of the independent variable. Great improvements in measurement precision are not likely to be available, since all good experimenters work to polish the measurement process before starting the actual experiments. The number of observations turns out to be relatively less important than the location of the observations. In the case study example of the Monod model, doubling the number of observations by making duplicate tests at the five selected settings of substrate concentration would not help very much. Often the best way to shrink the size of the confidence region is to be more intelligent about selecting the settings of the independent variable at which observations will be made. Examples are given in the next two sections to illustrate this.

HOW MUCH DO MORE OBSERVATIONS CONTRIBUTE TO PRECISION?

Figure 26.6 shows the Monod model fitted to the original data set augmented with two observations at large substrate concentrations, specifically at $S = 225$ and $S = 375$ mg/L. It also shows the resulting approximate 95% joint confidence region. The corresponding observed growth rates were 0.122 and 0.125/hr. The least squares parameter estimates are $\hat{\mu}_{max} = 0.146$/hr and $\hat{K}_s = 49.8$ mg/L, with a residual sum of squares $S_R = 0.000738$. These values compare closely with $\hat{\mu}_{max} = 0.153$, $\hat{K}_s = 55.4$, and $S_R = 0.00079$ obtained with the original five observations. The differences in numerical values are unimportant, and noting this alone one might wonder whether the two additional observations contributed any new knowledge at all. Actually, there has been a substantial gain in precision of the estimates, but this is only apparent when the joint confidence region is examined.

Both the size and the shape of the confidence region have become more favorable. Whereas before the joint confidence region was so large that the parameter estimates

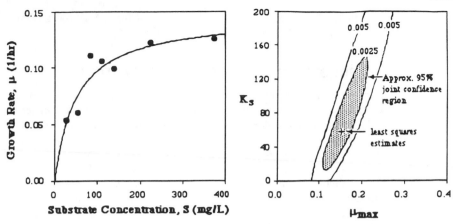

Figure 26.6 Monod model fitted to the original five points plus two more at higher substrate concentrations and the resulting joint confidence region of the parameters.

were virtually unbounded, now the joint confidence region is reasonably small. In particular, μ_{max} is estimated with good precision, since we are highly confident the true value lies between about 0.1 and 0.2, and K_s is seen to be between about 30 and 150 mg/L. The improvement is due mainly to having the new observations at large concentrations, because they contribute directly to estimating μ_{max}. The additional benefit is that any improvement in the precision of μ_{max} simultaneously improves the estimate of K_s.

One reason the confidence region is smaller is because n = 7. This gives a value of the F statistic of F = 5.79, compared with F = 9.94 for n = 5. The sum of squares value that bounds the confidence region when n = 7 is now $(1 + (2/5)5.79)S_R = 3.216S_R$ instead of $(1 + (2/3)9.94)S_R = 7.627S_R$ when n = 5. In order to see that the improvement in precision is not due to this alone, compare Figure 26.6 with Figure 26.7. Figure 26.7 is the result obtained from four of the five original data points (at S = 28, 55, 83, and 138) plus the new observation at S = 375. Indeed, the confidence region is about twice

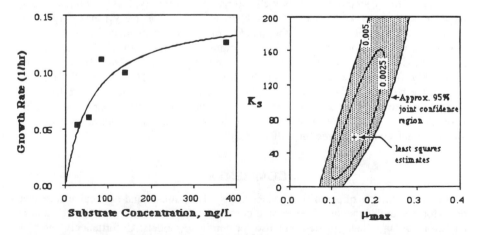

Figure 26.7 Monod model fitted to four of the original five points and one additional point at a higher substrate concentration and the resulting joint confidence region of the parameters.

as large as that in Figure 26.6 being bounded by $S_c = 0.0054$ instead of $S_c = 0.00245$, but it is much smaller than the region drawn in Figure 26.5.

THE PROBLEM OF PARAMETER CORRELATION

Here we look again at the importance of experimental design. Again it will be seen that the location of observations is crucial and, furthermore, that making a large number of observations at the wrong locations does not overcome the weaknesses of a bad experimental design. The weakness, which is obvious when the confidence region is plotted, has gone largely unrecognized. A great many articles on the effect of temperature, pH, metal concentration, etc. on reactions contain parameters estimated from such weak designs (for example, Berthouex and Szewczyk, 1983). The efficiency of aeration equipment was often studied using weak experimental designs (Boyle et al., 1974).

Asymptotic functions, so common in environmental modeling, present a particular problem in parameter estimation, which we will illustrate with the classical first-order model for long-term BOD:

$$y_i = L[1 - \exp(-k\ t_i)] + e_i$$

where y is the BOD measured at time t. The reaction is slow and t is measured in days. Each observation of y comes from incubating a test specimen for time t, and as a result, the y values and their errors, e, are independent. We further assume that the errors are normally distributed and have constant variance. The parameters to be estimated are the ultimate BOD; L, which is approached asymptotically as time goes on; and the first-order reaction rate coefficient, k.

Many published papers show estimates of L and k derived from measurements at just a few early times, say at day 1 through day 5. The experimenters may have reasoned something like this: "I am interested mainly in an estimate of the rate constant, and I want to get in in the shortest time possible. If I know k, I can always compute the ultimate BOD. Since the rate is high on days one through five, data over this range will give a good estimate of k and I can finish the experiment within a few days."

This may not sound too foolish, but it happens to be wrong. Such an experiment gives poor estimates of both parameters. The rate constant can only be precisely estimated if the asymptotic level is well estimated, and this typically requires having measurements at 15 to 20 days. The problem is that data at early times only allow estimation of the initial slope of the curve, which is dy/dt evaluated at $t = 0$. This is $dy/dt|_{t=0} = k\ L\ \exp(-k\ t)|_{t=0} = k\ L$ = constant. Thus, any values of L and k that have a product nearly equal to the slope of the curve over the first few days will reasonably fit the data collected on the first few days. The shape of the joint confidence region will be similar to Figure 26.1d. The hyperbolic shape reflects the correlation between the parameter estimates and shows that neither parameter is well known. Well-designed experiments will yield precise, uncorrelated parameter estimates, and the joint confidence region will be elliptical.

COMMENTS

Computing estimates of parameters values is easy. Even most hand-held calculators can do it for a straight line, and inexpensive commercial software for personal computers can do it for large linear models and also for nonlinear models. Unfortunately, most of these computing resources do not compute or plot the joint confidence region for the parameters. Let us hope that in the future this capability will be added, because our work is not complete until we know the precision with which the estimates have been made.

This chapter has shown by example how the precision of parameter estimates can be evaluated by examining the joint confidence region. For linear models, exact confidence regions can be developed. If the model is nonlinear, we must be content to construct an *approximate* joint confidence region. Even though the region is approximate, its size and shape still convey important information, and the effort required to determine these characteristics is usually rewarded.

The highly correlated parameter estimates that result from weak experimental designs can lead to serious disputes. Suppose that two laboratories made independent experiments to estimate the parameters in a model and, furthermore, that both used a poor design. One might report $\theta_1 = 750$ and $\theta_2 = 0.3$, while the other reports $\theta_1 = 13,000$ and $\theta_2 = 0.1$. They easily might suspect the other of doing careless work or of making a mistake in calculating the parameter values. If the experimental design produced an elongated confidence region with a high degree of correlation, both sets of parameters values could be within the confidence region. In a sense, both could be considered correct and both could be useless. If the joint confidence region were presented along with the estimated parameter values, it would be clear that the disagreement stemmed from the experimental design and not from flawed analytical procedures.

The examples presented here have, for graphical convenience, been based on two-parameter models. We can try to get a picture of the joint confidence region when there are three parameters by making contour maps of slices through *volume* which is bounded by the critical sum of squares value. The critical sum of squares value is determined exactly as illustrated in this chapter.

REFERENCES

Bailey, C. J., E. A. Cox, and J. A. Springer (1978). "High Pressure Liquid Chromatographic Determination of the Imediate/Side Reaction Products in FD & C Red No. 2 and FD & C Yellow No. 5: Statistical Analysis of Instrument Response," *J. Assoc. Off. Anal. Chem.*, 61, 1404–1414.

Berthouex, P. M. and J. E. Szewczyk (1983). "Discussion: Influence of Toxic Metal Ions on the Repression of Carbonaceous Oxygen Demand," *Water Res.*, 18, 385–386.

Boyle, W. C., P. M. Berthouex, and T. Rooney (1974). "Pitfalls in Parameter Estimation for Oxygen Transfer," *J. Sanit. Eng. Div. ASCE*, 100, 2, 391–408.

Draper, N. R. and H. Smith (1981). *Applied Regression Analysis*, 2nd ed., New York, John Wiley & Sons.

Chapter 27

Calibration

Key words: calibration, confidence intervals, inverse prediction, least squares, measurement error, precision, prediction, straight line, regression

---

Instrumental methods of chemical analysis (and also some "wet" analyses) are based on being able to construct a calibration curve which will translate the instrument's highly precise physical signal (light absorption, peak height, voltage, etc.) into an estimate of the true concentration of the specimen being tested. The usual procedure to construct a calibration curve is for the analyst to take a series of measurements on prepared specimens (normally, at least three or four and sometimes more) in which the concentration is *known*. It is important that these *calibration standards* cover the whole range of concentrations required in subsequent analyses. Concentrations of test specimens are to be determined by interpolation and not by extrapolation.

The calibration standards are measured under the same conditions as those that subsequently will be used to test the unknown specimens. Also, a *blank* specimen should be included in the calibration curve. The blank contains no deliberately added analyte, but it does contain the same solvents, reagents, etc. as the other calibration standards and test specimens and is subjected to the same sequence of preparatory steps. The instrumental signal given by the blank sample often will not be zero. It is, however, subject to error, and it should be treated in the same manner as any of the other points on the calibration plot. It is wrong to subtract the blank value from the other standard values before plotting the calibration curve.

One could construct a calibration curve with a good eye and a straight edge and use it without bothering with statements about the precision of the measured concentrations. No doubt, this is often done. Let us assume, however, that the measurements are important enough that the best calibration line is to be determined and that the measured concentrations will be accompanied by a statement of their precision. This raises several statistical questions (Miller and Miller, 1984):

1. Does the plot of the calibration curve seem to be well described by a straight line? If it seems to be curved, what is the mathematical form of the curve?
2. Bearing in mind that each point on the calibration curve is subject to error, what is the best straight line (or curved line) through these points?
3. If the calibration curve is actually linear, what are the estimated errors and confidence limits for the slope and intercept of the line?
4. When the calibration curve is used for the analysis of a test specimen, what are the error and confidence limits for the determined concentration?

The calibration curve is always plotted with the instrument response on the vertical (y) axis and the standard concentrations on the horizontal (x) axis. This is because the standard statistical procedures used to fit a calibration line to the data and to interpret the precision of estimated concentrations assume that the y values contain experimental errors, but the x values are error free. Anyone experienced in laboratory work will recognize instances where this last assumption is not strictly correct, but it is often true that standards can be prepared with an error of about 0.1% or less, while the measurements themselves have errors of 1% or larger. Another assumption implicit in the usual analysis of the calibration data is that the magnitude of the errors in the y values is independent of the analyte concentrations. Common sense and experience indicate that measurement error is often proportional to analyte concentration. The analyst is reminded that these

214

assumptions can be checked once data are available. If either assumption is violated, the statistical procedures can be modified.

CASE STUDY — HPLC CALIBRATION

The data in Table 27.1 relate the concentration of a dye to the peak area response of a high pressure liquid chromatographic (HPLC) instrument (Bailey et al., 1978; Hunter, 1981). The chemist needs a calibration curve for these data. A plot of the data, Figure 27.1, indicates that a straight-line calibration curve is suitable.

The calibration problem has two parts. The first is to construct the calibration curve. The second is to use that line to estimate a concentration from an observed instrument response. The statistical problem is to fit the straight line and provide some useful statements about the precision of the fitted line and the estimates of dye concentration that will be made from its use with future HPLC measurements of peak area (Hunter, 1981).

THEORY — CONSTRUCTING A STRAIGHT-LINE CALIBRATION CURVE

The calibration curve will relate the concentration of the standard solution, ξ, and the instrument response, η. Assume that the functional relationship between these two variables can be well approximated by a straight line of the form $\eta = \beta_0 + \beta_1 \xi$, where β_0 is the intercept and β_1 is the slope of the calibration line. In practice, the true values of η and ξ are not known. Instead, we have the values x and y, where $x = \xi + e_x$ and $y = \eta + e_y$. Here e_x is a random measurement error associated with the attempt to realize the true dye concentration, ξ, and e_y is another random measurement error associated with the response, η. Assuming this error structure, the model is

$$y = \beta_0 + \beta_1(x - e_x) + e_y = \beta_0 + \beta_1 x + (e_y - \beta_1 e_x)$$

In the usual straight-line model, it is assumed that the error in x is zero ($e_x = 0$) or,

Table 27.1 **Calibration Data for HPLC Measurement of Dye**

Dye conc.	0.18	0.35	0.055	0.022	0.29	0.15	0.044	0.028
HPLC peak area	26.666	50.651	9.628	4.634	40.206	21.369	5.948	4.245
Dye Conc.	0.044	0.073	0.13	0.088	0.26	0.16	0.10	
HPLC peak area	4.786	11.321	18.456	12.865	35.186	24.245	14.175	

Data from Bailey et al. (1978).

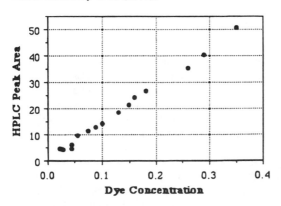

Figure 27.1 A plot of the calibration data.

if that is not literally true, that the error in x is much smaller than the error in y (i.e., $e_x \ll e_y$). In terms of the experiment, this means that the x value is controlled and that the experiment can be repeated at any desired x value. In most calibration problems, one of these conditions is reasonably accepted, and the random variability of the measurement system can be attributed solely to the y values. Denoting the errors in y simply as ε, the model becomes the familiar

$$y = \beta_0 + \beta_1 x + e$$

Usually, the errors, e, are assumed to be independent, with normal distribution having a mean equal to zero and a constant variance σ^2.

It is now clear that fitting the best straight-line calibration curve is just a problem of linear regression. The two parameters (the slope and intercept of the straight line) are estimated and their precision is evaluated using the procedures described in Chapters 25 and 26.

Using the Calibration Curve

In practice, our objective is not to construct a calibration curve. It is to estimate concentrations (x values) from measured peak areas (y values). An interesting problem arises in using the calibration curve in this inverse fashion. Of course, it is a simple matter to use the fitted equation, $y = b_0 + b_1 x$, to compute an x value for any given y value. Specifically, the estimate of x for any future observation y is simply $\hat{x} = (y - b_0)/b_1$. Since the calibration curve is to be used over and over again, what is required for a future value of y, in addition to the predicted value $\hat{x}$, is an interval estimate that has the property that at least 100P% of the intervals will contain the true concentration value ξ with $100(1 - \alpha)\%$ confidence. (For example, for P = 0.9 and $\alpha = 0.05$, we can assert with 95% confidence that 90% of the computed interval estimates of concentration will contain the true value.) This problem was resolved by Leiberman et al. (1967); Hunter (1981) gives a full example.

The error in $\hat{x}$ clearly will depend in some way on the error in measuring y and also on how closely b_0 and b_1 estimate the true slope and intercept of the calibration line. The uncertainties associated with estimating β_0 and β_1 imply that the regression line is not unique. Another set of standards and measurements would yield a different line. We should consider, instead of a regression line, that there is a regression band (Sharaf et al., 1986), which can be represented by a confidence band for the calibration line. The confidence band for the entire calibration line (i.e., valid for any value of x) can be used to translate the confidence interval for the true response η, based on a future observation of y, into a confidence interval for the abscissa value.

The confidence interval for the entire line was first proposed by Working and Hotelling (1927) and is computed from

$$b_0 + b_1 x \pm \left[2 F_{2,\nu,\alpha} \left(\frac{1}{n} + \frac{(x - \bar{x})^2}{\Sigma(x - \bar{x})^2} \right) s^2 \right]^{1/2}$$

in which

$$s^2 = \frac{\Sigma(y_i - \hat{y})^2}{(n - 2)}$$

Note that this confidence band is not the same as the confidence interval for y_f predicted from a future value of x_f, a case that was described in Chapter 26. This confidence region can be narrowed somewhat by limiting the range of the abscissa variable (Wynn and Bloomfield, 1971; Hunter, 1981).

Leiberman et al. (1967) proposed a practical approach to using the Working-Hotelling confidence region for the calibration line to obtain a confidence interval for the predicted concentration. They construct a 100P% confidence interval for the true response, η, based upon a future observation of y. This interval is computed using

$$y \pm z_p \left[\frac{\nu\, s^2}{\chi^2_{\nu,\alpha/2}} \right]^{1/2}$$

where y is the measured instrument response and z_p is the standard normal deviate for percentage P. The quantity in brackets is an estimate of the upper $100(1-\alpha)\%$ confidence limit on the standard deviation of the instrument response, y, where s^2 is the variance of replicate measurements of y, and $\chi^2_{\nu,\alpha/2}$ is the lower $\alpha/2$ percentile point of the χ^2_ν distribution, where $\nu = n - 2$ degrees of freedom associated with the estimate of s^2.

This interval for y and the Working-Hotelling confidence region for the calibration line are used to estimate a confidence interval about $\hat{x}$ for the true concentration ξ. This typically is done graphically. Hunter (1981) provides a full derivation and explanation of the procedure.

CASE STUDY — SOLUTION

Fitting the Calibration Line

The equation for the calibration line is

$$\hat{y} = b_0 + b_1 x = 0.566 + 139.759\ x$$

The calibration data and the fitted model are plotted in Figure 27.2. Also shown are the Working-Hotelling 95% confidence region plus the confidence bounds for 90% of future observations. The estimate of the variance[1] is $s^2 = 1.194$, which is estimated with

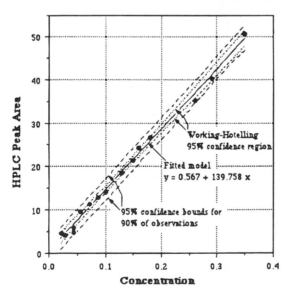

Figure 27.2 The calibration data and the fitted calibration curve with its Working-Hotelling 95% confidence region plus the confidence bounds for 90% of future observations.

---

[1] The residual mean square error is our estimate of σ^2.

$\nu = 15 - 2 = 13$ degrees of freedom, and the estimated variances of the parameters are Var(b_0) = 0.2625 and Var(b_1) = 2.8898. The appropriate value of the t statistic for estimation of the 95% confidence intervals of the parameters is $t_{\nu=13,\alpha/2=0.025} = 2.16$, giving 95% confidence intervals as follows:

$$\beta_0 = 0.567 \pm 1.023 \qquad \text{or} \qquad -0.456 < \beta_0 < 1.587$$
$$\beta_1 = 139.759 \pm 6.242 \qquad \text{or} \qquad 133.52 < \beta_1 < 146.00$$

Using the Calibration Curve to Predict Concentrations

Straightforward application of linear regression to fit a straight line provided the calibration curve, but it has not answered critical questions about using the curve to estimate concentration from an observation of HPLC peak area.

Specifically, suppose that the chemist measured peak area = 22.0. The predicted concentration is read from the calibration line or computed from the calibration equation $\hat{x} = (y - b_0)/b_1 = (22.0 - 0.566)/139.76 = 0.153$. To establish confidence limits for the true concentration, first determine the confidence limits for the value y = 22.0. Then, using the Working-Hotelling confidence limits for the calibration line, translate these limits into confidence limits on the abscissa. The graphical solution is shown in Figure 27.3.

The specific steps are as follows. The measured peak area of 22.0 yields an estimated concentration of 0.153. Use $z_{p=0.9} = 1.26$ for the confidence limits for the y value that will contain the true value of the peak area (ordinate) 90% of the time. To have a confidence level of $100(1 - \alpha)\% = 95\%$, specify $\alpha/2 = 0.025$, which gives $\chi^2_{13,0.025} = 5.01$. The corresponding interval for the peak area is

$$y \pm 1.26(13(1.194)/5.01)^{1/2} = y \pm 2.218$$

For the observed peak area y = 22.0, the relevant interval estimate of the ordinate variable is 22 ± 2.218. Or, we can state with 95% confidence that the bounds

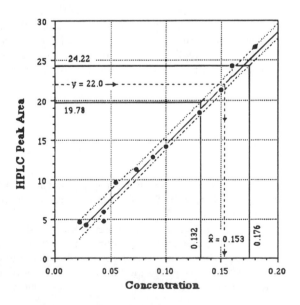

Figure 27.3 Calibration line with the 95% Working-Hotelling confidence region showing an example of predicting ξ from y = 22.0.

$$19.78 \leq \eta \leq 24.22$$

contain the true value of peak area (η) 90% of the time.

The bounds for the true concentration, read from the Working-Hotelling confidence bounds in Figure 27.3, are

$$0.132 \leq \xi \leq 0.176$$

These bounds contain the true value of the abscissa variable 90% of the time, a statement we can make with 95% confidence.

COMMENTS

Suppose that the calibration curve is constructed and the confidence interval for predicted concentration values is found to be undesirably large. How can it be reduced, aside from simply using more care to make the measurements?

The confidence limits of the calibration curve can be narrowed by increasing the number of calibration points on the straight line (this would include replication of points). Instead of trying to shrink the confidence interval of the calibration line, we could shrink the confidence interval of the measured y value that will be translated into a prediction of x. These possibilities will be examined. In order to explain the benefits of replication as simply as possible, we will use an approximate method (instead of Leiberman's method) to evaluate the error in the predicted concentrations (Miller and Miller, 1984).

Approximate confidence limits on ξ can be computed using $x_0 \pm t s_{x_0}$, where t is the appropriate value of the t statistic with $n - 2$ degrees of freedom, and

$$s_{x_0} = \frac{s}{b_1} \left[\frac{1}{m} + \frac{1}{n} + \frac{(y_0 - \bar{y})^2}{b_1^2 \Sigma (x_i - \bar{x})^2} \right]^{1/2}$$

All terms have been defined earlier except m, which is the number of replicate measures of y that are averaged to get y_0.

The two approaches to shrinking the confidence interval can be assessed by considering the terms inside the brackets in the equations above. If only one measurement of y is made, $m = 1$ and the first term will be large compared with the $1/n$ of the second term. Making m replicate measurements of y reduces the first term to $1/m$. Just a few replicates give a rapid gain. For example, suppose that we have $b_1 = 0.2$, $\Sigma(x_i - \bar{x})^2 = 100$, $\bar{y} = 20$, $s = 0.4$, and $n = 5$. The confidence interval on x is directly proportional to the standard deviation

$$s_{x_0} = \frac{0.4}{2} \left(\frac{1}{m} + \frac{1}{5} + \frac{y_0 - 20}{2(100)} \right)$$

Suppose we are interested in predicting concentration from a measured value of $y_0 = 30$. For $m = 1$, $s_{x_0} = 0.25$. If we had four measurements that average to give $y_0 = 30$, then $m = 4$, and $s_{x_0} = 0.1$.

This is a considerable gain in precision for a small amount of work.

The effect of the number of calibration points (n) is more complex because changing n also changes the value of the t statistic used to compute the confidence intervals. The motivation to reduce n is to limit the amount of work involved with preparing standards at many concentrations. It is clear, however, that small values of n will inflate the confidence interval by making the term 1/n large, while simultaneously increasing the corresponding t statistic. In most calibrations, six or so calibration points will be adequate, and if extra precision is needed, it can be gained most efficiently by making repeated measurements at some or all calibration levels.

There are a number of other interesting problems associated with calibrations. For example, in some analytical applications (photometric titrations), it is necessary to locate the intersection of two regression lines; in the standard addition method, the concentration is estimated by extrapolating a straight line down to the abscissa. Then there are nonlinear calibration curves and cases where the variance of the signal increases as the concentration increases. References listed at the end of the chapter provide some guidance on these problems.

A special case is where there are errors in both x and y, which could occur when no primary standard is available. As a rough but useful guideline, errors in the x variable can be ignored if $\sigma_x^2 < 3\sigma_y^2$. The effect of ignoring errors in x is to pull the regression line down so that the estimated slope is less than would be estimated if errors in x were taken into account. In going from y to x, this could badly overestimate x (Carroll and Spiegelman, 1986). This emphasizes the importance of using accurate standards in preparing calibration curves.

Many computer programs used to fit the straight line automatically provide values for the correlation coefficient, r, and the coefficient of determination, R^2. Often values of R^2 and r are cited as evidence that the calibration relation is strong and useful or that the calibration is in fact a straight line. These quantities do not prove these points, and in the context of calibration curves, they have little meaning of any kind. The correlation coefficient, r, is a measure of association between two *random* variables. In calibration, the concentration of the standard and the response of the instrument are functionally related. Random variation enters because the measurements are not perfect, but correlation in the mathematical-statistics sense does not exist (Hunter, 1981).

A discussion of the coefficient of determination, R^2, is found in Chapter 29. Here we note simply that values of R^2 close to -1 or $+1$ are to be expected in calibration curves. First, of all measurement situations, this is the one that is best under control. If the relation between standard and instrumental response is not clean and strong, there is simply no useful measurement method. Second, the value of R^2 value can be increased without increasing the precision of the measurements or of the predictions. This is done simply by expanding the range of concentrations covered by the standards. Third, R^2 can be large (>0.98), even though the curve deviates slightly from linearity.

REFERENCES

Bailey, C. J., E. A. Cox, and J. A. Springer (1978). "High Pressure Liquid Chromatographic Determination of the Immediate/Side Reaction Products in FD & C Red No. 2 and FD & C Yellow No. 5: Statistical Analysis of Instrument Response," *J. Assoc. Off. Anal. Chem.*, 61, 1404–1414.

Carroll, R. J. and C. H. Spiegelman (1986). "The Effect of Ignoring Small Measurement Errors in Precision Instrument Calibration," *J. Qual. Technol.*, 18, 3, 170–173.

Hunt, D. T. E. and A. L. Wilson (1986). *The Chemical Analysis of Water*, 2nd ed., London, Royal Society of Chemistry.

Hunter, J. S. (1981). "Calibration and the Straight Line: Current Statistical Practice," *J. Assoc. Off. Anal. Chem.*, 64, 3, 574–583.

Leiberman, G. J., R. G. Miller, Jr., and M. A. Hamilton (1967). "Unlimited Simultaneous Discrimination Intervals in Regression," *Biometrika*, 54, 122–145.

Mandel, J. (1964). *The Statistical Analysis of Experimental Data*, New York, Dover Publications.

Miller, J. C. and J. N. Miller (1984). *Statistics for Analytical Chemistry*, Chichester, England, Ellis Horwood Ltd.

Sharaf, M. A., D. L. Illman, and B. Kowalski (1986). *Chemometrics*, New York, John Wiley & Sons.

Working, H. and H. Hotelling (1929). "Application of the Theory of Error to the Interpretation of Trends," *J. Am. Stat. Assoc., Suppl. (Proceedings)*, 24, 73–85.

Wynn, H. P. and P. Bloomfield (1971). "Simultaneous Confidence Bands in Regression Analysis," *J. R. Stat. Soc. Ser. B*, 33, 202–217.

Chapter 28

Empirical Model Building by Linear Regression

Key words: diagnostic checking, empirical models, polynomials, linear regression, F tests, sedimentation, solids removal, all possible regressions, residual plots

---

Empirical models are widely used in engineering. They are useful when we lack knowledge or data needed to construct a mechanistic model. Sometimes the model is a straight line. Sometimes a mathematical French curve — a smooth interpolating function — is needed. Regression provides the means for selecting the complexity of the "French curve" that can be supported by the available data.

Regression begins with the specification of a model to be fitted. One goal is to find a parsimonious model — an adequate model with the fewest possible terms. Sometimes the proposed model turns out to be too simple and we need to augment it with additional terms. The much more common case, however, is to start with more terms than are needed or justified. This is called *overfitting*. Overfitting is harmful because the prediction error of the model is proportional to the number of parameters in the model.

A fitted model is always checked for inadequacies. The statistical output of regression programs is somewhat helpful for doing this, but a more satisfying and useful approach is to make diagnostic plots of the residuals. As a minimum, the residuals should be plotted against the predicted values of the fitted model. Plots of residuals against the independent variables are also useful. This chapter will illustrate how this kind of diagnosis suggests which terms should be investigated as potential improvements to a model. It will also illustrate a method for deciding whether one or more terms can be purged from a complicated model. If so, the simplified model is refitted and rechecked. The model builder thus works iteratively toward the simplest adequate model.

CASE STUDY — A MODEL OF SEDIMENTATION

Sedimentation is a process which removes solid particles from a liquid by allowing them to settle under quiescent conditions. An ideal sedimentation process can be created in the laboratory in the form of a batch column. The column is filled with the suspension (turbid river water, industrial wastewater, or sewage), and samples are taken over time from several sampling ports located at various depths along the column. The concentration of pollutant in these samples is measured, and the fraction that have settled below a certain depth in particular time is calculated. We will discuss the process as though the pollutant of interest is solids that are present in the form of small particles that will settle under the influence of gravity. The measure of sedimentation efficiency will be solids concentrations (or percent solids removed), which will be measured as a function of time and depth. The goal is to create an empirical model of the removal efficiency that might be expected of a real sedimentation process. Camp (1946) proposed a method for transforming the data into empirical models in the form of graphs. Here we seek to construct an empirical statistical model of the sedimentation process.

The data come from a quiescent batch settling test. At the beginning of the test, the concentration is uniform over the depth of the test settling column. The mass of solids in the column initially is Mass = C_0 Z A, where C_0 is the initial concentration (g/m^3), Z is the water depth in the settling column (m), and A is the cross-sectional area of the column (m^2). This is shown in the left-hand panel of Figure 28.1.

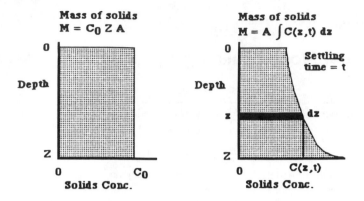

Figure 28.1 Solids concentration as a function of depth and time. The initial condition (t = 0) is shown on the left. The condition at time t is shown on the right.

After settling has progressed for time t, the concentration lower in the column has increased relative to the concentration at the top to give a solids concentration profile that is a simple function of depth at any time t. The mass of solids remaining above depth z is $M = A \int C(z,t)\, dz$. The total mass of solids in the column is still $C_0\, A\, Z$. This is shown in the right-hand panel of Figure 28.1.

The fraction of solids removed in a settling tank at any depth z, that has a detention time t, is estimated as

$$R(z, t) = \frac{A Z C_0 - A \int_0^z C(z,t)\, dz}{A Z C_0} = 1 - \frac{1}{Z C_0} \int_0^z C(z,t)\, dz$$

This integral could be calculated graphically, which is what Camp (1946) recommended, but it is also easy to derive an approximating polynomial function for the concentration curve and algebraically calculate the fraction, R, of solids removed.

The simplest possible model is $C(z,t) = b_0 + b_1 t$, and the most complicated that might be needed is

$$C(z,t) = b_0 + b_1\, t + b_2\, t^2 + b_3\, z + b_4\, z^2 + b_5\, z\, t$$

Intermediate models could be constructed by adding or dropping terms.

An empirical polynomial model of this kind can be used to estimate the solids removal efficiency. Suppose, for example, that the appropriate model is found to be

$$C(z,t) = 167 + 11.9\, z - 2.74\, t + 0.014\, t^2 - 0.08\, z\, t$$

We want to use this model to predict the removal that will be achieved with 60 min detention time for a depth of 8 ft and an initial concentration of 500 mg/L. The solids concentration profile as a function of depth at t = 60 min is

$$C(z, 60) = 167 + 11.9\, z - 2.74(60) + 0.014(60)^2 - 0.08(60)\, z = 53.0 + 7.1\, z$$

This is integrated over depth, Z = 8 ft, to give the fraction of solids that are expected to be removed.

$$R(z=8,t=60) = 1 - \frac{1}{8(500)}\int_{z=0}^{z=8} (53 + 71.1\ z)\ dz$$

$$= 1 - \frac{1}{8(500)} [53(8) - 3.55(8)^2] = 0.95$$

The regression problem is to determine the form of the polynomial function and to estimate the coefficients of the terms in the function.

METHOD — LINEAR REGRESSION

Suppose that the correct model for the process is $\eta_i = \beta_0 + \beta_1 x + \beta_2 x_i^2$. The value of the response observed at a setting of the independent variable, x_i, is y_i. Because of random measurement error, y_i differs from η_i by an amount e_i, and the measured response is $y_i = \beta_0 + \beta_1 x + \beta_2 x_i^2 + e_i$. The e_i are the residuals, i.e., the differences between the observed values (y_i) and values calculated from the proposed model. According to the least squares criterion, the best estimates of the values of η are obtained from the values of β that minimize the sum of the squared residuals

$$S(\beta) = \Sigma e_i^2 = \Sigma [y_i - \eta_i]^2$$

where the summation is over all observations. For the specific model $\eta_i = \beta_0 + \beta_1 x + \beta_2 x_i^2$, the sum of squares function is

$$S(\beta) = \Sigma [y_i - (\beta_0 + \beta_1 x + \beta_2 x_i^2)]^2$$

The best estimates (the least squares parameter estimates) of the parameters β_0, β_1, and β_2 are the values b_0, b_1, and b_2 that minimize the least squares function. When the model is linear, as in this example, the least squares estimates are particularly easy to compute (see Chapters 25 and 26).

The least squares criterion for fitting the model is based on an assumption that the residuals (the e_i) are normally distributed, independent, and have constant variance. This assumption always needs to be checked after the model has been fitted. This is done by computing and plotting the residuals $e_i = y_i - \hat{y}_i$, where $\hat{y}_i = b_0 + b_1 x + b_2 x_i^2$ is the value of the response predicted by the fitted model.

CASE STUDY SOLUTION

A column settling test was done on a suspension with an initial concentration of 560 mg/L. Samples were taken at depths of 2, 4, and 6 ft (measured from the water surface) at 20, 40, 60, and 120 min. The data are given in Table 28.1.

Table 28.1   Data from a Laboratory Settling Column Test

Depth (ft)	Suspended Solids Concentration at Time (min)			
	20	40	60	120
2	135	90	75	48
4	170	110	90	53
6	180	126	96	60

Starting from a Simple Model

Fitting the simplest possible model involving time and depth gives

$$\hat{y} = 132.32 + 7.125\ t - 0.968\ z$$

which has $R^2 = 0.844$ and residual mean square = 355.82. R^2 is the coefficient of determination and is equal to the percentage of the total variation in the data explained by the model (see next section and Chapter 29). The residual mean square is a measure of the total variation in the data that is *not* explained by the model. The residual mean square is computed as the ratio of minimum sum of squares, S(b), divided by its degrees of freedom; that is, $S(b)/(n - p)$, where n = number of observations and p = number of parameters estimated.

Figure 28.2a shows the diagnostic residual plots for the model. When plotted against the predicted values, the residuals do not appear random. This suggests an inadequacy in the model, but it does not tell us how the model might be improved. The pattern of the residuals plotted against time (Figure 28.2b) suggests that adding a quadratic term, t^2, may be helpful. This was done to obtain

$$\hat{y} = 185.97 + 7.125\ t + 0.014\ t^2 - 3.057\ z$$

which has $R^2 = 0.97$ and residual mean square = 81.5. A diagnostic plot of the residuals, Figure 28.3, reveals no inadequacies. Similar plots of residuals against the independent variables also support the model. This simple model is adequate to describe the data.

Starting from a Complicated Model

Here we start with a complete quadratic function. Linear regression was used to fit the following six parameter model to the data:

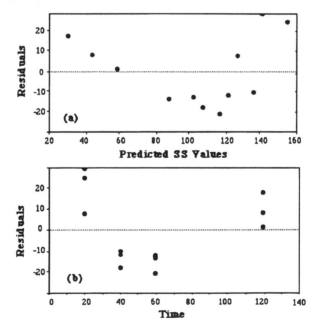

Figure 28.2 (a) Residuals plotted against the predicted suspended solids concentrations are not random. (b) Residuals plotted against settling time suggest that a quadratic term is needed in the model.

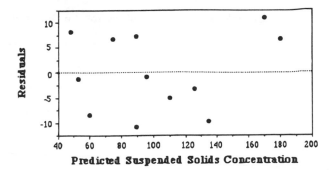

Figure 28.3 Plot of residuals against the predicted values of the regression model y = 185.97 + 7.125 t + 0.014 t² - 3.057 z.

$$\hat{y} = 152 + 20.9\ z - 2.74\ t - 1.13\ z^2 - 0.014\ t^2 - 0.08\ t\ z$$

Depth and time are denoted by z and t. The model contains quadratic terms for each of these and the interaction of depth and time (zt). The analysis of variance of this regression model is given in Table 28.2. This information is produced by computer programs that do linear regression, so we need to understand how it is interpreted.

The sum of squares "due to regression" is called the *regression sum of squares* (Reg. SS = 20,555.5); it shows how much of the total variation (i.e., total SS) has been explained by the fitted equation. The sum of squares "due to residuals" is called the *residual sum of squares* (RSS = 308.8); it represents the variation that is not explained by the model. The mean square (MS) values are the sums of squares divided by the corresponding degrees of freedom. If the model fits the data, the mean square error for the residuals estimates the variance due to random experimental error. The residual sum of squares (RSS = 308.8) and its residual mean square (RSS/df = 308.8/6 = 51.5) are the key statistics in comparing this model with simpler models.

Table 28.3 summarizes parameter estimates, the coefficient of determination R^2, and the regression sum of squares for all eight possible linear models. The total sum of squares, of course, is the same in all eight cases, since it depends on the data and not on the form of the model.

The coefficient of determination, commonly denoted as R^2, is the regression sum of squares expressed as a percentage of the total sum of squares. For the complete six-parameter model (Model A in Table 28.3), R^2 = (20556/20564)100 = 98.5, so it can be said that this model explains 98.5% of the total variation in the data.

We are fascinated by high R^2 values,[1] and this may tempt us to think that the goal of regression is to make this value as high as possible. Obviously, this is can be done

Table 28.2 Analysis of Variance for the Settling Linear Regression Model

Due to	df	SS	MS = SS/df
Regression	5	20255.5	4051.1
Residuals	6	308.8	51.5
Total	11	20564.2	

[1] Chapter 29 discusses the interpretation of R^2 values.

Table 28.3 **Summary of All Possible Regressions for the Quadratic Settling Test Model**

Model	$b_0 +$	$b_1z +$	$b_2t +$	$b_3z^2 +$	$b_4t^2 +$	$b_5t\ z$	R^2 (%)	Reg. SS	Decrease in Reg. SS
A	152	20.9	−2.74	−1.13	0.014	−0.08	98.5	20256	
B	167	11.9	−2.74		0.014	−0.08	98.2	20202	54
C	171	16.1	−3.06	−1.13	−0.014		97.1	19966	289
D	186	7.1	−3.06		0.143		96.8	19912	343
E	98	20.9	−0.65	−1.13		−0.08	86.4	17705	2550
F	113	11.9	−0.65			−0.80	85.8	17651	2605
G	117	16.1	−0.97	−1.13			84.9	17416	2840
H	132	7.1	−0.97				84.4	17362	2894

by putting more high-order terms into a model, but it should be obvious that this does not necessarily improve the predictions that will be made using the model. Increasing R^2 is the wrong goal. Instead of worrying about R^2 values, we should seek the simplest adequate model.

We need a method to select this model from among the eight candidates. One approach is to examine the confidence intervals of the estimated parameters. If this interval includes zero, the variable associated with the parameter can be dropped from the model. For example, in Model A, the coefficient of z^2 is −1.13 with a 95% confidence interval of [−3.81 to 1.56]. This confidence includes zero, indicating that the true value of the coefficient is likely to be zero, and therefore, the term z^2 can be tentatively dropped from the model. Refitting the simplified model (without z^2) to the data gives Model B in Table 28.3. After doing this, an additional test should be made to compare the regression sum of squares of the two models. Details of this test are given in texts on regression analysis (e.g., Draper and Smith, 1981). Here the test will be illustrated by example.

The regression sum of squares for the complete model, Model A, is 20,256. Dropping the z^2 term to get Model B reduced the regression sum of squares by only 54.[2] This reduction can be thought of as a variance associated with the z^2 term. We need to consider that a reduction of 54 in the regression sum of squares may not be a statistically significant difference. In order to test its significance, an estimate is needed of the variance due to "pure experimental error." There were no repeated measurements in this experiment, so an independent estimate of this quantity cannot be computed. The best that can be done under the circumstances is to use the mean residual sum of squares of the complete model as an estimate of the pure error variance. The residual mean square for the complete model, Model A, is 51.5. This is compared with the difference in regression sum of squares of the two models of interest. The difference in regression sum of squares between Models A and B is 54. The ratio of the two variances is F = 54/51.5 = 1.06. This value is compared against the upper 95% point of the F distribution for 1.6 degrees of freedom or 6.61. The degrees of freedom are 1 for the numerator (1 degree of freedom for the one parameter that was dropped from the model) and 6 for the denominator (the mean residual sum of squares). Since $F = 1.06 < 6.61 = F_{1,6,\alpha=0.05}$,

---

[2] Note that the residual sum of squares is increased by the amount that the regression sum of squares is reduced.

we conclude that removing the z^2 term does not result in a significant reduction in the regression sum of squares. Therefore, the z^2 term is not needed in the model.

In contrast, consider Models A and E. Omitting t^2 decreases the regression sum of squares by $20256 - 17705 = 2551$. The F statistic is $2551/51.5 = 49.5$. Since $49.5 \gg 6.61$ (the upper 95% point of the F distribution with 1 and 6 degrees of freedom), this change is significant, and t^2 needs to be included in the model.

A similar analysis shows that the zt term is not needed; Model C is as good as Model A. Dropping both z^2 and zt gives Model D, which is the simplest adequate model.

$$\text{Model D} \qquad \hat{y} = 186 + 7.12\,t - 3.06\,z + 0.143\,t^2$$

Note that this is the same simple model that was obtained by adding terms to the simplest original linear function.

COMMENTS

Regression is used to estimate the parameters and to determine the simplest adequate model. Parameter estimation is easily accomplished by linear regression. The more difficult part of the analysis is determining how many terms to put into the model. The goal is not to maximize the value of R^2, because maximizing R^2 leads to putting unneeded high-order terms into the polynomial model. The best model should have the fewest possible parameters because this will minimize the prediction error of the model. One approach to finding the simplest adequate model is to start with a simple tentative model and use diagnostic checks such as residuals plots for guidance. The alternate approach is to start by overfitting the data with a highly parameterized model and to then find appropriate simplifications. Each time a term is added or deleted from the model, a check is made on whether the difference in the regression sum of squares of the two models is large enough to justify modification of the model.

REFERENCES

Berthouex, P. M. and D. K. Stevens (1982). "Computer Analysis of Settling Data," *J. Environ. Eng. Div. ASCE*, 108, 1065–1069.

Camp, T. R. (1946). "Sedimentation and Design of Settling Tanks," *Trans. Am. Soc. Civil Eng.*, 3, 895–936.

Chatterjee, S. and B. Price (1977). *Regression Analysis by Example*, New York, John Wiley & Sons.

Draper, N. R. and H. Smith (1981). *Applied Regression Analysis*, 2nd ed., New York, John Wiley & Sons.

Chapter 29

The Coefficient of Determination, R^2

Key words: coefficient of determination, coefficient of multiple correlation, linear regression, R^2, spurious correlation

---

Regression analysis is so easy to do — many hand-held calculators can do linear regression — that one of the best-known statistics is the coefficient of determination, R^2. But, what does it really reveal about how well the model fits the data? And, what important information may be overlooked if too much reliance is placed in the interpretation of R^2?

The coefficient of determination is that proportion of the total variability in the dependent variable that is accounted for by the regression equation. A value of $R^2 = 1$ indicates that the fitted equation accounts for all the variability of the values of the dependent variables in the sample data. At the other extreme, $R^2 = 0$ indicates that the regression equation explains none of the variability. This idea is so simple that we naturally tend to assume that a high R^2 assures a statistically significant regression equation and that a low R^2 proves the opposite. A "statistically significant equation" would mean that we conclude there is some true relationship between the independent and dependent variables and that this relationship could be used to predict new conditions. Life is not this simple, as will be shown by some examples.

A HIGH R^2 DOES NOT ASSURE A VALID RELATION

Figure 29.1 shows a regression with $R^2 = 0.746$, which is statistically significant at almost the 1% level of confidence (a 1% chance of concluding significance when there is no true relation). This might be impressive until one knows the source of the "data." x is the first six digits of pi, and y is the first six Fibonocci numbers. There is no true relation between x and y. The linear regression equation has no predictive value (the seventh digit of pi does not predict the next Fibonocci number).

Anscombe (1973) published a famous and fascinating example of how R^2 and other statistics that are routinely computed as part of a regression analysis can fail to reveal the important features of the data. Table 29.1 gives Anscombe's four data sets. Each data set has n = 11, $\bar{x} = 9.0$, $\bar{y} = 7.5$, equation of the regression line = $\hat{y} = 3 + 0.5x$, standard error of estimate of the slope = 0.118, t statistic = 4.24, regression sum of squares (corrected for mean) = 110.0, residual sum of squares = 13.75, correlation coefficient = 0.82, and $R^2 = 0.67$. All four data sets appear to be described equally well by exactly the same linear model, at least until the data are plotted (or until the residuals are examined). Figure 29.2 shows how vividly they differ. The example is a persuasive argument for always plotting the data.

A LOW R^2 DOES NOT MEAN THE MODEL IS USELESS

Hahn (1973) explains that the chances are one in ten of getting R^2 as high as 0.9756 in fitting a simple linear regression equation to the relationship between an independent variable x and a normally distributed variable y based on only three observations, even if x and y are totally unrelated. On the other hand, with 100 observations, a value of $R^2 = 0.07$ is sufficient to establish statistical significance at the 1% level.

Table 29.2 lists the values of R^2 required to establish statistical significance for a simple linear regression equation. This tabulation gives values at the 10, 5, and 1%

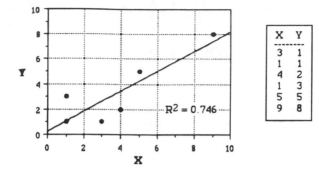

Figure 29.1 An example of nonsense in regression. X is the first six digits of pi, and Y is the first six Fibonocci numbers. $R^2 = 0.746$ is high, even though there is no actual relation between x and y.

Table 29.1 **Anscombe's Four Data Sets**

A		B		C		D	
x	y	x	y	x	y	x	y
10.0	8.04	10.0	9.14	10.0	7.46	8.0	6.58
8.0	6.95	8.0	8.14	8.0	6.77	8.0	5.76
13.0	7.58	13.0	8.74	13.0	12.74	8.0	7.71
9.0	8.81	9.0	8.77	9.0	7.11	8.0	8.84
11.0	8.33	11.0	9.26	11.0	7.81	8.0	8.47
14.0	9.96	14.0	8.10	14.0	8.84	8.0	7.04
6.0	7.24	6.0	6.13	6.0	6.08	8.0	5.25
4.0	4.26	4.0	3.10	4.0	5.39	19.0	12.50
12.0	10.84	12.0	9.13	12.0	8.15	8.0	5.56
7.0	4.82	7.0	7.26	7.0	6.42	8.0	7.91
5.0	5.68	5.0	4.74	5.0	5.73	8.0	6.89

Note: Each data set has $n = 11$, mean of $\bar{x} = 9.0$, mean of $\bar{y} = 7.5$, equation of the regression line $= \hat{y} = 3 + 0.5x$, standard error of estimate of the slope $= 0.118$, t statistic $= 4.24$, regression sum of squares (corrected for mean) $= 110.0$, residual sum of squares $= 13.75$, correlation coefficient $= 0.82$, and $R^2 = 0.67$.
Data from Anscombe (1973).

significance level. These correspond, respectively, to the situations where one is ready to take 1 chance in 10, 1 chance in 20, and 1 chance in 100 of *incorrectly* concluding there is evidence of a statistically significant linear regression when in fact x and y are unrelated. Table 29.2 applies only for the straight-line model $y = a + bx$; for multivariable regression models, statistical significance must be determined by other means.

A SIGNIFICANT R² DOES NOT MEAN THE MODEL IS USEFUL

Practical significance and statistical significance are not equivalent. Significance and importance are not equivalent. A regression based on a modest and unimportant true relationship may be established as statistically significant if a sufficiently large number of observations are available. On the other hand, with a small sample, it may be difficult to obtain statistical evidence of a strong relation.

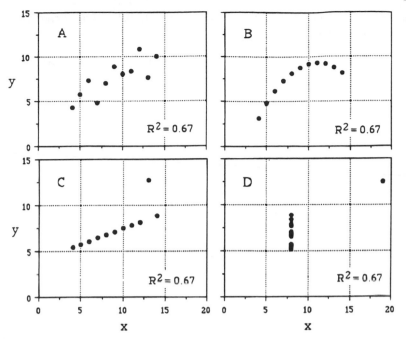

Figure 29.2 Plot of Anscombe's four data sets, which all have R^2 = 0.67 and identical results from simple linear regression analysis.

It generally is good news if we find R^2 large and also statistically significant, but it does not assure a useful equation, especially if the equation is to be used for prediction. One reason is that the coefficient of determination is not expressed on the same scale as the dependent variable. A particular equation may explain a large proportion of the variability in the dependent variable and thus have a high R^2, yet the total variability may be so large that the unexplained variability is still too large for useful prediction. It is not possible to tell from the magnitude of R^2 how accurate the predictions will be.

THE MAGNITUDE OF R^2 DEPENDS ON THE RANGE OF VARIATION IN X

The value of R^2 decreases with a decrease in the range of variation of the independent variable, other things being equal, assuming the correct model is being fitted to the data. Figure 29.3 shows a set of 50 data points that has R^2 = 0.77. Suppose, however, that the range of x that could be investigated is only from 14 to 16 (for example, because a process is carefully constrained within narrow operating limits) and the available data are those shown in the upper right-hand panel of Figure 29.3. The underlying relationship is the same, but R^2 is now only 0.12. This dramatic reduction in R^2 occurs mainly because the range of x is restricted and not because the number of observations is reduced. This is shown by the two lower panels. Fifteen points (the same number as found in the range of x = 14–16), located at x = 10, 15, and 20, give R^2 = 0.88. Just ten points, at x = 10 and 20, gives an even larger value, R^2 = 0.93.

These examples show that a large value of R^2 might reflect the fact that data were collected over an unrealistically large range of the independent variable x. This can happen especially when x is time. Conversely, a small value might be due to a limited range of x, such as when x is a variable that is carefully controlled within a narrow

Table 29.2   Values of R² Required to Establish Statistical Significance of a
Simple Linear Regression Equation for Various Sample Sizes

Sample Size	Statistical Significance Level		
	10%	5%	1%
3	0.9756	0.9938	0.998
4	0.810	0.903	0.980
5	0.65	0.77	0.92
6	0.53	0.66	0.84
7	0.45	0.57	0.77
8	0.39	0.50	0.70
9	0.34	0.44	0.64
10	0.30	0.40	0.59
11	0.27	0.36	0.54
12	0.25	0.33	0.50
13	0.23	0.31	0.47
14	0.21	0.28	0.44
15	0.19	0.26	0.41
20	0.14	0.20	0.31
25	0.11	0.16	0.26
30	0.09	0.13	0.22
40	0.07	0.10	0.16
50	0.05	0.08	0.13
100	0.03	0.04	0.07

Data from Hahn (1973).

limit in order to control a process. In this case, x is carefully controlled because it is known to be highly important, yet this importance will not be revealed by doing regression on typical data from the process.

OTHER WAYS TO EXAMINE A MODEL

If R^2 does not tell all that is needed about how well a model fits the data and how good the model may be for prediction, what else could be examined?

Graphics reveal information in data (Tufte, 1983). Always examine the data and the proposed model graphically. How sad that this advice is often forgotten in a rush to compute some statistic (like R^2).

A more useful single measure of the prediction capability of a model (including a k variate regression model) is the *standard error of the estimate*. The standard error of the estimate is computed from the variance of the predicted value, $\hat{y}$, and it indicates the precision with which the model estimates the value of the dependent variable. This statistic is used to compute intervals which have the following meanings (Hahn, 1973):

- The *confidence interval for the dependent variable* is an interval that one expects, with a specified level of confidence, to contain the *average value* of the dependent variable at a set of specified values for the independent variables.

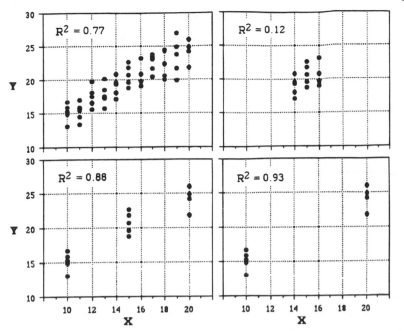

Figure 29.3 The full data set of 50 observations (upper left panel) has $R^2 = 0.77$. The other three panels show how R^2 depends on the range of variation in the independent variable.

- A *prediction interval* for the *dependent variable* is an interval that one expects, with a specified degree of confidence, to contain a *single future value* of the dependent variable from the sampled population at a set of specified values of the independent variables.
- A *confidence interval around a regression coefficient* (i.e., parameter in a model) is an interval that one expects, with a specified degree of confidence, to contain the *true regression coefficient*.

The exact method of obtaining these intervals is explained in texts such as Draper and Smith (1981), Rawlings (1988), and Meyers (1986). They are computed by most statistics software packages. Confidence intervals for parameter estimates are discussed in Chapter 25, and confidence intervals of the dependent variable are discussed in Chapter 26.

WHAT DOES "EXPLAINED" MEAN?

One often heard definition of R^2 is that it "explains" a certain proportion of the variability in the observed response. Caution is recommended in interpreting the phrase "R^2 explains the variation in the dependent variable." If the data are from a well-designed controlled experiment, with proper replication and randomization, it is reasonable to infer that a significant association of the variation in y with variation in the level of x is a causal effect of x. If the data had been observational, what Box (1966) calls *happenstance data*, there is a high risk of a causal interpretation being wrong. With observational data, there can be many reasons for associations among variables, only one of which is causality.

COMMENTS

It is a pity, but the most widely used tools and methods have the greatest potential to be the most frequently misused tools. Of all statistical methods, linear regression is probably the most widely used, and there are risks of it being misused or misinterpreted, especially if one relies too much on R^2 as a characterization of how well a model fits. This is why studying examples of a few traps opened by its use help us remember what R^2 means and what it does not mean.

Totally spurious correlations, often with high R^2 values, can arise when unrelated variables are combined. Two examples of particular interest to environmental engineers are presented by Sherwood (1974) and Rowe (1974). Both emphasize graphical analysis to stimulate and support any regression analysis. Rowe discusses the particular dangers that arise when sets of variables are combined to create new variables (such as dimensional numbers, Froude number, etc.). Benson (1965) points out the same kinds of dangers in the context of hydraulics and hydrology.

REFERENCES

Anscombe, F. J. (1973). "Graphs in Statistical Analysis," *Am. Statistician*, 27, 17–21.

Benson, M. A. (1965). "Spurious Correlation in Hydraulics and Hydrology," *J. Hydraul. Div. ASCE*, 91, HY4, 35–45.

Box, G. E. P. (1966). "The Use and Abuse of Regression," *Technometrics*, 8, 4, 625–629.

Draper, N. R. and H. Smith (1981). *Applied Regression Analysis*, 2nd ed., New York, John Wiley & Sons.

Hahn, G. J. (1973). "The Coefficient of Determination Exposed!," *Chemtech*, October, 609–611.

Meyers, R. H. (1986). *Classical and Modern Regression with Applications*, Boston, Duxbury Press.

Rawlings, J. O. (1988). *Applied Regression Analysis: A Research Tool*, Pacific Grove, CA, Wadsworth and Brooks/Cole.

Rowe, P. N. (1974). "Correlating Data," *Chemtech*, January, 9–14.

Sherwood, T. K. (1974). "The Treatment and Mistreatment of Data," *Chemtech*, December, 736–738.

Tufte, E. R. (1983). *The Visual Display of Quantitative Information*, Cheshire, CT, Graphics Press.

Chapter 30

Regression Analysis with Categorical Variables

Key words: acid rain, pH, categorical variables, F test, linear model, regression, dummy variables, qualitative variables, regression sum of squares

---

Qualitative variables can be used as explanatory variables in regression models. A typical case would be when several sets of data have been collected that are similar except that each set was measured by a different chemist (or a different instrument or laboratory), each set comes from a different location, or each set was measured on a different day. The qualitative variables — chemist, location, or day — typically take on discrete values (i.e., chemist Smith or chemist Jones). For convenience, they are usually represented numerically by a combination of zeros and ones to signify that an observation belongs in one of two possible categories, hence the name *categorical variables*.

One task in the analysis of such data is to determine whether the same model structure holds for each data set and, if the structure seems to be the same, whether the values of the model parameters are also the same. One way to do this evaluation would be to fit the proposed model to each individual data set and then try to make some assessment of the similarities and differences in the goodness of fit. Another way would be to fit the proposed model to all the data as though they were one data set instead of several, assuming that each data set had the same pattern, and then to look for inadequacies in the fitted model.

Neither of these approaches is as convenient as using categorical variables to create a collective data set that can be fitted to a single model, while retaining the distinction between the smaller individual data sets. This technique allows both the model structure and the model parameters to be evaluated using statistical methods like those discussed in the previous chapter.

CASE STUDY — ACIDIFICATION OF A STREAM DURING STORMS

Cosby Creek, in the southern Appalachian Mountains, was monitored during three storms in order to study how pH and other measures of acidification were affected by the rainfall in that region. Samples were taken every 30 min, and 19 characteristics of the stream water chemistry were measured (Meinert et al., 1982). Weak acidity (WA) and pH will be examined in this case study.

There were 17 observations for Storm 1, 14 for Storm 2, and 13 for Storm 3, giving a total of 44 observations. If they are plotted without distinguishing between storms, the result is Figure 30.1, which shows WA and pH to be correlated (r = –0.7). This might suggest a model of the form pH = β_0 + β_1 WA. A plot that does distinguish between storms, Figure 30.2, still shows a relation between pH and WA, but suggests that this relation is not the same for all three storms. It is clear that Storm 3 does not have the same slope and intercept as Storms 1 and 2, but Storms 1 and 2 might have the same intercept and equal slopes. This can be checked by using categorical variables to estimate a different slope and intercept for each storm.

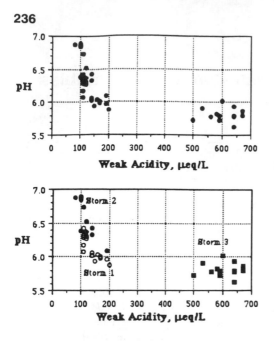

Figure 30.1 pH and weak acidity data for Cosby Creek.

Figure 30.2 Cosby Creek data plotted to show each storm.

METHOD — REGRESSION WITH CATEGORICAL VARIABLES

Suppose that a model needs to include an effect due to the category (storm event, farm plot, treatment, truckload, operator, laboratory, etc.) from which the data came. This effect is included in the model in the form of categorical variables (also called dummy or indicator variables). In general $n - 1$ categorical variables are needed to specify n categories.

Begin by considering data from a single category. The quantitative predictor variable is x_1, which can predict the independent variable y_1 using the linear model

$$y_i = \beta_0 + \beta_1 x_{1i} + e_i$$

where β_0 and β_1 are parameters to be estimated by least squares.

If there are data from two categories, for example, data produced at two different laboratories, one approach would be to model the two sets of data separately as

$$y_{1i} = \alpha_0 + \alpha_1 x_{1i} + e_i$$

and

$$y_{2i} = \beta_0 + \beta_1 x_{2i} + e_i$$

and then to compare the estimated intercepts, α_0 and β_0, and the estimated slopes, α_1 and β_1, using confidence intervals and/or t-tests.

A second, and often better, method is to simultaneously fit a single augmented model to all the data. To construct this model, define a categorical variable Z as follows: $Z = 0$ if the data are in the first category and $Z = 1$ if the data are in the second category.

The augmented model is

$$y_i = \alpha_0 + \alpha_1 x_i + Z(\beta_0 + \beta_1 x_i) + e_i$$

or, with some rearrangement,

$$y_i = \alpha_0 + Z\beta_0 + \alpha_1 x_i + \beta_1 Z x_i + e_i$$

In this last form, it is apparent that the regression is done as though there are three independent variables, x, Z, and Zx. The vectors of Z and Zx have to be created from the categorical variables as defined above. The four parameters α_0, β_0, α_1, and β_1 are estimated by linear regression.

A model for each category can be obtained by substituting the defined values. For the first category, which was defined by Z = 0,

$$y_i = \alpha_0 + \alpha_1 x_i$$

and for the second category, which has Z = 1,

$$y_i = (\alpha_0 + \beta_0) + (\alpha_1 + \beta_1)x_i$$

The regression might estimate either β_0 or β_1 as zero or both as zero. If $\beta_1 = 0$, the two lines have the same intercept. If $\beta_0 = 0$, the two lines have the same slope. If both β_1 and β_0 equal zero, a single straight line fits all the data. Figure 30.3 shows the four possible outcomes. Figure 30.4 shows the particular case where the slopes are equal and the intercepts are different.

If any of the simplifications seems justified from the results obtained by fitting the full model (the four-parameter model), the model is modified and the simplified version is fitted to the data. We show later how the full model and simplified model are compared to check whether the simplification is justified.

To deal with three categories, define two categorical variables:

	Slopes different	Slopes equal
Intercepts different	$y_i = (\alpha_0 + \beta_0) + (\alpha_1 + \beta_1)x_i + e$	$y_i = (\alpha_0 + \beta_0) + \alpha_1 x_i + e_i$
Intercepts equal	$y_i = \alpha_0 + (\alpha_1 + \beta_1)x_i + e_i$	$y_i = \alpha_0 + \alpha_1 x_i + e_i$

Figure 30.3 Four possible models to fit a straight line to data in two categories.

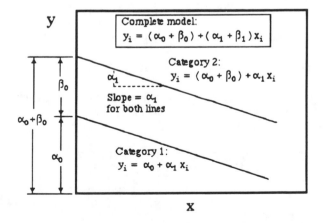

Figure 30.4 Model with two categories having different intercepts, but equal slopes.

$$\text{Category 1:} \quad Z_1 = 1 \quad \text{and} \quad Z_2 = 0$$
$$\text{Category 2:} \quad Z_1 = 0 \quad \text{and} \quad Z_2 = 1$$

and construct the model

$$y_i = (\alpha_0 + \alpha_1 x_i) + Z_1(\beta_0 + \beta_1 x_i) + Z_2(\gamma_0 + \gamma_1 x_i) + e_i$$

This can be rearranged to give

$$y_i = (\alpha_0 + \beta_0 Z_1 + \gamma_0 Z_2) + \alpha_1 x_i + \beta_1 Z_1 x_i + \gamma_1 Z_2 x_i + e_i$$

which shows that the six parameters are estimated by fitting the original independent variable x_i plus the four created variables Z_1, Z_2, $Z_1 x_i$, and $Z_2 x_i$. The parameters with subscript 0 estimate the intercept, and those with subscript 1 estimate the slope. This is more apparent when the model is written as

$$y_i = (\alpha_0 + \beta_0 Z_1 + \gamma_0 Z_2) + (\alpha_1 + \beta_1 Z_1 + \gamma_1 Z_2)x_i + e_i$$

Any of the parameters might be estimated as zero by the regression analysis. A couple of examples may explain how the simpler models may be identified. In the simplest possible case, the regression would estimate β_0 and $\gamma_0 = 0$ and β_1 and $\gamma_1 = 0$, and the same slope (α_1) and intercept (α_0) apply to all three categories. If the intercepts are different for the three categories, but the slopes are the same, the regression would estimate β_1 and $\gamma_1 = 0$, and the model becomes

$$y_i = (\alpha_0 + \beta_0 Z_1 + \gamma_0 Z_2) + \alpha_1 x_i$$

CASE STUDY — SOLUTION

The model under consideration allows a different slope and intercept for each storm. Two dummy variables are needed: $Z_1 = 1$ for Storm 1 and zero otherwise and $Z_2 = 1$ for Storm 2 and zero otherwise.
The model is

$$pH = \alpha_0 + \alpha_1 \, WA + Z_1(\beta_0 + \beta_1 \, WA) + Z_2(\gamma_0 + \gamma_1 \, WA)$$

where the α's, β's, and γ's are estimated by regression. The model can be rewritten as

$$pH = (\alpha_0 + \beta_0 Z_1 + \gamma_0 Z_2) + \alpha_1 \, WA + \beta_1 Z_1 \, WA + \gamma_1 Z_2 \, WA$$

The dummy variables are incorporated into the model by creating the "new" variables $Z_1 \cdot WA$ and $Z_2 \cdot WA$. Table 30.1 shows how this is done.
Fitting the full six-parameter model gave

Model A: pH = 5.77 − 0.00008 WA + 0.998 Z_1 + 1.65 Z_2
             (0.11)   (2.14)   (3.51)

+ 0.005 $Z_1 \cdot$ WA − 0.008 $Z_2 \cdot$ WA
   (3.63)   (4.90)

which is also shown as Model A in Table 30.2 (top row). The numerical coefficients

Table 30.1 **Weak Acidity (WA), pH, and Categorical Variables for Three Storms on Cosby Creek**

Storm	WA	Z_1	Z_2	Z_1WA	Z_2WA	pH
1	190	1	0	190	0	5.96
1	110	1	0	110	0	6.08
1	150	1	0	150	0	5.93
1	170	1	0	170	0	5.99
1	170	1	0	170	0	6.01
1	170	1	0	170	0	5.97
1	200	1	0	200	0	5.88
1	140	1	0	140	0	6.06
1	140	1	0	140	0	6.06
1	160	1	0	160	0	6.03
1	140	1	0	140	0	6.02
1	110	1	0	110	0	6.17
1	110	1	0	110	0	6.31
1	120	1	0	120	0	6.27
1	110	1	0	110	0	6.42
1	110	1	0	110	0	6.28
1	110	1	0	110	0	6.43
2	140	0	1	140	0	6.33
2	140	0	1	140	0	6.43
2	120	0	1	120	0	6.37
2	190	0	1	190	0	6.09
2	120	0	1	120	0	6.32
2	110	0	1	110	0	6.37
2	110	0	1	110	0	6.73
2	100	0	1	100	0	6.89
2	100	0	1	100	0	6.87
2	120	0	1	120	0	6.30
2	120	0	1	120	0	6.52
2	100	0	1	100	0	6.39
2	80	0	1	80	0	6.87
2	100	0	1	100	0	6.85
3	580	0	0	0	0	5.82
3	640	0	0	0	0	5.94
3	500	0	0	0	0	5.73
3	530	0	0	0	0	5.91
3	670	0	0	0	0	5.87
3	670	0	0	0	0	5.80
3	640	0	0	0	0	5.80
3	640	0	0	0	0	5.78
3	560	0	0	0	0	5.78
3	590	0	0	0	0	5.73
3	640	0	0	0	0	5.63
3	590	0	0	0	0	5.79
3	600	0	0	0	0	6.02

Data provided by D. Meinart et al. (1982).

Table 30.2 **Alternate Models for pH at Cosby Creek**

Model	Parameter Estimates for the Fitted Model	Reg. SS	R^2
A	pH $= 5.77 - 0.00008$ WA $+ 0.998$ $Z_1 + 1.65$ Z_2	4.278	0.866
	$+ 0.005$ Z_1 WA $- 0.008$ Z_2 WA		
B	pH $= 5.82 + 0.95$ $Z_1 + 1.603$ $Z_2 - 0.005$ Z_1 WA $- 0.008$ Z_2 WA	4.278	0.866
C	pH $= 5.82 + $ **1.11 Z_1** $+ 1.38$ $Z_2 - 0.006$ Z WA	4.229	0.866

are the least squares estimates of the parameters. The small numbers in parentheses beneath the coefficients are the t statistics for the parameter values. A parameter with $t \geq 2$ is almost certain to be significant. Terms with $t < 2$ are candidates for elimination from the model because they are almost certainly not significant.

The term WA appears to be insignificant. If this term is dropped and we refit the simplified model, the result is Model B, in which all coefficients are significant:

Model B: pH $= 5.82 + 0.95$ $Z_1 + 1.60$ Z_2 $- 0.005$ $Z_1 \cdot$WA $- 0.008$ $Z_2 \cdot$WA
(t statistic of parameter)   (6.01)   (2.23)   (4.53)   (5.54)
(95% conf. int.)   [0.63 to 1.27]   [1.26 to 1.95]   [-0.007 to -0.002]   [$-0.01 - 0.005$]

The regression sum of squares, listed in Table 30.2, is 4.278 for Models A and B. Clearly, dropping the WA term caused no decrease in the regression sum of squares. Model A is equivalent to Model B.

Is any further simplification possible? Notice that the 95% confidence intervals overlap for the terms -0.005 Z_1WA and -0.008 Z_2WA. Therefore, the coefficients of these two terms might be the same. To check this, we fit Model C, which has the same slope but different intercepts for Storms 1 and 2. This is done by defining $Z_1 = 1$ for Storm 1 and zero otherwise, $Z_2 = 1$ for Storm 2 and zero otherwise, and a third categorical variable $Z = 1$ for Storms 1 and 2 and zero otherwise.

The fitted model is

Model C: pH $= 5.82 + 1.11$ $Z_1 + 1.38$ $Z_2 - 0.006$ Z $\cdot$ WA
(t statistics)   (8.43)   (12.19)   (6.67)

This simplification of the model can be checked in a more formal way by comparing regression sums of squares of the simplified model with the more complicated one. The regression sums of squares is a measure of how well the model fits the data; a lower value indicates a better fit. Dropping an important term will cause the regression sum of squares to decrease by a noteworthy amount, whereas dropping an unimportant term will change the regression sum of squares very little. An example shows how we decide whether a change is "noteworthy" (i.e., statistically significant).

If two models are equivalent, the difference of their regression sums of squares will be small, within an allowance for variation due to random experimental error. (See also Chapter 28.) The variance due to experimental error can be estimated by the mean residual sum of squares of the full model (Model A). The variance due to the deleted term is estimated by the difference between the regression sums of squares of Models A and C, with an adjustment for their respective degrees of freedom. The ratio of the variance due to the deleted term is compared with the variance due to experimental error by computing the F statistic, as follows:

$$F = \frac{(\text{Reg. SS}_A - \text{Reg. SS}_C)/(\text{Reg. df}_A - \text{Reg. df}_C)}{\text{Res. SS}_A/\text{Res. df}_A}$$

where

Reg. SS = regression sum of squares

Reg. df = degrees of freedom associated with the appropriate regression sum of squares

Res. SS = residual sum of squares

Res. df = degrees of freedom associated with the residual sum of squares

Model A has 5 degrees of freedom associated with the regression sum of squares (Reg. df = 5, one for each of the six parameters in the model minus one for computing the mean). Model C has 3 degrees of freedom. Thus,

$$F = \frac{(4.278 - 4.229)/(5 - 3)}{0.66/38} = \frac{0.0245}{0.017} = 1.44$$

For a test of significance at the 95% confidence level, this value of F is compared with the upper 5% point of the F distribution with the appropriate degrees of freedom (5 − 3 = 2 in the numerator and 38 in the denominator): $F_{2,38,0.05} = 3.25$. The computed value, F = 1.44, is smaller than the critical value $F_{2,38,0.05} = 3.25$, which confirms that omitting WA from the model and forcing Storms 1 and 2 to have the same slope has not significantly worsened the fit of the model. In short, Model C describes the data as well as Models A or B. Since it is simpler, it is preferred.

Models for the individual storms are derived by substituting the values of Z_1, Z_2, and Z into Model C:

Storm 1 $Z_1 = 1, Z_2 = 0$ pH = 6.94 − 0.006 WA

Storm 2 $Z_1 = 0, Z_2 = 1$ pH = 7.21 − 0.006 WA

Storm 3 $Z_1 = 0, Z_2 = 0$ pH = 5.82

The model indicates a different intercept for each storm, a common slope for Storms 1 and 2 and a slope of zero for Storm 3, as shown by Figure 30.5. In Storm 3, the variation in pH was random about a mean of 5.82, which is indicated by the zero slope. For Storms 1 and 2, increasing WA was associated with a lowering of the pH. It is not difficult to imagine conditions that would lead to two different storms having the same slope, but different intercepts. It is more difficult to understand how the same stream could respond so differently to Storm 3, which had a range of WA that was much higher than either Storms 1 or 2, a lower pH, and no change of pH over the observed range of WA. Apparently, high WA depresses the pH and also buffers the stream against

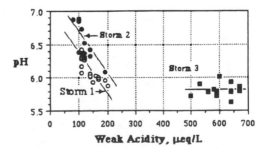

Figure 30.5 Stream acidification data fitted to Model C (Table 30.2). Storms 1 and 2 have the same slope.

changes in pH. But, why was the WA so much different during Storm 3? The data alone and the statistical analysis do not answer this question. They do, however, serve the investigator by raising the question.

COMMENTS

The variables considered in regression equations usually take numerical values over a continuous range, but occasionally it is advantageous to introduce a factor which has two or more discrete levels or categories. For example, data may arise from three storms or three operators. In such a case, we cannot set up a continuous measurement scale for the variable "storm" or "operator." We must create categorical variables (dummy variables) that account for the possible different effects of separate storms or operators. The levels assigned to the categorical variables are unrelated to any physical level that might exist in the factors themselves. Draper and Smith (1981) provide several examples on creating efficient patterns for assigning categorical variables. This method was used to model the disappearance of polychlorinated biphenyls (PCBs) from soil by Berthouex and Gan (1991).

REFERENCES

Berthouex, P. M. and D. R. Gan (1991). "Fate of PCBs in Soil Treated with Contaminated Municipal Sludge," *J. Environ. Eng. Div. ASCE*, 116, 1, 1–18.

Chatterjee, S. and B. Price (1977). *Regression Analysis by Example,* New York, John Wiley & Sons.

Daniel, C. and F. S. Wood (1980). *Fitting Equations to Data: Computer Analysis of Multifactor Data*, 2nd ed., New York, John Wiley & Sons.

Draper, N. R. and H. Smith (1981). *Applied Regression Analysis*, 2nd ed., New York, John Wiley & Sons.

Meinert, D. L., S. A. Miller, R. J. Ruane, and H. Olem (1982). *A Review of Water Quality Data in Acid Sensitive Watersheds in the Tennessee Valley,*" Report No. TVA.ONR/WR-82/10, TVA, Chattanooga, TN.

Chapter 31

The Effect of Autocorrelation on Regression

Key words: autocorrelation, Durbin-Watson statistic, serial dependence, regression, variance (inflation)

Many environmental data exist as sequences of measurements over time or space. The time sequence is obvious in some data series, like daily measurements on river quality. There can also be a time sequence in the way experimental runs are done in the laboratory. A characteristic of such data can be that neighboring observations tend to be somewhat alike. High values are followed by high values, and low values are followed by low values. The same tendency could arise from chemists on different shifts having slightly different results. This tendency is called autocorrelation or serial correlation. It is generally good practice to recognize that autocorrelation is likely to be present and to look for evidence that it exists in a set of data. Here we explain why it is a problem if overlooked. We hope the examples will convince engineers that experiments should be designed and conducted in a way that reduces the possibility for autocorrelation to enter and that data from unplanned experiments should be analyzed with an eye open to detect autocorrelation.

Most statistical methods, such as estimation of confidence intervals, ordinary least squares regression, etc., depend on the errors being independent, having constant variance, and being normally distributed. "Independent" means that the errors are not autocorrelated. The errors in statistical conclusions caused by violating the condition of independence can be more serious errors than those caused by not having normality.

Unrecognized (or ignored) autocorrelation is likely to seriously bias estimates of variances and any statistics, such as confidence intervals, calculated using variances. All statements about probabilities will be wrong. Parameter estimates may or may not be seriously affected by autocorrelation. One possibility is that a value is estimated with what appears to be great precision when in fact there is nothing that can be estimated with statistical confidence.

This chapter explains why ignoring existing autocorrelation can lead to serious errors and broadly describes the methods for detecting autocorrelation. Checking for autocorrelation is relatively easy (although in small data sets it may be undetected even when present). No particular knowledge of time series analysis or modeling is needed. Making suitable provisions to incorporate existing autocorrelation into the data analysis can be difficult. Explaining how to do it requires more space than can be afforded in this book. Some useful references are given, but the best approach may be to consult with a statistician.

CASE STUDY — A SUSPICIOUS LABORATORY EXPERIMENT

A laboratory experiment was done to demonstrate to students that increasing factor x by one unit should cause factor y to increase by one half unit. Preliminary experiments indicated that the standard deviation of repeated measurements on y was about one unit. To make measurement error small relative to the signal, the experiment was designed to produce 20 to 25 units of y. The data obtained, which are plotted in Figure 31.1, were

x =	0	1	2	3	4	5	6	7	8	9	10
y =	21.0	21.8	21.3	22.1	22.5	20.6	19.6	20.9	21.7	22.8	23.6

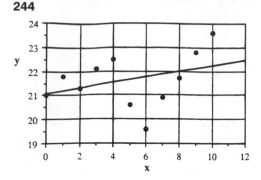

Figure 31.1 The original data from a suspicious laboratory experiment.

Linear regression gave the equation $\hat{y} = 21.04 + 0.12x$, with $R^2 = 0.12$. The 95% confidence interval of the slope was -0.12 to 0.31. This interval includes zero, so we cannot even be sure that x and y are related. And, it does not include the theoretical slope of 0.5, which the experiment was designed to reveal. This was an unpleasant surprise, since the experiment was designed to demonstrate that x and y are related.

One might be tempted to blame the peculiar result entirely on the low value measured at $x = 6$, but the experimenters did not leap to premature conclusions. Discussion of the experimental procedure revealed that the students had done the tests starting with $x = 0$ first, then with $x = 1$, and so on up through $x = 10$. The measurements of y were also done in order of increasing concentration. It was also discovered that the injection port of the instrument used to measure y might not have been thoroughly cleaned between each run. Yes, they had heard about randomization, but time was short and they could complete the experiment faster by not randomizing. The penalty was serial correlation and a wasted experiment.

They were asked to repeat the experiment, this time randomizing the order of the runs and taking more care with cleaning the injection port. This time the data were as shown in Figure 31.2. The regression equation was $\hat{y} = 19.75 + 0.48x$, with $R^2 = 0.87$. The confidence interval of the slope was 0.27–0.64, showing that x and y are related. Also, this confidence interval includes the expected theoretical slope of 0.5.

Can the dramatic difference in the outcome of the first and second experiments possibly be due to the presence of autocorrelation in the experimental data? It is both possible and likely, in view of the lack of randomization in the order of running the tests.

THE CONSEQUENCES OF AUTOCORRELATION ON REGRESSION

When doing regression, we often focus too much on obtaining estimates of the parameters and not enough on the equally important task of obtaining a valid statement about the precision of the estimates. The main reason for doing statistics is to make statements about the precision of estimated values and not merely to estimate parameters. Unfortunately,

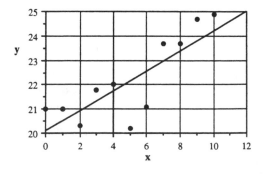

Figure 31.2 Data obtained from a repeated experiment with randomization to eliminate autocorrelation.

autocorrelation acts to destroy our ability to make such statements. If the error terms are positively autocorrelated, the usual confidence intervals and tests using t and F distributions are no longer strictly applicable because the variance estimates are distorted (Neter et al., 1983).

WHY AUTOCORRELATION DISTORTS THE VARIANCE ESTIMATES

Suppose that the system generating the data has the true underlying relation $\eta = \beta_0 + \beta_1 x$, where x could be any independent variable, including time as in a times series of data. What we observe is $y_i = \eta_1 + e_i = \beta_0 + \beta_1 x_i + e_i$. The usual assumption is that the residuals, e_i, are independent, meaning that the value of e_i is not related to e_{i-1}, e_{i-2}, etc. Let us examine what happens when this is not true.

Suppose that the residuals, instead of being random and independent, are dependent in a simple way that is described by $e_i = \rho \, e_{i-1} + a_i$, in which the errors, a_i, are independent and normally distributed with constant variance σ^2. The strength of the autocorrelation depends on the value of ρ, the *autocorrelation coefficient*. If $\rho = 0$, the e_i are independent. If $0 \leq \rho \leq 1$, the autocorrelation is said to be positive, meaning that successive values of e_i are similar to each other.

More specifically it means that

$$e_i = \rho e_{i-1} + a_i, \qquad e_{i-1} = \rho e_{i-2} + a_{i-1}, \qquad e_{i-2} = \rho e_{i-3} + a_{i-2}$$

and so on. By recursive substitution, we can show that

$$e_i = \rho(\rho e_{i-2} + a_{i-1}) + a_i = \rho^2 e_{i-2} + \rho a_{i-1} + a_i$$

and

$$= \rho^2(\rho e_{i-3} + a_{i-2}) + \rho a_{i-1} + a_i = \rho^3 e_{i-3} + \rho^2 a_{i-2} + \rho a_{i-1} + a_i$$

This shows that the process is "remembering" past conditions to some extent, and the strength of this memory is reflected in the value of ρ.

Reversing the order of the terms and continuing the recursive substitution gives

$$e_i = a_i + \rho a_{i-1} + \rho^2 a_{i-2} + \rho^3 a_{i-3} + \rho^4 a_{i-4} + \cdots$$

The expected values of a_i, a_{i-1}, ... are zero and so are the expected values of e_i, e_{i-1}, ... The variance of e_i and the variance of a_i, however, are not the same. The variance of the a's, is σ_a^2. The variance of e_i is

$$\sigma_e^2 = \mathrm{Var}(a_i) + \rho^2 \mathrm{Var}(a_{i-1}) + \rho^4 \mathrm{Var}(a_{i-2}) + \rho^6 \mathrm{Var}(a_{i-3}) + \rho^8 \mathrm{Var}(a_{i-4}) +$$

$$\sigma_e^2 = \sigma_a^2(1 + \rho^2 + \rho^4 + \rho^6 + \rho^8 + \ldots + \rho^\infty)$$

For positive correlation $(0 \leq \rho \leq 1)$, the power series of ρ equals $1/(1 - \rho^2)$ and

$$\sigma_e^2 = \sigma_a^2/(1 - \rho^2)$$

This means that when there is positive autocorrelation, and we do not recognize and account for it, the estimated variance σ_e^2 will be larger than the true variance of the random independent errors, σ_a^2, by the factor $1/(1 - \rho^2)$ (Box et al., 1977).

AN EXAMPLE OF AUTOCORRELATED ERRORS

The "laboratory data" presented as the case study were created to illustrate the consequences of autocorrelation on regression. The true model of the experiment is $\eta = 20 + 0.5x$. The data structure is shown in Table 31.1. If there were no autocorrelation, the observed values would be as shown in Figure 31.2. These are the fourth column in Table 31.1, which is computed as $y_i = 20 + 0.5x + a_i$, where the a_i's are independent values drawn randomly from a normal distribution with mean zero and variance of one (the variance of the actual 11 a_i's selected was 1.00; their mean was –0.28).

In the flawed experiment, hidden factors in the experiment were assumed to operate on the random errors a_i to make them serially correlated. This could happen because the injection port on an instrument was not perfectly cleaned out between samples so each measurements was affected by the previous sample injected. If the order of injections happened to be from high concentrations to low concentrations, the serial correlation might be strengthened. The "data" were computed assuming that the experiment generated errors having first-order autocorrelation with $\rho = 0.8$. The last three columns in Table 31.1 show how the independent random errors are converted to correlated errors. The function producing the data for the flawed experiment is

$$y_i = \eta + e_i = (20 + 0.5\ x_i) + (0.8\ e_{i-1} + a_i)$$

If the data were produced by the above model, but we were unaware of the autocorrelation and fit the simpler model $\eta = \beta_0 + \beta_1 x$, the estimates of β_0 and β_1 would reflect this misspecification of the model. Perhaps, more serious is the fact that t-tests and F tests on the regression results will be wrong, so we may be badly misled as to the significance or precision of estimated values. Fitting the data produced from the autocorrelation model of the process gives $\hat{y}_i = 21.0 + 0.12\ x_i$. The 95% confidence interval of the slope is [–0.12 to 0.36] and the t statistic for the slope is 1.11. Both of these results indicate the the slope is not significantly different from zero. Even though the result is statistically significant, it is wrong, since the true slope is 0.5.

This is in contrast to what would have been obtained if the experiment had been conducted in a way that prevented autocorrelation from entering. The data for this case

Table 31.1 **Data Created for Using True Values of $y_i = 20 + 0.5\ x_i + a_i$, with $a_i = N(0,1)$**

	No Autocorrelation			Autocorrelation, $\rho = 0.8$			
x	η	a_i	$y_i = \eta_i + a_i$	$0.8\ e_{i-1} + a_i$	=	e_i	$y_i = \eta_i + e_i$
0	20.0	1.0	21.0	0.00 + 1.0	=	1.0	21.0
1	20.5	0.5	21.0	0.80 + 0.5	=	1.3	21.8
2	21.0	–0.7	20.3	1.04 + –0.7	=	0.3	21.3
3	21.5	0.3	21.8	0.27 + 0.3	=	0.6	22.1
4	22.0	0.0	22.0	0.46 + 0.0	=	0.5	22.5
5	22.5	–2.3	20.2	0.37 + –2.3	=	–1.9	20.6
6	23.0	–1.9	21.1	–1.55 + –1.9	=	–3.4	19.6
7	23.5	0.2	23.7	–2.76 + 0.2	=	–2.6	20.9
8	24.0	–0.3	23.7	–2.05 + –0.3	=	–2.3	21.7
9	24.5	0.2	24.7	–1.88 + 0.2	=	–1.7	22.8
10	25.0	–0.1	24.9	–1.34 + –0.1	=	–1.4	23.6

are listed in the "no autocorrelation" section of Table 31.2, and the results are shown in Table 31.2. The fitted model is $\hat{y}_i = 20.1 + 0.43\,x_i$, the confidence interval of the slope is [0.21 to 0.65], and the t statistic for the slope is 4.4. The slope is statistically significant, and the true value of the slope ($\beta = 0.5$) falls within the confidence interval.

Table 31.2 summarizes the results of these two regression examples ($\rho = 0$ and $\rho = 0.8$). The Durbin-Watson statistic (explained in the next section) provided by the regression program indicates independence in the case where $\rho = 0$, and they show serial correlation in the other case.

A STATISTIC TO INDICATE POSSIBLE AUTOCORRELATION

Detecting autocorrelation in a small sample is difficult. Sometimes it is not possible. In view of this, it is better to design and conduct experiments to exclude autocorrelated errors. In undesigned experiments, there is always the possibility of autocorrelation in the errors, and this can be disastrous in regression. Because of this, most computer programs that do regression can compute the Durbin-Watson statistic, which is based on an examination of the residual errors for autocorrelation. Unfortunately, the Durbin-Watson statistic cannot always detect correlation when it exists.

The Durbin-Watson test assumes a first-order model of autocorrelation. Higher-order autocorrelation structure is possible, but less likely than first order, and verifying higher-order correlation would be more difficult. Even testing for the first-order effect is hard when the number of observations is small.

The test examines whether the first-order autocorrelation parameter ρ is zero. In the case where $\rho = 0$, the errors are independent. The test statistic is

$$D = \sum_{i=2}^{n} (e_i - e_{i-1})^2 \Big/ \sum_{i=1}^{n} (e_i)^2$$

where the e_t is the residuals determined by fitting a model using least squares.

Durbin and Watson (1971) obtained approximate upper and lower bounds, d_L and d_U, on the statistic D. If $d_L \leq D \leq d_U$, the test is inconclusive. However, if $D > d_U$ conclude $\rho = 0$, and if $D < d_L$ conclude $\rho > 0$. A few Durbin-Watson test bounds for the 0.05 level of significance are given in Table 31.3. Note that this test is for positive ρ. If a test for negative correlation, $\rho < 0$, is required, the test statistic to be used is $4 - D$, where D is calculated as before.

Table 31.2 **Summary of the Regression "Experiments"**

Estimated Statistics	No Autocorrelation $\rho = 0$	Autocorrelation $\rho = 0.8$
Intercept	20.1	21.0
Slope	0.43	0.12
Standard error of slope	0.10	0.1
Confidence interval of slope	0.21 to 0.65	−0.12 to 0.3
R^2	0.68	0.1
Mean square error	1.06	1.2
Durbin-Watson D	1.38	0.91

Table 31.3 **Some Durbin-Watson Test Bounds for the 0.05 Level of Significance**

n	p = 2		p = 3		p = 4	
	d_L	d_U	d_L	d_U	d_L	d_U
15	1.08	1.36	0.95	1.54	0.82	1.75
20	1.20	1.41	1.10	1.54	1.00	1.68
25	1.29	1.45	1.21	1.55	1.12	1.66
30	1.35	1.49	1.28	1.57	1.21	1.65
50	1.50	1.59	1.46	1.63	1.42	1.67

Note: n = number of observations and p = number of parameters estimated in the model.

AUTOCORRELATION AND TREND ANALYSIS

Sometimes we are tempted to take an existing record of environmental data (pH, temperature, etc.) and analyze it for a trend by doing linear regression to estimate a slope. A slope statistically different from zero being taken as evidence shows that some long-term change has been occurring. Resist the temptation, because such data are almost always serially correlated. *Serial correlation* is autocorrelation between data that constitute a time series. An example, similar to the regression example, helps make the point.

Figure 31.3 shows two time series of simulated environmental data. There are 50 values in each series. The model used to construct data set in Figure 31.3a was $y_i = 10 + a_i$, where a_t is a random, independent variable with N(0, 1). The model used to construct data set in Figure 31.3b was $y_i = 10 + 0.8\ e_{i-1} + a_i$. The a_i are the same as in data set Figure 31.3a; the e_i variates are serially correlated with $\rho = 0.8$. The influence of the serial correlation on the pattern of the time series is evident. Naturally, it affects the statistics estimated from the series as well.

For both data sets, the true underlying trend is zero (the models contain no term for slope). If trend is examined by fitting a model of the form $\eta_i = \beta_0 + \beta_1\ t_i$, where t is time, the results are

Data set (Figure 31.3a): $\hat{y}_i = 9.98 + 0.005\ t_i$
mean square error = 0.798
confidence interval of the slope [–0.012 to 0.023]

Data set (Figure 31.3b): $\hat{y}_i = 9.71 + 0.033\ t_i$
mean square error = 1.996
confidence interval of the slope [0.005 to 0.061]

For the data set in Figure 31.3a, the slope is estimated to be zero, but for the data set in Figure 31.3b, it is estimated to be positive. This is caused by the serial correlation. The inflation of the mean square error is also caused by the serial correlation. The Durbin-Watson statistic does give the correct warning about the serial correlation, but the data analyst is left to decide how to deal with it.

COMMENTS

We have seen that autocorrelation can cause serious problems in regression. The Durbin-Watson statistic *might* indicate when there is cause to worry about autocorrelation. It will not always detect autocorrelation, and it is especially likely to fail when the data

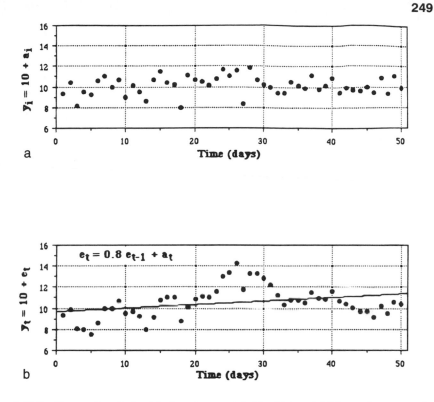

Figure 31.3 Time series of simulated environmental data. Data set (a) is random, normally distributed values with mean zero and variance of one added to a constant value of 10. Data set (b) was constructed using the random variates of data set (a) to compute serially correlated values which were added to a constant value of 10.

set is small. Even when autocorrelation is revealed as a problem, it is too late to eliminate it from the data, and one faces the task of deciding how to model it.

The pitfalls inherent with autocorrelated errors provide a strong incentive to plan experiments to include proper randomization whenever possible. If an experiment is intended to define a relationship between x and y, the experiments should not be conducted by gradually increasing (or decreasing) the x's. Randomize over the settings of x to eliminate autocorrelation due to time effects in the experiments.

REFERENCES

Box, G. E. P., W. G. Hunter, and J. S. Hunter (1977). *Statistics for Experimenters*, New York, Wiley Interscience.

Durbin, J. and G. S. Watson (1951). "Testing for Serial Correlation in Least Squares Regression, II," *Biometrika*, 38, 159–178.

Durbin, J. and G. S. Watson (1971). "Testing for Serial Correlation in Least Squares Regression, III," *Biometrika*, 58, 1–19.

Neter, J., W. Wasserman, and M. H. Kutner (1983). *Applied Regression Models*, Homewood, IL, Richard D. Irwin Co.

Rawlings, J. O. (1988). *Applied Regression Analysis: A Research Tool*, Pacific Grove, CA, Wadsworth and Brooks/Cole.

Chapter 32

The Iterative Approach to Modeling

Key words: biokinetics, chemostat, dilution rate, experimental design, factorial designs, iterative design, model building, Monod model, parameter estimation, sequential design

---

The dilemma of model building is that what needs to be known in order to design good experiments is exactly what the experiments are supposed to discover. We could be easily frustrated by this if we imagined that success depended on one grand experiment that would lead directly to the desired results. Life, science, and statistics do not work this way. Knowledge is gained in small steps, each step guiding us to the next. We begin with a modest initial experiment which produces information we use to design the second experiment, which in turn guides design of the third, and so on. Between each step there is need for reflection, study, and creative thinking. Experimental design, then, is a philosophy as much as a technique.

The iterative (or sequential) philosophy of experimental investigation diagrammed in Figure 32.1 applies to empirical exploration of operating conditions (Chapter 33) and to mechanistic model building. It includes checks on the adequacy of the model, because we are never certain the model is correct. The iterative approach is illustrated for an experiment in which each observation requires a considerable investment.

CASE STUDY — BACTERIAL GROWTH

The material balance equations for substrate (S) and bacterial solids (X) in a completely mixed reactor operated without recycle are

$$V \frac{dX}{dt} = 0 - QX + \left(\frac{\theta_1 S}{\theta_2 + S} \right) X \, V \qquad \text{material balance on bacterial solids}$$

$$V \frac{dS}{dt} = Q \, S_0 - QS - \frac{1}{\theta_3} \left(\frac{\theta_1 S}{\theta_2 + S} \right) X \, V \qquad \text{material balance on substrate}$$

This assumes there are no bacterial solids in the influent and uses Monod kinetics for bacterial growth. After dividing by V, these equations are written more conveniently as

$$\frac{dX}{dt} = \frac{\theta_1 S \, X}{\theta_2 + S} - D \, X$$

and

$$\frac{dS}{dt} = DS_0 - DS - \left(\frac{\theta_1 X}{\theta_3} \right) \left(\frac{S}{\theta_2 + S} \right)$$

where
  Q = liquid flow rate
  V = reactor volume
  D = Q/V = dilution rate

251

252

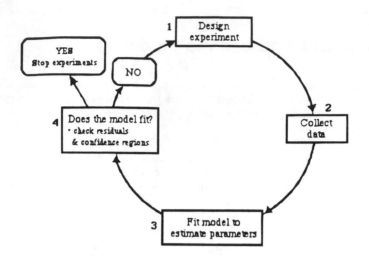

Figure 32.1 The iterative cycle of experimentation. (From Box and Hunter, 1965.)

S_0 = influent substrate concentration
X = organism concentration in the reactor and in the effluent
S = substrate concentration in the reactor and also in the effluent
θ_3 = the yield coefficient

   θ_1 and θ_2 are parameters of the Monod model for bacterial growth (θ_1 is the maximum growth rate and θ_2 is the half saturation constant)

The steady-state solutions of the equations are

$$S = \frac{\theta_2 D}{\theta_1 - D}$$

and

$$X = \theta_3(S_0 - S) = \theta_3\left[S_0 - \frac{\theta_2 D}{\theta_1 - D} \right]$$

   If the dilution rate is sufficiently large, the organisms will be washed out of the reactor faster than they can grow. If all the organisms are washed out, the effluent concentration will equal the influent concentration, $S = S_0$. The lowest dilution rate at which washout occurs is called the critical dilution rate, which is derived by substituting $S = S_0$ into the substrate model above.

$$D_c = \frac{\theta_1 S_0}{\theta_2 + S_0}$$

When $S_0 \gg \theta_2$, which is often the case, $D_c \approx \theta_1$.

   Experiments will be performed at several dilution rates (i.e., flow rates), while keeping the influent substrate concentration constant. When the reactor has been brought to steady state at the selected dilution rate, X and S will be measured, and these data can be used to estimate θ_1, θ_2, and θ_3. Let us assume that the experimenter who wishes to do this has only two reactors that can be used to test two different dilution rates simultaneously. Note that because two responses (X and S) are measured, the two experimental runs provide four data points (X_1 and S_1 at D_1; X_2 and S_2 at D_2), and this provides enough information to estimate the three parameters in the model. Since several weeks may be needed to start a reactor and bring it to steady-state conditions, the experimenter naturally wants to get as much information as possible from each run. Here is how the iterative approach can be used to do this.

Three iterations of the experimental cycle are shown in Figure 32.2. The bottom line represents the data analysis phase of work, where the parameters are estimated and the adequacy of the model is checked. The values shown along this line are the parameter estimates at each iteration. At the beginning, before any experiments are done, an initial guess of parameter values is used to get started. The top line represents the experimental design phase. Knowledge gained in previous experiments is used to plan the next iteration. Two experimental runs are done in each iteration. The values of X and S measured in these runs are shown in the middle of the diagram.

The initial guesses of the parameter values were $\theta_3 = 0.50$, $\theta_1 = 0.70$, and $\theta_2 = 200$. This led to selecting flow rate $D_1 = 0.64$ for one run and $D_2 = 0.35$ for the other. The method of choosing efficient experimental settings of D is ignored for now, since our purpose is merely to show the efficiency of iterative experimentation. Let it suffice to simply state that the experimental design criterion advises doing two runs, one with the dilution rate set as near the critical value D_c as the experimenter dares to operate, and the other at about half this value. At any stage in the experimental cycle, the best current estimate of the critical flow rate is $D_c = \theta_1$.

The experimenter must be cautious in using this advice because operating conditions become unstable as D_c is approached. If the actual critical dilution rate is exceeded, the experiment fails entirely and the reactor has to be restarted, at a considerable loss of time. On the other hand, staying too far on the safe side (keeping the dilution rate too low) will yield poor estimates of the parameters, especially of θ_1.

The parameter values estimated using data from the first pair of experiments were $k_3 = 0.60$, $k_1 = 0.55$, and $k_2 = 140$. (see Table 32.1). These values were used to design two new experiments. A second experimental iteration was done and the parameters

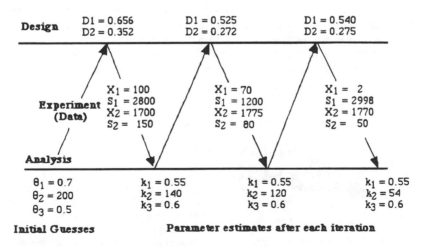

Figure 32.2 Three iterations of the Design-Experiment-Analyze experimental cycle converge to precise estimates of the parameters θ_1, θ_2, and θ_3 that predict cell growth and substrate removal in a steady-state completely mixed biological reactor.

Table 32.1 **Three Iterations of the Experiment
to Estimate Biokinetic Parameters**

Exp. Run	Best Current Estimates of Parameter Values			Dilution Rate D = V/Q	Observed Values		Parameter Values Estimated from New Data		
	θ_3	θ_1	θ_2		S	X	k_3	k_1	k_2
Iteration 1									
1	0.50	0.70	200	0.656	2800	100			
2				0.352	150	1700	0.60	0.55	140
Iteration 2									
3	0.60	0.55	140	0.525	1200	70			
4				0.275	80	1775	0.60	0.55	120
Iteration 3									
5	0.60	0.55	120	0.540	2998	2			
6				0.275	50	1770	0.60	0.55	54

Data from Johnson and Berthouex (1975).

were reestimated using all available data. The values of k_1 and k_3 (estimates of θ and θ_3 respectively) did not change from the first iteration, but k_2 (our estimate of θ_2) was reduced from 140 to 120. Since θ_1 and θ_3 seem to be estimated quite well (because k_1 and k_3 did not change from one iteration to the next), the third iteration essentially focused on improving the estimate of θ (i.e., k_2). At the end of three iterative steps — a total of only six experiments — the joint confidence region of the three parameters was small and the experiment was ended.

Figure 32.3 shows how the approximate 95% joint confidence region for θ_1 and θ_2 decreased in size from the first to the second to the third set of experiments. The large shaded region is the approximate joint 95% confidence region for the parameters after the first set of n = 2 experiments. Neither θ_1 or θ_2 was estimated very precisely. At the end of the second iteration, there were n = 4 observations at four settings of the dilution rate. The resulting joint confidence region (the unshaded area) is horizontal, but elongated, showing that θ_1 was estimated with good precision, but θ_2 was not. The third iteration invested in data that would more precisely define the value of θ_2. Fitting the model to the n = 6 tests gives the estimates $k_1 = 0.55$ and $k_2 = 54$ and the small joint confidence

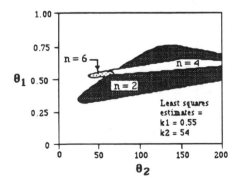

Figure 32.3 Approximate joint 95% confidence regions for θ_1 and θ_2 estimated after the first, second, and third experimental iterations. Each iteration consisted of experiments at two dilution rates, giving n = 2 after the first iteration, n = 4 after the second, and n = 6 after the third.

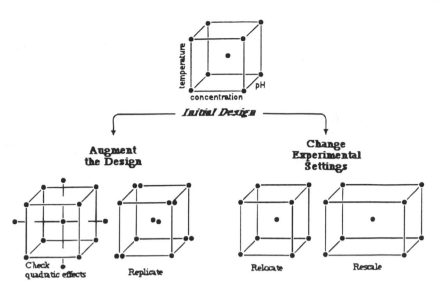

Figure 32.4 Illustration of some sequential modifications of an experimentation.

region shown in Figure 32.3. Note that this confidence region falls inside the region determined after n = 4 tests.

COMMENTS

The iterative experimental approach is very efficient. It is especially useful when measurements are difficult or expensive. It is recommended in almost all model-building situations, whether the model is linear or nonlinear, simple or complicated.

The example described in this case study was able to obtain precise estimates of the three parameters in the model with experimental runs at only six experimental conditions. Six runs is not many in this kind of experiment, so it should be obvious that the success was not due to making a large number of experimental runs. It was due to selecting experimental conditions (dilution rates) that produced a lot of information about the parameter values. Chapters 34, 36, and 37 will show that making a large number of runs can yield poorly estimated parameters if the experiments are run at the wrong conditions. Chapter 26 explains how the precision of parameter estimates is determined.

Factorial and fractional factorial experimental designs (Chapters 20–22) are especially well suited to the iterative approach since they can be modified in many ways to suit the experimenters' need for additional information. Figure 32.4 shows six ways that a three-variable design (a 2^{3-1} fractional factorial design with a center point) might be modified, depending on the results of the initial design and subsequent designs. Some of the various modifications might be to (a) move the design to a new location (i.e., run the same design at new settings of the independent variables); (b) add the other half fraction to create a full 2^3 factorial experiment; (c) keep the experiment centered on the original location, but expand the range of one or more independent variables; (d) change, add, or drop independent variables; (e) do replicate tests at some (or all) of the experimental settings; or (f) augment the design by adding "star points" in order to investigate second-order effects (Box, 1991). The example in Chapter 33 illustrates optimizing process conditions by moving to explore a new location and augmenting the design to investigate a curvature at a peak.

REFERENCES

Box, G. E. P. (1991). "Quality Improvement: The New Industrial Revolution," Tech. Report No. 74, Center for Quality and Productivity Improvement, University of Wiscosin-Madison.

Box, G. E. P. and W. G. Hunter (1965). "The Experimental Study of Physical Mechanisms," *Technometrics*, 7, 23.

Johnson, D. B. and P. M. Berthouex (1975). "Efficient Biokinetic Designs," *Biotechnol. Bioengr.*, 18, 557–570.

Chapter 33

Seeking Optimum Conditions by Response Surface Methodology

Key words: factorial designs, iterative design, optimization, quadratic effects, response surface methodology, regression, steepest ascent

---

The response of a system or process may depend on many variables such as temperature, organic carbon concentration, air flow rate, etc. An important problem is to discover the settings for the critical variables that give the best system performance. Response surface analysis is a powerful experimental approach to optimizing the performance of systems for which no detailed process model is known. It is the ultimate application of the iterative approach to experimentation.

The method was first demonstrated by Box and Wilson (1951) in a paper that Margolin (1985) describes as follows:

> The paper . . . is one of those rare, truly pioneering papers that completely shatters the existing paradigm of how to solve a particular problem. . . . The design strategy . . . to attain an optimum operating condition is brilliant in both its logic and its simplicity. Rather than exploring the entire continuous experimental region in one fell swoop, one explores a sequence of subregions. Two distinct phases of such a study are discernible. First, in each subregion a classical two-level fractional factorial design is employed . . . . This first-phase process is iterative and is terminated when a region of near stationarity is reached. At this point a new phase is begun, one necessitating radically new designs for the successful culmination of the research effort.
>
> It is in the exploration of the nearly stationary region that Box most clearly exhibits his superb geometrical talents and insights. Box's approach involves contour plots, reductions to canonical form, and composite designs (e.g. designs consisting of two-level factorials augmented with axial points to permit estimation of quadratic effects).

RESPONSE SURFACE METHODOLOGY

Response surface methodology uses a strategy of sequential study of small parts of the experimental space. The strategy is to explore, analyze what has been learned, and then move to a promising new location, where the learning cycle is repeated. Each exploration points the way to a new location, where conditions are expected to be better. Eventually, a set of optimal operating conditions may be determined. We visualize these as the peak of a hill[1] that has been climbed after stopping periodically to explore and locate the most locally promising path to the summit. When we are starting the climb, we imagine the shape of the hillside is relatively smooth and we worry mainly about its steepness. Figure 33.1 sketches the progress of an iterative search for the optimum conditions in a process that has two active independent variables. The early phases use two-level factorial experimental designs, perhaps augmented with a center point. The main effects estimated from these designs point the way of steepest ascent toward the optimum. Near

---

[1] Or, alternately, a crater if the optimum is a minimum. Sometimes we find that the surface is a saddle between two hills instead of a hill with a peak. The quadratic model will also describe this kind of surface.

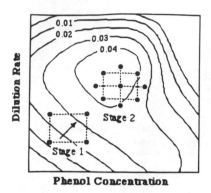

Figure 33.1 Two stages of a response surface optimization. The second stage is a two-level factorial augmented to define quadratic effects.

Phenol Concentration

the optimum, the curvature is more extreme and the simple designs do not capture it entirely. Indeed, a simple factorial design might be located astride the optimum in a way that makes the main effects appear to be zero. When we are at the peak and exploring on all sides of it, a quadratic model will be needed, and the experimental design to fit the model is a two-level factorial augmented with star points, as in the Stage 2 design shown in Figure 33.1. Of course, in general, we do not know how many experimental stages will be required to locate the optimum.

CASE STUDY — INHIBITION OF MICROBIAL GROWTH BY PHENOL

Wastewater from a coke plant contains phenol, which is known to be biodegradable at low concentrations and inhibitory at high concentrations. Hobson and Mills (1990) used a laboratory scale treatment system to vary the influent phenol concentration and the flow rate in order to determine how these two variables affect the phenol oxidation rate and whether there is an operating condition at which the removal rate is a maximum. This case study is based on their data, which we used to create a response surface by drawing contours. The "data" given in the following sections were interpolated from this surface, and a small experimental error was added.

To some extent, a treatment process operated at a low dilution rate can tolerate high phenol concentrations better than a process operating at a high dilution rate. We need to define "high" and "low" for a particular wastewater and a particular biological treatment process and find the operating conditions that give the most rapid phenol oxidation rate (R). The experiment is arranged so that the rate of biological oxidation of phenol depends on only two factors: (1) the concentration of phenol in the reactor and (2) the dilution rate. Dilution rate is defined as the reactor volume divided by the wastewater flow rate through the reactor. These factors will be denoted by C and D, respectively. Other factors, such as temperature, have been held constant.

The iterative approach of experimentation, as embodied in response surface methodology, will be illustrated. The steps in each iteration are *design*, *data collection*, and *data analysis*. Here only the design and data analysis steps are discussed.

First Iteration

Design — Because each experiment takes several days to complete, the experiment was run in small sequential stages. The first was a two-level two-factor experiment — a 2^2 factorial design. The two experimental factors are dilution rate (D) and residual phenol concentration (C). Each factor was investigated at two levels, and the observed phenol removal rates are given in Table 33.1.

Analysis — The data can be analyzed by calculating the effects, as done in Chapter 20, or by linear regression. Here we will illustrate the regression method. There are four

Table 33.1   **Experimental Design and Results for Iteration 1**

C (g/L)	D (1/hr)	R (g/hr)
0.5	0.14	0.018
0.5	0.16	0.025
1.0	0.14	0.030
1.0	0.16	0.040

observations, so the most complicated model we can fit would have four parameters. Since we expect the surface to be relatively smooth, a reasonable model has the form $R = b_0 + b_1C + b_2D + b_{12}CD$. The terms b_1C and b_2D represent the main effects of concentration and dilution rate; the last term, $b_{12}CD$, is the interaction between the two factors.

The fitted model is $R = -0.022 - 0.018C + 0.2D + 0.3 \, CD$. The level of the response R as a function of the two experimental factors is depicted as a contour map in Figure 33.2. The contours are values of R in units of grams per hour. The approximation is good only in the neighborhood of the 2^2 experiment, which is indicated by the four dots at the corner of the rectangle. The direction in which the experiment should be moved to seek higher reaction rates is clear from the contour lines; this is indicated by an arrow. Mathematically this is the direction of steepest ascent and is perpendicular to the contour line at the point of interest.

Second Iteration

Design — The results of the first iteration indicate the direction to move, but do not tell us how much each setting should be increased. Boldly make a big step and you risk going over the peak. Make a cowardly small step and progress toward the peak will be slow. How bold — or how cowardly — should we be? This may seem like a serious dilemma, but it really is not, for two reasons. First, the experimenter usually has prior experience and special knowledge about the experimental conditions. We know, for example, that there is a practical upper limit on the dilution rate because at some level all the bacteria will wash out of the reactor. We also know from previously published results that phenol becomes inhibitory at some level. We may have a fairly good idea of the concentration at which this should be observed. In short, the experimenter knows something about limiting conditions at the start of the experiment (and will quickly learn more). To a large extent, we trust our judgment about how far to move.

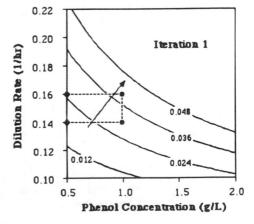

Figure 33.2 Response surface computed from the data collected in exploratory Stage 1 of the optimizing experiment.

The second factor is that doing iterative factorial experiments is so extremely efficient that the total number of experiments will be small regardless of how boldly we proceed. Even if we make what seems in hindsight to be a mistake, whether in direction or distance, this will be quickly discovered, and the same experiments that reveal it will help put us onto a better path toward the optimum.

In this case of phenol degradation, published experience indicates that inhibitory effects will probably become evident with the range of 1–2 g/L. This suggests that an increase of 0.5 g/L in concentration should be a suitable next step, so we decide to try C = 1.0 and C = 1.5 as the low and high settings. Going roughly along the line of steepest ascent, this would give dilution rates of 0.16 and 0.18 as the low and high settings of D. This leads to the second stage experiment, which is the 2^2 factorial design shown in Table 33.2. Table 33.2 also gives the measured phenol removal rates at the four experimental settings.

Analysis — The average performance has improved, and two of the response values are larger than the maximum observed in the first iteration. The fitted model is R = 0.047 – 0.014 C + 0.05 D. The estimated coefficient for the C*D term was zero, so no interaction term is included in the model for this experiment. Figure 33.3 shows the response surface.

In the first iteration, the sign of the coefficient of C was positive and we wanted to increase C. Here it is negative, indicating C should be reduced. The positive sign on D indicates that further increase in D may be beneficial. Before jumping in a new direction, consider the model more carefully. The fitted model describes a plane and the plane is almost horizontal, as indicated by the small coefficients of both C and D. This indicates that we may be close to the optimum conditions. To check on this and to be able to model the curvature at the peak, we need to use a more elaborate experimental design in the next iteration.

Table 33.2 **Experimental Design and Results for Iteration 2**

C (g/L)	D (1/hr)	R (g/hr)
1.0	0.16	0.041
1.0	0.18	0.042
1.5	0.16	0.034
1.5	0.18	0.035

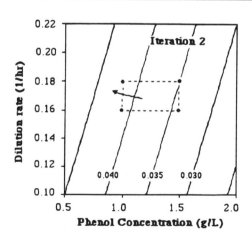

Figure 33.3 Approximation of the response surface estimated from the second-stage exploratory experiment.

Third Iteration — Exploring for Optimum Conditions

Design — One way we can observe a nearly zero effect for both variables is if the four corners of the 2^2 experimental design straddle the peak of the response surface. If there is substantial curvature in the true response surface, a two-level experiment lacks the power to define it. Nevertheless, the fundamental efficiency of the two-level design still works to our favor. The design is easy to augment so that curvature can be detected and modeled.

We design a more ambitious experiment in anticipation of needing to fit a model that contains some quadratic terms, such as $R = b_0 + b_1C + b_2D + b_{11}C^2 + b_{22}D^2 + b_{12}CD$. The basic experimental design is still a two-level factorial, but it will be augmented by adding "star" points. The easiest way to picture this design is to imagine a circle (or ellipse, depending on the scaling of our sketch) that passes through the four corners of the two-level design. It should be clear that we could rotate the four points of the two-level experiment to any position on the circle and have experiments that produce equivalent information about the surface. There are already four points on this "circle," and four more will be added in a way that maintains the symmetry of the original design. The augmented design has eight equally distributed points, each equidistant from the center of the design. Adding one more point at the center of the design will provide a better estimate of the curvature, while maintaining the symmetric design. The experimental design and the results are shown in Figure 33.4 and Table 33.3.

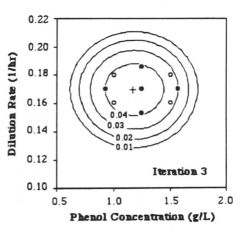

Figure 33.4 Contour plot of the quadratic response surface model fitted to an augmented two-level factorial experimental design. The open symbols are the two-level design from Iteration 2; the solid symbols are the center and star points added to investigate curvature near the peak. The cross (+) locates the optimum computed from the model.

Table 33.3 **Experimental Results for Iteration 3**

C (g/L)	D (1/hr)	R (g/hr)	Notes
1.0	0.16	0.041	Iteration 2 design
1.0	0.18	0.042	Iteration 2 design
1.5	0.16	0.034	Iteration 2 design
1.5	0.18	0.035	Iteration 2 design
0.9	0.17	0.038	Augmented "star" point
1.25	0.156	0.043	Augmented "star" point
1.25	0.17	0.047	Center point
1.25	0.184	0.041	Augmented "star" point
1.6	0.17	0.026	Augmented "star" point

Analysis — These data were fitted to a quadratic model to get

$$R = -0.76 + 0.28\,C + 7.54\,D - 0.12\,C^2 - 22.2\,D^2$$

Once again, the interaction term has a very small coefficient and can be omitted.

Contours computed from this model are plotted in Figure 33.4. The location of the nine experimental runs are also shown on the plot. The open circles are the two-level design from Iteration 2; the solid circles are the center and star points added to investigate curvature near the peak.

The maximum predicted phenol oxidation rate is 0.047 g/hr, which is obtained at $C = 1.17$ g/L and $D = 0.170$/hr. These values are obtained by taking derivatives of the response surface model and simultaneously solving $dR/dC = 0$ and $dR/dD = 0$ for C and D.

Iteration 4 — Is It Needed?

The question now is whether a fourth iteration is needed. One possibility is to declare that enough is known and to stop. Another possibility would be to replicate some of the previous experimental settings to confirm the quadratic model. And yet another would be to contract the experimental region around the predicted optimum conditions.

HOW EFFECTIVELY WAS THE OPTIMUM LOCATED?

Let us see how efficient the method was in this case. Figure 33.5 (top) shows the contour plot from which the experimental "data" were obtained. This plot was constructed by interpolating the Hobson and Mills data with a simple contour plotting routine; no equations were fitted to the data to generate the surface. An experiment was "run" by interpolating a value of R from the graph and adding to it an "experimental error."

Even though the first 2^2 design was not very close to the peak, the maximum was located with a total of only 13 experimental runs (4 in Iteration 1, 4 in Iteration 2, plus 5 in Iteration 3). The predicted optimum is very close to the peak of the contour map (Figure 33.5) from which the "data" were taken. Furthermore, the region near the optimum is nicely approximated by the contours derived from the fitted model, as can be seen by comparing Figures 33.4 and 33.5.

Hobson and Mills made 14 observations covering an area of roughly $C = 0.5$–0.15 and $D = 0.125$–0.205. The location of their 14 runs is shown in Figure 33.5, which also shows the three-dimensional response surface from two points of view. (These surfaces were plotted using empirical data smoothing methods; no simple model fitted to the data will produce these surface shapes.) Their model predicted an optimum at about $D = 0.15$/hr and $C = 1.1$ g/L, whereas the largest removal rate they observed was at $D = 0.178$/hr and $C = 1.37$ g/L. Their predicted optimum differs from experimental observation because they tried to describe the entire experimental region (i.e., all their data) with a simple quadratic model. A quadratic model gives a poor fit and a poor estimate of the optimum's location because it is not adequate to describe the irregular response surface. Observations that are far from the optimum can be useful in pointing us in a profitable direction, but they may provide no useful information about its location or value. Obviously, such observations can be omitted when the region near the optimum is modeled.

COMMENTS

Response surfaces are effective ways to empirically study the effect of explanatory variables on the response of a system and can help guide experimentation to obtain

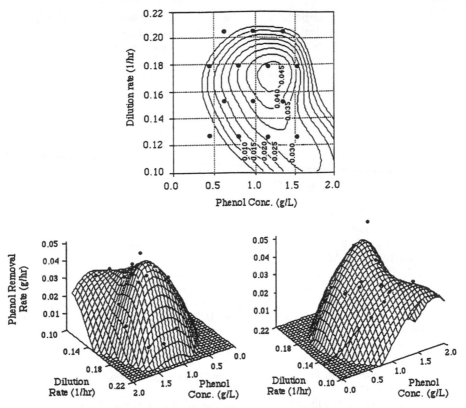

Figure 33.5 The location of the Hobson and Mills experimental runs are shown on the contour plot (top). Their data are shown from two perspectives in the three-dimensional plots.

further information. The approach should have tremendous natural appeal to environmental engineers because (1) their experiments often take a long time to complete and (2) only a few experiments at a time can be conducted. Both characteristics make it attractive to do a few runs at a time and to intelligently use the early results to guide the design of additional experiments. This strategy is also powerful in process control. In most processes, the optimal settings of control variables change over time, and factorial designs can be used iteratively to follow shifts in the response surface. This is a wonderful application of the iterative approach to experimentation (Chapter 32).

The experimenter should keep in mind that response surface methods are not designed to faithfully describe large regions in the possible experimental space. The goal is to explore and describe the most promising regions as efficiently as possible. Indeed, large parts of the experimental space may be ignored. In this example, the direction of steepest ascent was found graphically. Early in the search for the optimum, we expect that the surface can be approximated by a plane; then the direction of steepest ascent is the direction of greatest slope on the plane. If there are more than two variables, this is not convenient, so the direction is found either by using derivatives of the regression equation or the main effects computed directly from the factorial experiment (see Chapter 20) to find the gradient of the surface. Engineers are familiar with these calculations, and good explanations may be found in several of the books and papers referenced at the end of this chapter.

The wonderful paper of Box and Wilson (1951) is recommended for study. Davies (1956) contains an excellent chapter on this topic. The approach has been applied to seeking optimum conditions in full-scale manufacturing plants under the name of Evolutionary Operation (Box, 1957; Box and Draper, 1969). An interesting application to biological processing is given by Kargi et al. (1980). Springer et al. (1984) applied these ideas to wastewater treatment plant operation.

REFERENCES

Box, G. E. P. (1954). "The Exploration and Exploitation of Response Surfaces: Some General Considerations and Examples," *Biometrics*, 10, 1, 16–60.

Box, G. E. P. (1957). "Evolutionary Operation: A Method for Increasing Industrial Productivity," *Appl. Stat.*, 6, 2, 3–23.

Box, G. E. P. and N. R. Draper (1969). *Evolutionary Operation — A Statistical Method for Process Improvement*, New York, John Wiley & Sons.

Box, G. E. P. and N. R. Draper (1989). *Empirical Model Building and Response Surfaces*, New York, John Wiley & Sons.

Box, G. E. P. and J. S. Hunter (1957). "Multi-Factor Experimental Designs for Exploring Response Surfaces," *Ann. Math. Stat.*, 28, 1, 195–241.

Box, G. E. P., W. G. Hunter, and J. S. Hunter (1978). *Statistics for Experimenters: An Introduction to Design, Data Analysis, and Model Building*, New York, Wiley Interscience.

Box, G. E. P. and K. B. Wilson (1951). "On the Experimental Attainment of Optimum Conditions," *J. R. Stat. Soc. Ser. B*, 13, 1, 1–45.

Davies, O. L. (1956). *Design and Analysis of Industrial Experiments*, New York, Hafner Publishing Co.

Hobson, M. J. and N. F. Mills (1990). "Chemostat Studies of Mixed Culture Growing of Phenolics," *J. Water Pollut. Control Fed.*, 62, 684–691.

Kargi, F., M. L. Shuler, R. Vashon, J. W. Seeley, A. Henry, and R. E. Austic (1980). "Continuous Aerobic Conversion of Poultry Waste into Single Cell Protein Using A Single Reactor: Kinetic Analysis and Determination of Optimal Conditions," *Biotechnol. Bioeng.*, 22, 1567–1600.

Margolin, B. H. (1985). "Experimental Design and Response Surface Methodology — Introduction," *The Collected Works of George E. P. Box*, Vol. 1, pp. 271–276, Ed. G. Tiao, Belmont, CA, Wadsworth Books.

Springer, A. M., R. Schaefer, and M. Profitt (1984). "Optimization of a Wastewater Treatment Plant by the Employee Involvement Optimizations System (EIOS)," *J. Water Pollut. Control Fed.*, 56, 1080–1092.

Chapter 34

Designing Experiments to Estimate Parameters in Nonlinear Models

Key words: Box-Lucas designs, derivative matrix, experimental design, nonlinear least squares, nonlinear models, parameter estimation, joint confidence region

The goal is to design experiments that will yield precise estimates of the parameters in a nonlinear model with a minimum of work and expense. Design means specifying what settings of the independent variable will be used and how many observations will be made. The design should recognize that each observation, even though measured with equal accuracy and precision, will not contribute an equal amount of information about parameter values. In fact, the size and shape of the joint confidence region often depends more on where observations are located in the experimental space than on how many measurements are made.

CASE STUDY — A FIRST-ORDER MODEL

A great number of environmental models have the general form $\eta = \theta_1[1 - \exp(-\theta_2 t)]$. For example, oxygen transfer from air to water according to a first-order mass transfer has this model, in which case η is dissolved oxygen concentration, θ_2 is the first-order overall mass transfer coefficient, and θ_1 is the effective dissolved oxygen equilibrium concentration in the system. Experience has shown that θ_1 should be estimated experimentally because the equilibrium concentration achieved in real systems is not the handbook saturation concentration (Boyle et al., 1974).

The biochemical oxygen demand (BOD) model is another familiar example in which θ_1 is the ultimate BOD and θ_2 is the reaction rate coefficient. Figure 34.1 shows some BOD data obtained from analysis of a dairy wastewater specimen (Berthouex and Hunter, 1971). Figure 34.2 shows two joint confidence regions for θ_1 and θ_2 estimated by fitting the model to the entire data set (n = 59) and to a much smaller subset of the data (n = 12). An 80% reduction in the number of measurements has barely changed the size or shape of the joint confidence region. We wish to discover the efficient smaller design in advance of doing the experiment. This is possible if we know the form of the model to be fitted.

METHOD — A CRITERION TO MINIMIZE THE JOINT CONFIDENCE REGION

A model contains p parameters that are to be estimated by fitting the model to observations located at n settings of the independent variables (time, temperature, dose, etc.). The model is of the form $\eta = f(\theta_i, x_j)$, where the θ_i are the parameters and x_j the independent variables. If we assume that *the form of the model is correct*, it is possible to determine settings of the independent variables that will yield precise estimates of the parameters with a small number of experiments. The parameters will be estimated by linear or nonlinear least squares, depending on whether the model is linear or nonlinear in the parameters. Our interest lies mainly in nonlinear models because, as shall be explained shortly, finding an efficient design for a linear model is intuitive.

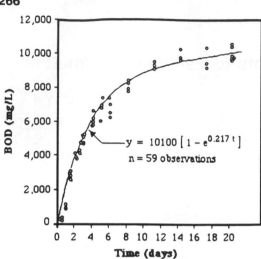

$$y = 10100 \left[1 - e^{0.217\,t} \right]$$

n = 59 observations

Figure 34.1 A BOD experiment with 59 observations covering the range of 1–20 days, with three to six replicates at each time of measurement. The curve is the fitted model with nonlinear least squares parameter estimates $\hat{\theta}_1 = 10,100$ mg/L and $\hat{\theta}_2 = 0.217$/day.

The minimum number of observations that will yield p parameter estimates is n = p. The fitted nonlinear model generally will not pass perfectly through these points, unlike a linear model with n = p which will fit each observation exactly.[1] The regression analysis will yield a residual sum of squares and a joint confidence region for the parameters. The goal is to have the joint confidence region small (see Chapter 26). The joint confidence region for the parameters is small when their variances and covariances are small.

We will develop the regression model and the derivation of the variance of the parameter estimates in matrix notation. Our explanation is necessarily brief; for more details, one may consult almost any modern reference on regression analysis (e.g., Draper and Smith, 1981; Rawlings, 1988; and Bates and Watts, 1988).

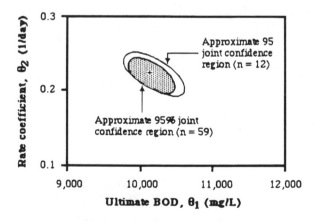

Figure 34.2 The shaded ellipse is the approximate 95% joint confidence region for parameters estimated using all n = 59 observations. The cross locates the nonlinear least squares parameter estimates for n = 59. The unshaded ellipse, which encloses the shaded ellipse, is for parameters estimated using only n = 12 observations (6 on day 4 and 6 on day 20).

---

[1] If p = 2 and the two (x,y) data pairs are, for example, (1,1) and (2,1), the nonlinear model y = a*exp(–bx) cannot fit both points, but the linear model y = a + bx can (in this case a = 1 and b = 0).

In matrix notation, a linear model is

$$\mathbf{y} = \mathbf{X}\boldsymbol{\beta} + \mathbf{e}$$

where $\mathbf{y}$ is a $n \times 1$ column vector of the observations, $\mathbf{X}$ is an $n \times p$ matrix of the independent variables (or combinations of them), $\boldsymbol{\beta}$ is a $p \times 1$ column vector of the parameters, $\mathbf{e}$ is an $n \times 1$ column vector of the residual errors, which are assumed to have constant variance (n is the number of observations and p is the number of parameters in the model).

The least squares parameter estimates and their variances and covariances are given by

$$\mathbf{b} = (\mathbf{X}'\mathbf{X})^{-1}\mathbf{X}'\mathbf{y}$$

and

$$V(\mathbf{b}) = (\mathbf{X}'\mathbf{X})^{-1}\sigma^2$$

The same equations apply for nonlinear models, except that the definition of the $\mathbf{X}$ matrix changes. A nonlinear model cannot be written as a matrix product of $\mathbf{X}$ and $\boldsymbol{\beta}$, but we can circumvent this difficulty by using a linear approximation (Taylor series expansion) to the model. When this is done, the $\mathbf{X}$ matrix, instead of being just the independent variables, becomes a derivative matrix, which is a function of the independent variables. This is shown below.

The variances and covariances of the parameters are given exactly by $\{\mathbf{X}'\mathbf{X}\}^{-1}\sigma^2$ when the model is linear. This expression is approximate when the model is nonlinear in the parameters. The minimum sized joint confidence region corresponds to the minimum of the quantity $\{\mathbf{X}'\mathbf{X}\}^{-1}\sigma^2$. Since the variance of random measurement error, σ^2, is a constant (even though its value may be unknown), it is only the $\{\mathbf{X}'\mathbf{X}\}^{-1}$ matrix that must be considered.

It is not necessary to compare entire variance-covariance matrices for different experimental designs. All we need to do is minimize the determinant of the $\{\mathbf{X}'\mathbf{X}\}^{-1}$ matrix or the equivalent of this, which is to maximize the determinant of $\{\mathbf{X}'\mathbf{X}\}$. This latter design criterion, presented by Box and Lucas (1959), is written as

$$\max \Delta = \max|\mathbf{X}'\mathbf{X}|$$

where the vertical bars indicate the determinant. Maximizing Δ results in minimizing the approximate joint confidence region. The size of the approximate confidence region is inversely proportional to the square root of the determinant, i.e., proportional to $\Delta^{-1/2}$.

$\{\mathbf{X}'\mathbf{X}\}^{-1}$ is called the variance-covariance matrix. For nonlinear models, it is obtained from $\mathbf{X}$, an n row by p column ($n \times p$) matrix, called the derivative matrix

$$\mathbf{X} = \mathbf{X}_{ij} = \begin{bmatrix} X_{11} & X_{21} & \cdots & X_{p1} \\ X_{12} & X_{22} & \cdots & X_{p2} \\ \cdots & \cdots & \cdots & \cdots \\ X_{1n} & X_{2n} & \cdots & X_{pr} \end{bmatrix} \qquad i = 1, 2, \ldots, p \qquad j = 1, 2, \ldots, n$$

where p and n are the number of parameters and observations as defined earlier.

The elements of the **X** matrix are partial derivatives of the model with respect to the parameters.

$$X_{ij} = \frac{\partial f(\theta_i, x_j)}{\partial \theta_i} \qquad i = 1, 2, \ldots, p; \quad j = 1, 2, \ldots, n$$

For nonlinear models, however, the elements X_{ij} are functions of both the independent variables x_j and the unknown parameters θ_i. Thus, some preliminary work is required to compute the elements of the matrix in preparation for maximizing $|\mathbf{X'X}|$.

For linear models, the elements X_{ij} do not involve the parameters of the model. They are functions only of the independent variables, x_j, or combinations of them. (This is the characteristic that defines a model as linear in the parameters.) It is easily shown that the minimum variance design for a linear model spaces observations as far apart as possible. This result is intuitive in the case of fitting $\eta = \beta_0 + \beta_1 x$; the estimate of β_0 is enhanced by making an observation near the origin, and the estimate of β_1 is enhanced by making the second observation at the largest feasible value of x. This simple example also points out the importance of the qualifier "if the model is assumed to be correct." This design with measurements at two widely spaced settings of x, which is ideal for fitting a straight line, has terrible deficiencies if the correct model is quadratic. Obviously, the design strategy is different when we know the form of the model and when we are seeking to discover the form of the model. In this chapter, the correct form of the model is assumed to be known.

Returning now to the design of experiments to estimate parameters in nonlinear models, we see a difficulty in going forward. In order to find the settings of x_j that maximize $|\mathbf{X'X}|$, the values of the elements X_{ij} must be expressed in terms of numerical values for the parameters. The experimenter's problem then is how to provide these numerical values.

At first, this seems an insurmountable problem, since we are planning the experiment because the values of θ are unknown. Is it necessary, in order to design efficient experiments, to know in advance the answer that those experiments will give? Never say die — there is a way. The experimenter, based on experience and perhaps on previous similar experiments, always has some prior knowledge from which to "guess" parameter values that are not too remote from the true values. These *a priori* estimates, being the best available information about the parameter values, are used to evaluate the elements of the derivative matrix and design the first experiments.

The experimental design based on maximizing $|\mathbf{X'X}|$ is *optimal*, in the mathematical sense, with respect to the *a priori* parameter values, and based on the critical assumption that the model is correct. This does not mean the experiment will be perfect or even that its results will satisfy the experimenter. If the initial parameter "guess" is not close to the true underlying value, the confidence region will be large and more experiments will be needed. If the model is incorrect, the experiments planned using this criterion will not reveal it. The so-called "optimal" design, then, should be considered as advice that should make experimentation more economical and rewarding. It is not a prescription for getting perfect results with the first set of experiments. Because of these caveats, an iterative approach to experimentation is often planned. (See Chapter 32.)

If the parameter values provided are very near the true values, the experiment designed by this criterion will give precise parameter estimates. If they are distant from the true values, the estimated parameters will have a large joint confidence region. In either case, the first experiments provide useful information about the parameter values, and the estimates from the first experiment are used to design a second set of tests, which may be used to design a third set, and so on until the parameter estimates have been estimated with the desired precision. This iterative approach to experimentation, discussed and illustrated in Chapter 32, can be extremely efficient. Even if the initial design is poor, knowledge increases in steps, sometimes large ones, and the joint confidence region is reduced with each additional iteration. Checks on the structural adequacy of the model can be made at each iteration.

CASE STUDY SOLUTION

The model is $\eta = \theta_1[1 - \exp(-\theta_2 t)]$. There are $p = 2$ parameters, and we will plan an experiment with $n = 2$ observations placed at locations that are optimal with respect to our best informed initial estimates of the parameter.

The partial derivatives of the model with respect to each parameter are

$$X_{1j} = \frac{\partial[\theta_1(1 - e^{-\theta_2 t_j})]}{\partial\theta_1} = 1 - e^{-\theta_2 t_j} \qquad j = 1, 2$$

$$X_{2j} = \frac{\partial[\theta_1(1 - e^{-\theta_2 t_j})]}{\partial\theta_2} = \theta_1 t_j e^{-\theta_2 t_j} \qquad j = 1, 2$$

The derivative matrix X for $n = 2$ experiments is 2×2:

$$X = \begin{bmatrix} X_{11} & X_{21} \\ X_{12} & X_{22} \end{bmatrix} = \begin{bmatrix} 1 - e^{-\theta_2 t_1} & \theta_1 t_1 e^{-\theta_2 t_1} \\ 1 - e^{-\theta_2 t_2} & \theta_1 t_2 e^{-\theta_2 t_2} \end{bmatrix}$$

where t_1 and t_2 are the times at which observations will be made.

Premultiplying X by its transpose gives

$$X'X = \begin{bmatrix} X_{11}^2 + X_{12}^2 & X_{11}X_{21} + X_{12}X_{22} \\ X_{11}X_{21} + X_{12}X_{22} & X_{21}^2 + X_{22}^2 \end{bmatrix}$$

The objective now is to maximize the determinant of the $X'X$ matrix.

$$\max \Delta = |X'X|$$

The vertical bars indicate a determinant. In a complicated problem, the matrix multiplication and the minimization of the determinant of the matrix would be done using numerical methods. The analytical solution for a number of interesting models (Box 1971) and for this example can be derived rather easily. For the case where $n = p = 2$,

$$\Delta = (X_{11}X_{22} - X_{12}X_{21})^2$$

This expression is maximized when the absolute value of the quantity $(X_{11}X_{22} - X_{12}X_{21})$ is maximized. Therefore, the design criterion becomes

$$\Delta^* = \max|X_{11}X_{22} - X_{12}X_{21}|$$

where here the vertical bars designate the absolute value. The asterisk on Δ merely indicates this redefinition of the criterion Δ.

Substituting the appropriate derivative elements gives

$$\Delta^* = (1 - e^{-\theta_2 t_1})(\theta_1 t_2 e^{-\theta_2 t_2}) - (1 - e^{-\theta_2 t_2})(\theta_1 t_1 e^{-\theta_2 t_1})$$

$$= \theta_1[(1 - e^{-\theta_2 t_1})(e^{-\theta_2 t_2}) - (1 - e^{-\theta_2 t_2})(e^{\theta_2 - t_1})]$$

The factor θ_1 is a numerical constant that can be ignored. Thus, we only need to maximize the quantity in brackets, which is independent of the value of θ_1. This can be done by taking derivatives and solving for the roots t_1 and t_2. The algebra is omitted. The solution is

$$t_1 = 1/\theta_2$$

and

$$t_2 = \infty$$

The value $t_2 = \infty$ is interpreted as advice to collect data with t_2 set at the largest feasible level. A measurement at this time will provide a direct estimate of θ_1, since η approaches the asymptote θ_1 as t becomes large (i.e., 20 days or more). This estimate of θ_1 will be essentially independent of θ_2. Notice that the value of θ_1 is irrelevant in setting the level of both t_2 and t_1 (which is a function only of θ_2).

The other observation at $t_1 = 1/\theta_2$ is on the rising part of the curve. If we estimate $\theta_2 = 0.23$/day, the optimal setting for t_1 is $1/0.23 = 4.3$ days. As a practical matter, we might say that values of t_1 should be in the range of 4–5 days (since $\theta_2 = 0.2$ gives $t_1 = 5$ days and $\theta_2 = 0.25$ gives $t_1 = 4$ days). Notice that the optimal settings of t_1 depend only on the value of θ_2; the value of θ_1 is irrelevant. Likewise, t_2 is set at a large value regardless of the value of θ_2.

Table 34.1 compares the three arbitrary experimental designs and the optimal design shown in Figure 34.3. The insets in Figure 34.3 suggest the shape and relative size of the confidence regions one expects from the designs. A smaller value of $\Delta^{-1/2}$ in Table 34.1 indicates a smaller joint confidence region. The absolute value of $\Delta^{-1/2}$ has no meaning, it depends upon the magnitude of the parameter values used, the units, the time scale, etc., but the relative magnitude of $\Delta^{-1/2}$ for different designs indicates relative precision of the parameter estimates. The values of $\Delta^{-1/2}$ do not indicate the shape of the confidence region, but it happens that the best designs give well-conditioned regions because parameter correlation has been reduced.

Table 34.1 **Relative Size of Joint Confidence Region for Several Experimental Designs**

Design	Total Number of Observations	$\Delta^{-1/2}$ (10^{-4})
Designs shown in Figure 34.3		
Optimal design	2	27
Design A	6	28
Design B	10	7
Design C	20	3.5
Replicated optimal designs ($t_1 = 4.5$, $t_2 = 20$)		
2 Replicates	4	14
3 Replicates	6	9
4 Replicates	8	7
5 Replicates	10	5

Note: The optimal designs ($t_1 = 4.5$, $t_2 = 20$) are based on the model $\eta = \theta_1[1 - \exp(-\theta_2 t)]$ and $\theta_1 = 250$ and $\theta_2 = 0.23$.

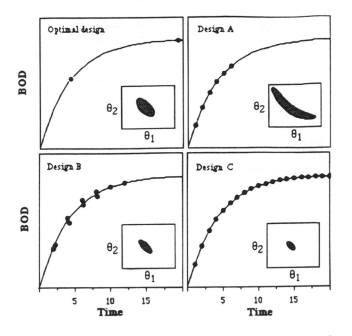

Figure 34.3 The optimal design and three arbitrary designs. Insets suggest the relative size and shape of the confidence regions.

The optimal design and design A have confidence regions that are about the same size. What is not indicated by the $\Delta^{-1/2}$ values is that the region for optimal design will be elliptical, while design A will be elongated. Design A has been used very often in kinetic studies. It is inefficient. It does not give independent estimates of the parameters because all the observations are in the early part of the experiment and none are on the asymptote. It tends to estimate the product $\theta_1\theta_2$ rather than θ_1 and θ_2 independently. In fact, θ_1 and θ_2 are estimated so poorly as to be virtually useless; the joint confidence region is hyperbolic and elongated (banana shaped) instead of elliptic as we would like. This weakness in the design cannot be overcome merely by putting more observations into the same region. To improve, the observations must be made at times that yield more information.

Design B, with ten observations, is similar to design A, but the observations cover a wider range of time and four points are duplicated. This almost doubles the experimental work, and it reduces the size of the confidence region by a factor of four. This reduction is due mainly to putting some observations nearer the asymptote. Adding five observations will do nothing if they are made in the wrong places. For example, duplicating the five points of design A will not do much to improve the shape of the confidence region or to reduce its size.

Design C has 20 observations and yields a confidence region that is half the size obtained by design B. This design will tend to give precise and uncorrelated estimates. Also, it has the advantage of providing a check on the model over the range of the experimental space. If the first-order model is wrong, this design experiment should reveal it. On the other hand, if the model is correct, the same result can be attained with less work by replicating the optimal design.

The simplest possible optimal design has only two observations, one at $t_1 = 1/\theta_2$ and one at large t_2 (for $\theta_2 = 0.23$, $t_1 = 5$ days, and $t_2 = 30$ days.) Two well-placed observations are better than six badly located ones (design A). The confidence region is smaller, and the parameter estimates will be independent (the confidence region will tend to be elliptic rather than elongated). Replication of the optimal design quickly reduces the joint confidence region. Five replicates, giving a total of 10 observations, is about equal to design C with 20 observations. This shows that design C has about half its observations made at times that contribute little information about the parameter values.

COMMENTS

The approach given here for designing experiments that are efficient for estimating parameters in nonlinear models depends on the experimenter assuming that the form of the model is correct. The goal is to estimate parameters in a known model and not to discover the correct form of the model.

The most efficient experimental strategy is to start with very simple designs, even as small as n = p observations, and then to work iteratively. The first experiment provides parameter estimates that can be used to plan a few additional experiments that will refine the parameter estimates.

In many cases, the experimenter will not want to make measurements at only the p locations that are "optimal" based on the criterion of from maximizing Δ. If setting up the experiment is costly, but each measurement is inexpensive, it may be preferable to use several observations at near-"optimal" locations. Or, some observations may be made at nonoptimal locations to check the adequacy of the model. Augmenting the "optimal" design is sensible. The design criterion, after all, provides advice, not orders.

REFERENCES

Atkinson, A. C. and W. G. Hunter (1968). "Design of Experiments for Parameter Estimation," *Technometrics*, 10, 271.

Bates, D. M. and D. G. Watts (1988). *Nonlinear Regression Analysis and Its Applications*, New York, John Wiley & Sons.

Berthouex, P. M. and W. G. Hunter (1971). "Problems Associated with Planning BOD Experiments," *J. Sanit. Eng. Div. ASCE*, 97, 333–344.

Berthouex, P. M. and W. G. Hunter (1971). "Statistical Experimental Design: BOD Tests," *J. Sanit. Eng. Div. ASCE*, 97, 393–407.

Box, G. E. P. and W. G. Hunter (1965). "The Experimental Study of Physical Mechanisms," *Technometrics*, 7, 23.

Box, G. E. P. and H. L. Lucas (1959). "Design of Experiments in Nonlinear Situations," *Biometrika*, 45, 77–90.

Box, M. J. (1971). "Simplified Experimental Design," *Technometrics*, 13, 19–31.

Boyle, W. C., P. M. Berthouex, and T. C. Rooney (1974). "Pitfalls in Parameter Estimation for Oxygen Transfer Data," *J. Environ. Eng. Div. ASCE*, 100, 391–408.

Draper, N. R. and H. Smith (1981). *Applied Regression Analysis*, New York, John Wiley & Sons.

Rawlings, J. O. (1988). *Applied Regression Analysis: A Research Tool*, Pacific Grove, CA, Wadsworth and Brooks/Cole.

Chapter 35

Why Linearization Can Bias Parameter Estimates

Key words: bias, biological kinetics, linear model, linearization, Lineweaver-Burke, Michaelis-Menten, nonlinear least squares, nonlinear model, parameter estimation, precision, regression, transformations, Thomas slope method

---

An experimenter, having invested considerable care, time, and money to obtain data, should want to extract all the information they contain. If the purpose is to estimate parameters in a nonlinear model, we should insist that the parameter estimation method gives estimates that are (1) unbiased and (2) precise. Generally, the best method of estimating the parameters will be nonlinear least squares, in which variables are used in their original form and units. Some experimenters transform the model so it can be fitted by linear regression. This can, and often does, give biased or imprecise estimates. The dangers of linearization will be shown by examples.

CASE STUDY — BACTERIAL GROWTH

Linearization may be helpful, as shown in Figure 35.1. The plotted data show the geometric growth of bacteria; x is time and y is bacterial population. The measurements are more variable at higher populations. Taking logarithms gives constant variance at all population levels, as shown in the right-hand panel of Figure 35.1. Fitting a straight line to the log-transformed data is appropriate and correct.

CASE STUDY — A FIRST-ORDER KINETIC MODEL

The model for biochemical oxygen demand (BOD) exertion as a function of time, assuming the usual first-order model, is $y_i = \theta_1[1 - \exp(-\theta_2 t_1)] + e_i$. The rate coefficient, θ_2, and the ultimate BOD, θ_1, are to be estimated by the method of least squares from observations of y_i and various times t_i. The residual errors are assumed to be independent, normally distributed, and with constant variance. Constant variance means that the magnitude of the residual errors is the same over the range of observed values of y. It is this property that can be altered, either beneficially or harmfully, by linearizing a model.

One linearization of this model is the Thomas slope method (TSM). It has been used by many engineers over many years and still appears in some textbooks. The TSM may have been useful before nonlinear least squares estimates could be done easily. It should never be used today *unless* it can be shown that the transformation better fulfills the assumptions on which regression is based, in particular the assumption of constant variance. Our experience with many data sets is that it usually makes things worse rather than better. This example shows how poor the estimates are and explains why the method does poorly.

Why is the TSM so bad? The method involves plotting $Y_i = y_i^{-1/3}$ on the ordinate against $X_i = y_i/t_i$ on the abscissa, thus distorting both X and Y. In linear regression, the X variable is supposed to be free of error or at least have an error that is very small compared to the errors in Y. The transformed abscissa, y_i/t_i, now contains measurement error, and because it is scaled by t_i, each X_i contains a different amount of error. Furthermore, the ordinate $Y_i = y_i^{-1/3}$ is badly distorted, first by the reciprocal and then by the cube root. Suppose that the measured values $y_1 = 10$ and $y_2 = 20$ have constant variance, $Var(y_1) = Var(y_2) = 1$. The variance of the transformed variable Y is $Var(Y) =$

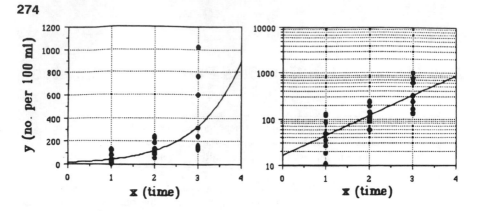

Figure 35.1 A log transformation of bacteria data linearizes the model and gives constant variance.

$1/9\ y^{8/3}$. Now, instead of the variance of the two observations being equal, the transformation makes $\text{Var}(Y_1) = 6.4\ \text{Var}(Y_2)$. The transformation has obliterated the condition of constant variance.

Figure 35.2 shows an example. The original data, shown on the left, have constant variance. The model is $y = 30(1 - \exp(-0.25\ t))$. The linearization, on the right, is not very linear, but it gives a high coefficient of determination ($R^2 = 0.72$), so the straight line is not entirely implausible. The disturbing thing about this method is that eliminating data at the longer times would make the plot of $y^{-1/3}$ vs y/t more nearly linear. This is a tragedy because the observations on the asymptote contain almost all the information about the parameter θ_1 and failing to use the observations in this region will make the estimates imprecise as well as biased.

Several linearization methods have been developed for transforming BOD data so they can be fitted to a straight line using linear regression. They should not be used because they all carry the danger of distortion illustrated with the TSM. This was shown by Marske and Polkowski (1972) and Berthouex and Szewczyk (1984) using a large number of data sets for which nonlinear least squares was the proper method of parameter estimation. The TSM estimates often were so badly biased that they did not fall within

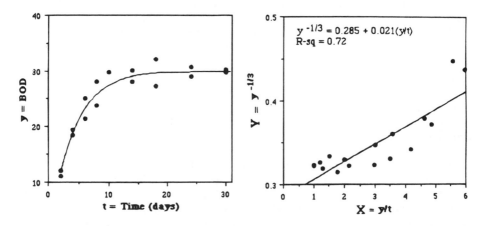

Figure 35.2 The Thomas slope method of linearizing the BOD model distorts the errors and gives biased parameter estimates.

CASE STUDY — MICHAELIS-MENTEN MODEL

The Michaelis-Menten model states, in biochemical terms, that an enzy_____ _____ reaction has a rate $\mu = \theta_1 x/(\theta_2 + x)$, where x is the concentration of substrate (the independent variable). The maximum reaction rate, θ_1, is approached as x gets large. The saturation constant, θ_2, is the substrate concentration at which $x = \theta_1/2$. The observed values are

$$y_i = \mu + e_i = \frac{\theta_1 x}{\theta_2 + x} + e_i$$

There are three ways of estimating the two parameters in the model.

1. Nonlinear least squares fit the original form of the model.

$$\min S = \Sigma \left(y_i - \frac{\theta_1 x_i}{\theta_2 + x_i} \right)^2$$

2. Linearization using a Lineweaver-Burke plot (double reciprocal plot). The model is rearranged to give

$$\frac{1}{y_i} = \frac{1}{\theta_1} + \frac{\theta_2}{\theta_1 x_i}$$

A straight line typically is fitted by ordinary linear regression to estimate the intercept $1/\theta_1$ and slope θ_2/θ_1.

3. Linearization using y against y/x gives

$$y_i = \theta_1 - \theta_2 \frac{y_i}{x_i}$$

A straight line typically is fitted by ordinary linear regression to estimate the intercept θ_1 and slope $-\theta_2$.

Assuming there is constant variance in the original measurements of y, only the method of nonlinear least squares gives unbiased parameter estimates. The Lineweaver-Burke plot will give the most biased estimates, and the y vs y/x linearization will have somewhat less bias.

The effectiveness of the three methods is demonstrated with simulated data. The simulated data in Table 35.1 were generated using the known parameter values $\theta_1 = 30.00$ and $\theta_2 = 15.00$. The "observed" y's are the true values plus random error (with mean = 0 and variance = 1). The nonlinear least squares parameter estimates are $\theta_1 = 31.45$ and $\theta_2 = 15.89$, which are close to the underlying true values. The two linearization methods give estimates (Table 35.2) that are distant from the true values. Figure 35.3 shows why.

Figure 35.3a was drawn using the values from Table 35.1 with five additional replicate observations that were arbitrarily chosen to make the spread the same between "dupli-cates" at each setting of x. Real data would not look like this, but this simplicity will

Substrate Conc. x	Observed Rate y	"True" Rate $\mu = \dfrac{30x}{15 + x}$
2.5	5.6	4.28
5.0	7.3	7.50
10.0	12.5	12.00
20.0	16.1	17.14
40.0	23.2	21.82

Table 35.2 Michaelis-Menten Parameter Estimates

Estimation Method	θ_1	θ_2
"True" parameter values	30.00	15.00
Nonlinear least squares	31.45	15.89
Lineweaver-Burke	22.58	8.16
y against y/x linearization	25.76	10.13

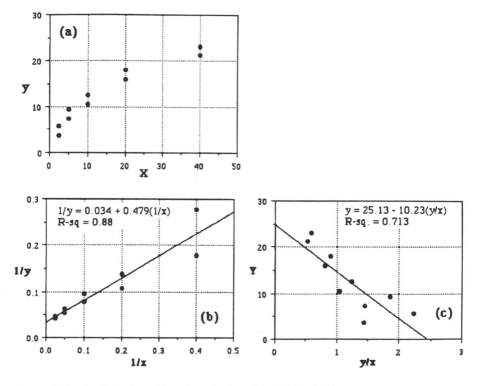

Figure 35.3 An illustration of how linearization of the Michaelis-Menten equation distorts the error structure.

help to illustrate how linearization distorts the errors. Figure 35.3b shows the Lineweaver-Burke plot. Values at large x (small 1/x) are squeezed together in the plot, making these values appear more precise than the others and literally fixing one end of the regression line. The values associated with small x (large 1/x) seem to be greatly in error. The consequence of this is that the least squares fit of a straight line will be strongly influenced by the large values of x, whereas according to the true error structure they should not be given any more weight than the other values. The plot of y against y/x, Figure 35.3c, shows some distortion, but less than the Lineweaver-Burke plot.

One simulated experiment may not be convincing, but the 750 simulated experiments of Coloquhoun (1971) are dramatic proof. An "experiment" was performed by selecting at random an observation from a normally distributed population known to have a mean equal to the "true" value shown in Table 35.1 and also having standard deviation $\sigma = 1.0$ at every concentration. Thus, unlike what happens in a real experiment, the distribution, mean, and standard deviations of the errors are known. Each set of data was analyzed to estimate the parameters by nonlinear least squares and by the two linearization methods.

The resulting 750 estimates of θ_1 were grouped to form histograms and are shown in Figure 35.4. The distributions for θ_2 (not reproduced here) were similarly biased because the estimates of θ_1 and θ_2 were highly correlated, i.e., experiments that yield an estimate of θ_1 that is too high tend to give an estimate of θ_2 that is too high also, whichever method of estimation is used. This parameter correlation is a result of the model structure and the settings of the independent variables.

The nonlinear least squares (NLLS) estimates are more closely grouped around the true value ($\theta_1 = 30.00$) than the estimates found by the other methods. The average value of the NLLS estimates is 30.4, close to the true value. They have little bias. The

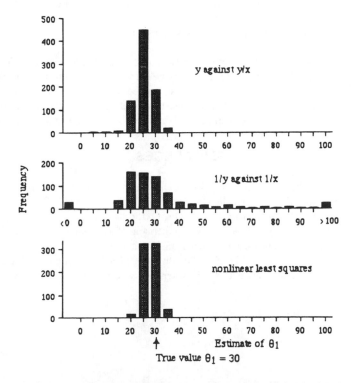

Figure 35.4 Distributions of 750 estimates of θ_1 (true value = 30) obtained using three methods in 750 simulated experiments. (Adapted from Coloquhoun [1971].)

NLLS have the smallest variance. By comparison, the Lineweaver-Burke method gives terrible estimates, including some that were negative and some greater than 100. The scatter is much greater. Near infinite estimates are obtained when the plot goes nearly through the origin, giving $1/\theta_1 = 0$. These distort the average of the estimated values so much that no realistic estimate of the bias is possible.

The method of plotting of y against y/x gives estimates falling between these extremes. Their standard deviation is only about 28% greater than that of the NLLS estimates. About 73% of the estimates were too low (below 30) and the average was 28.0. They have a negative bias. This bias is purely a consequence of the estimation.

The results of Coloquhoun's simulation should convince us to be careful about linearization for the purpose of parameter estimation, but they are not intended to prove that linearization is always bad. It can be helpful in the case of nonconstant errors. Dowd and Riggs (1965) found the y against y/x method better than NLLS for the case where the coefficient of variation (standard deviation/mean) was the same at each setting of x, but even in this case the Lineweaver-Burke method was still awful.

COMMENTS

Transformation always distorts the error structure of the original measurements. Sometimes this is beneficial (see Chapter 7). For example, a transformation may be necessary to make the variance constant over the range of settings of the independent variable.

Another reason to transform the data is to make it possible to estimate the parameters in a linearized model by ordinary linear regression. The linearization distorts the error structure of the data. This will be beneficial only if the transformation that linearizes the model *also* causes the data to have constant variance. Note that this can only happen when the original variances are not constant. If the original errors have constant variance, the linearization will make things worse instead of better and the parameter estimates will be biased.

The general lesson of this chapter is **beware of model linearization**. When it is wrong, it can be terrible. But, under the right circumstances, it can be helpful, but we should never assume it will be beneficial. This must be checked. Do not linearize merely to facilitate using linear regression. Learn to use nonlinear least squares. It is more natural, more likely to be the appropriate method, and easy to do with readily available software.

REFERENCES

Berthouex, P. M. and J. E. Szewczyk (1984). "Discussion of Influence of Toxic Metals on the Repression of Carbonaceous Oxygen Demand," *Water Res.*, 18, 385–386.

Coloquhoun, D. (1971). *Lectures in Biostatistics*, Oxford, Clarendon Press.

Dowd, J. E. and D. S. Riggs (1965). "A Comparison of Estimates of Michaelis-Menten Kinetic Constants from Various Linear Transformations," *J. Biol. Chem.*, 210, 863–872.

Marske, D. M. and L. B. Polkowski (1972). "Evaluation of Methods for Estimating Biochemical Oxygen Demand Parameters," *J. Water Pollut. Control Fed.*, 44, 1987–1992.

Chapter 36

Fitting Models to Multiresponse Data

Key words: biokinetics, Box-Draper criterion, determinant criterion, joint confidence regions, Monod model, multiresponse experiments, nonlinear least squares, parameter estimation

---

Frequently, data can be obtained simultaneously on two or more responses at a given experimental setting. Consider, for example, two sequential first-order reactions by which species A is converted to B which in turn is converted to C: A $\rightarrow$ B $\rightarrow$ C. The concentrations of species A, B, and C can be measured at any time of interest. If only B is measured, the rate constants for each step of the reaction can be estimated from the single equation that describes the concentration of B as a function of time. Better parameter estimates can be obtained if all three species (A, B, and C) are measured, and the three equations describing A, B, and C are fitted simultaneously to all the data. To do this, it is necessary to have a criterion for simultaneously fitting all three measured responses.

CASE STUDY — BACTERIAL GROWTH MODEL

The data in Table 36.1 were collected on a continuous-flow completely mixed biological reactor with no recycle (Ramanathan and Gaudy, 1969). At steady-state operating conditions, the effluent organism and substrate concentrations will be

$$X = \theta_3 \left[S_o - \frac{\theta_2 D}{\theta_1 - D} \right]$$

and

$$S = \frac{\theta_2 D}{\theta_1 - D}$$

These equations are derived from a mass balance assuming constant liquid volume V in the reactor, constant liquid feed rate Q, and Monod growth kinetics. The liquid detention time in the reactor is V/Q. The reciprocal of detention time is the dilution rate D = Q/V. The rate at which organisms initially present in the reactor would be washed out if growth ceased, but flow at rate Q continued is dX/dt = $-$DX, where X is the concentration of biomass (milligrams per liter). The feed contains substrate at concentration $S_o = 3000$ mg/L. The reactor contents and effluent have substrate concentration S. According to the Monod model, biomass grows at rate k = $\theta_1 S/(\theta_2 + S)$, where θ_1 and θ_2 are parameters representing the maximum growth rate and the half saturation constant (the substrate concentration at which k = $\theta_1/2$). Substrate consumption and net biomass production are related by the yield coefficient, θ_3; that is, dX/dt = θ_3dS/dt.

Experiments are performed at varied settings of D to obtain measurements of X and S in order to estimate θ_1, θ_2, and θ_3. One approach would be to fit the model for X to the X data and to independently fit the model for S to the S data. The disadvantage of this approach is that θ_1 and θ_2 appear in both equations, and two estimates for each would be obtained. These estimates might differ substantially. The alternative is to fit

Table 36.1 **Data from an Experiment on Bacterial Growth in a Continuous-Flow, Steady-State Reactor**

Dilution Rate (D) (1/hr)	Substrate Conc. (S) (mg/L COD)	Biomass Conc. (X) (mg/L)
0.042	221	1589
0.056	87	2010
0.083	112	1993
0.167	120	1917
0.333	113	1731
0.500	224	1787
0.667	1569	676
1.000	2745	122

Source: Ramanathan, M. and A. F. Gaudy (1969). "Effect of High Substrate Concentration and Cell Feedback on Kinetic Behavior of Heterogeneous Populations in Completely Mixed Systems," *Biotechnol. Bioeng.*, 11, 207.

both equations simultaneously to data on both X and S and obtain one estimate of each parameter. This makes better use of the data and will yield more precise parameter estimates.

METHOD — A MULTIRESPONSE LEAST SQUARES CRITERION

A logical criterion for simultaneously fitting three measured responses y_A, y_B, and y_C would be a simple extension of the least squares criterion to minimize the combined sums of squares for all three responses:

$$\min \Sigma(y_A - \hat{y}_A)^2 + \Sigma(y_B - \hat{y}_B)^2 + \Sigma(y_C - \hat{y}_C)^2$$

where y_A, y_B, and y_C are the measured responses and $\hat{y}_A$, $\hat{y}_B$, and $\hat{y}_C$ are values computed from the model. This criterion holds only if three fairly restrictive assumptions are satisfied, namely, (1) the errors of each response are normally distributed and all data points for a particular response are independent of one another, (2) the variances of all responses are equal, and (3) there is no correlation between data for each response for a particular experiment (Hunter, 1967).

Assumption 2 is violated when certain responses are measured more precisely than others. This condition is probably more common than all responses being measured with equal precision. If Assumption 2 is violated, the appropriate criterion is to minimize the weighted sums of squares, where weights are inversely proportional to the variance of each response. If both Assumptions 2 and 3 are violated, the analysis must account for variances and covariances of the responses. In this case, Box and Draper (1965) have shown that minimizing the determinant of the variance-covariance matrix of the residuals gives the maximum likelihood estimates of the model parameters. The Box-Draper *determinant criterion* is especially attractive because it is not restricted to linear models, and the estimates are invariant to scaling or linear transformations of the observations (Bates and Watts, 1985).

The criterion is written as a combined sum of squares function augmented by the covariances between responses. For the example reaction A → B → C, with three responses measured, the Box-Draper determinant criterion has the following form:

$$\min|V| = \begin{vmatrix} \Sigma(y_A - \hat{y}_A)^2 & \Sigma(y_A - \hat{y}_A)(y_B - \hat{y}_B) & \Sigma(y_A - \hat{y}_A)(y_C - \hat{y}_C) \\ \Sigma(y_A - \hat{y}_A)(y_B - \hat{y}_B) & \Sigma(y_B - \hat{y}_B)^2 & \Sigma(y_B - \hat{y}_B)(y_C - \hat{y}_C) \\ \Sigma(y_A - \hat{y}_A)(y_C - \hat{y}_C) & \Sigma(y_B - \hat{y}_B)(y_C - \hat{y}_C) & \Sigma(y_C - \hat{y}_C)^2 \end{vmatrix}$$

where y_A, y_B, y_C = the observed values of the ith response, $\hat{y}_A, \hat{y}_B, \hat{y}_C$ = the corresponding predicted values from the model, and Σ = the summations over all observations.

We assume that each response has been measured the same number of times. The vertical lines indicate the determinant of the matrix. The best parameter estimates, analogous to the least squares estimates, are those which minimize the determinant of this matrix.

The diagonal elements correspond to the residual (error) sum of squares for each of the three responses. The off-diagonal terms account for measurements on the different responses being correlated. If the residual errors of the three responses are independent of each other, the expected values of the off-diagonal terms will be zero, and the parameter estimation criterion simplifies to

$$\text{minimize } S = \Sigma(y_A - \hat{y}_A) + \Sigma(y_B - \hat{y}_B)^2 + \Sigma(y_C - \hat{y}_C)^2$$

For the special case of a single response, the determinant criterion simplifies to the method of least squares: minimize the sum of the squared residuals for the response of interest

$$\min \Sigma(y_A - \hat{y}_A)^2$$

or

$$\min \Sigma(y_B - \hat{y}_B)^2$$

or

$$\min \Sigma(y_C - \hat{y}_C)^2$$

Three caveats are in order when using the determinant criterion (Box et al., 1973). First is the problem of linear dependence in the data. The Box-Draper criterion requires that all responses be measured independently. If we know, for example, from a mass balance that the reaction was started with 1 mol of A and that the total amount of A, B, and C must equal 1 mol throughout the course of the experiment, it is tempting to reduce the experimental burden by measuring only A and B and then computing C by the difference, i.e., $y_C = 1 - y_A - y_B$. If we do this and proceed with the statistical analysis as though all three concentrations had been measured independently, we invite trouble. The computed concentration of y_C is not new information. It is redundant because it is linearly dependent on the sum of $y_A + y_B$. The mathematical consequence of using linearly dependent values is that the V matrix will be singular, and severe computational problems will arise in trying to complete the regression analysis. The practical consequence is a trade-off of dubious value. In using a mass balance constraint to reduce data collection (information gathering), the experimenter has given up the opportunity to gather data that could demonstrate model adequacy by verifying the mass balance constraint.

The second problem is linear dependence among the expected values of the responses, i.e., that one of the model equations is a linear combination of one or more of the other model equations. This situation is more subtle than the similar one described above with data, but its consequences are the same, namely, a singular **V** matrix.

A third problem is always present and it involves model lack of fit. The analysis assumes the model (the set of equations) is correct. Model inadequacy will inflate the residual errors which will cause the off-diagonal elements in the **V** matrix to be large even when correlation among the measurement errors is quite small.

How is the experimenter to guard against these potential difficulties? Our advice is the following:

1. Always check for model adequacy by plotting the residuals of the fitted model against the predicted values and against the independent variables. Inadequacies revealed by the residual plots usually provide clues about how the model can be improved.
2. Pay particular attention to the **V** matrix for indications of singularity. If there is such a problem, remove the linearly dependent data or the linearly dependent equation from the matrix.

In general, the variance-covariance matrix **V** is square with dimensions determined by the number of responses being fitted. Since the number of responses is usually small (usually two to four), it is easy to create the proper pattern of diagonal and off-diagonal terms in the matrix by following the three-response example given above. Any nonlinear optimization program should be able to handle the minimization calculations, but they will not provide any diagnostic statistical information. The GREG software program developed by Stewart et al. (1992) offers powerful methods for multiresponse parameter estimation.

The precision of the parameter estimates can be quantified by determining the approximate joint confidence region. The minimum value of the determinant criterion |V| takes the place of the minimum residual sum of squares value in the usual single-response parameter estimation problem (see Chapter 26). The approximate $100(1 - \alpha)\%$ joint confidence region is bounded by

$$|V|_{1-\alpha} = |V|_{min} \exp\left(\frac{\chi^2_{p(1-\alpha)}}{n}\right)$$

where p = the number of parameters estimated, n = the number of observations, and $\chi^2_{p(1-\alpha)}$ = chi-square value for p degrees of freedom and $(1 - \alpha)100\%$ probability level.

CASE STUDY SOLUTION

The data were fitted simultaneously to the two equations (for substrate and biomass) that constitute the model (Johnson and Berthouex, 1975). In this example, p = 3 and n = 8. The determinant criterion is

$$\min|V| =$$

$$\begin{vmatrix} \sum_{i=1}^{8}\left(x_i - \theta_3\left[S_o - \frac{\theta_2 D_i}{\theta_1 - D_i}\right]\right)^2 & \sum_{i=1}^{8}\left(x_i - \theta_3\left[S_o - \frac{\theta_2 D_i}{\theta_1 - D_i}\right]\right)\left(S_i - \frac{\theta_2 D_i}{\theta_1 - D_i}\right) \\ \sum_{i=1}^{8}\left(x_i - \theta_3\left[S_o - \frac{\theta_2 D_i}{\theta_1 - D_i}\right]\right)\left(S_i - \frac{\theta_2 D_i}{\theta_1 - D_i}\right) & \sum_{i=1}^{8}\left(S_i - \frac{\theta_2 D_i}{\theta_1 - D_i}\right)^2 \end{vmatrix}$$

This calculation is troublesome since the term $(\theta_1 - D)$ in the denominator can become zero during the numerical search for the minimum. A better formulation is to eliminate the denominator altogether. We can also simplify it by substituting $X_i = \theta_3(S_o - S_i)$. The rewritten criterion is

$$
\min \left|
\begin{array}{cc}
\sum_{i=1}^{8}(x_i - \theta_3(S_o - S_i))^2 & \sum_{i=1}^{8}(x_i - \theta_3(S_o - S_i))(S_i - \theta_2 D_i) \\
\sum_{i=1}^{8}(x_i - \theta_3(S_o - S_i))(S_i - \theta_2 D_i) & \sum_{i=1}^{8}(S_i - \theta_2 D_i)^2
\end{array}
\right|
$$

This was minimized to find least squares parameter estimates $\hat{\theta}_1 = 0.71$, $\hat{\theta}_2 = 114$, and $\hat{\theta}_3 = 0.61$. The data, the fitted model, and the residuals are plotted in Figure 36.1.[1]

The next step is to evaluate the precision of the estimated values. This involves understanding the shape of the $|V|$ function, which is a function of θ_1, θ_2, and θ_3 and is three dimensional. Figure 36.2 shows the general shape of $|V|$ for this set of data. Figure 36.3 shows contours of $|V|$ at the cross section where $\theta_3 = \hat{\theta}_3 = 0.61$. The shape of the contours show that θ_1 is estimated more precisely than θ_2 and there is slight correlation between the parameters. The approximate 95% joint confidence region is shaded.

The estimated parameters from fitting response X alone were $\theta_1 = 0.69$, $\theta_2 = 60$, and $\theta_3 = 0.62$. Fitting response S alone gave $\theta_1 = 0.71$ and $\theta_2 = 112$. The precision of the parameters estimated using the single responses is so poor that no segment of the 95% joint confidence regions falls within the borders of Figure 36.3. This is shown by Figure 36.4. Also, the correlation between the parameters is high. The imprecise parameter estimates are not caused simply by the individual models fitting the data very badly.

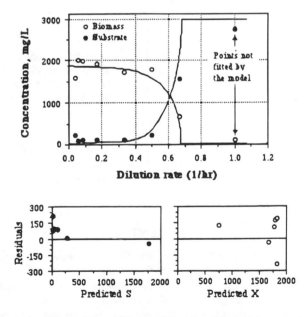

Figure 36.1 Fitted model and residual plots.

---

[1] There is a trend in the residuals for substrate, suggesting that a model of the form $S = \theta_4 + [\theta_2 D/(\theta_1 - D)]$, where θ_4 is a refractory substrate, might fit better.

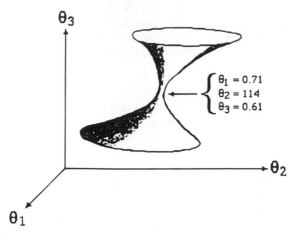

Figure 36.2 Three-dimensional representation of the variance-covariance matrix for the two-response three-parameter bacterial growth model.

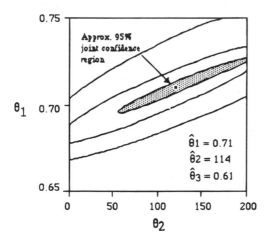

Figure 36.3 Contours of the determinant of the variance-covariance matrix at $\theta_3 = 0.61$. The shaded region is the approximate 95% joint confidence region.

Actually, the curves calculated using the individually estimated values fit nicely and are very near those obtained from the multiresponse estimates. The dramatic gain in the precision of the parameter estimates comes from simultaneously using both equations and both responses.

COMMENTS

When multiple responses have been measured in an experiment, parameters will be estimated more precisely if all responses are fitted simultaneously. This is accomplished by minimizing the determinant of a matrix that contains the residual sums of squares of each response along the diagonal and off-diagonal terms that account for correlation between the residuals. If the measurement errors of each response are independent of other responses, the determinant criterion simplifies to minimizing the sum of the individual sums of squares. In the case of a single response, the criterion simplifies to the usual least squares calculation. A comprehensive recent article of multiresponse estimation, with several excellent discussions, is Bates and Watts (1985).

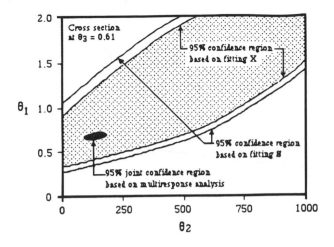

Figure 36.4 The 95% joint confidence regions at $\theta_3 = 0.61$ for the model fitted to both X and S simultaneously (dark shaded region), the model fitted to X only (light shaded region), and the model fitted to S only (unshaded region).

REFERENCES

Bates, D. M. and D. G. Watts (1985). "Multiresponse Estimation with Special Application to Linear Systems of Differential Equations," *Technometrics*, 27, 329–339 (with discussion and response, pp. 340–360).

Bates, D. M. and D. G. Watts (1988). *Nonlinear Regression Analysis and Its Applications*, New York, John Wiley & Sons.

Bard, Y. (1974). *Nonlinear Parameter Estimation*, New York, Academic Press.

Box, G. E. P. and N. R. Draper (1965). "Bayesian Estimation of Common Parameters from Several Responses," *Biometrika*, 52, 355.

Box, G. E. P., W. G. Hunter, J. F. MacGregor, and J. Erjavac (1973). "Some Problems Associated with the Analysis of Multiresponse Data," *Technometrics*, 15, 33.

Hunter, W. G. (1967). "Estimation of Unknown Constants from Multiresponse Data," *Ind. Eng. Chem. Fundam.*, 6, 461.

Johnson, D. B. and P. M. Berthouex (1975). "Using Multiresponse Data to Estimate Biokinetic Parameters," *Biotechnol. Bioeng.*, 17, 571–583.

Ramanathan, M. and A. F. Gaudy (1969). "Effect of High Substrate Concentration and Cell Feedback on Kinetic Behavior of Heterogeneous Populations in Completely Mixed Systems," *Biotechnol. Bioeng.*, 11, 207.

Stewart, W. E., M. Caracotsios, and J. P. Sorensen (1992). GREG Software Package Documentation, Department of Chemical Engineering, University of Wisconsin-Madison.

Chapter 37

A Problem in Model Discrimination

Key words: biological kinetics, least squares, mechanistic models, model building, model discrimination, Monod kinetics, posterior odds, posterior odds ratio, residual sum of squares, Tiessier model

---

When several rival models are tentatively considered, it is not uncommon to find that more than one model fits the data and gives residuals errors that are acceptable. For some applications, this inability to select one model as "the best" is of no consequence because, over the range of interest, either model will serve the intended purpose. In other situations, knowing which model is better may throw light on fundamental questions about reaction mechanisms, catalysis, inhibition, and other phenomena under investigation.

A fundamental concept in model discrimination is that rival models often diverge noticeably only at extreme conditions. It follows that extremes must be studied or discrimination on a statistical basis is impossible. This gives the researcher the choice of restricting the experiment to a limited range of conditions and accepting any plausible model that fits the data over this range or including a test at more stressful conditions in order to obtain more convincing evidence with which to select one among several rival models as the most plausible. In order to discriminate between models (and between mechanisms), the models must be put in jeopardy of failing.

Figure 37.1 shows two examples of this. The two first-order reactions in series, with a reversible second step, cannot be discriminated from the simple model A $\rightarrow$ B $\rightarrow$ C unless observations are carried out at long times. The Haldane inhibition model cannot be distinguished from the simpler Monod model unless studies are made at high concentrations. One might expect the problem of model discrimination to be fairly obvious with models such as these, but examples of weak experimental designs that were useless for discrimination do exist even for these models.

Given models that seem adequate on statistical grounds, we might try to select the "best" model on the basis of (1) minimum sum of squares, (2) lack of fit tests (F tests), (3) fewest parameters (parsimony), (4) simplest functional form, and (5) estimated parameter values consistent with the mechanistic premise of the model. In practice, all of these criteria come into consideration.

Of course, at any particular time, an investigator is not forced to choose a single model. There is also the choice of making more experimental measurements in order to clarify the situation. Sometimes, in order to be honest, an investigator will have to report that two or more models are equivalent, based on available data.

CASE STUDY — TWO RIVAL BIOLOGICAL MODELS

Bacterial growth rate is to be studied by operating a completely mixed reactor of volume V operated at constant flow rate Q without recycle. The influent to the reactor has bacterial solids concentration $X_0 = 0$ and constant influent substrate concentration S_0. At each experimental setting of dilution rate ($D = Q/V$), the reactor effluent will come

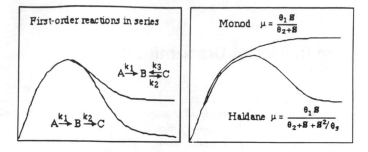

Figure 37.1 Examples of two models where discrimination is impossible unless observations are made at high levels of the independent variable.

to a steady-state bacterial concentration X and substrate concentration S. The specific bacterial growth rate, μ, is equal to the dilution rate (Q/V) of the system: $\mu = D = Q/V$.[1]

Therefore, each experimental setting of dilution rate produces a pair of values of growth rate, μ_i, and substrate concentration, S_i. Several models for bacterial growth in industrial biotechnology and biological wastewater treatment have been studied. We examine two that assume that the specific growth rate of bacterial solids is limited by the concentration of a single substrate. The two models are by Monod (1942) and Tiessier (1936):

$$\text{Monod} \qquad \mu_i = \frac{\theta_1 S_i}{\theta_2 + S_i} + e_i$$

$$\text{Tiessier} \qquad \mu_i = \theta_1[1 - \exp(\theta_3 S_i)] + e_i$$

where μ_i is the growth rate observed at substrate concentration S_i.

The models would be fitted, using nonlinear least squares, to estimate the parameters θ_1, θ_2, and θ_3. The parameter θ_1, which appears in both models, is an asymptotic value that is approached when the system is operated at the maximum bacterial growth rate and also the maximum dilution rate of the reactor. Operating the reactor at this asymptotic condition would provide a direct estimate of the parameter θ_1. The practical difficulty in doing this is that the system tends to become unstable at high dilution rates. At some critical dilution rate, bacteria will be "washed out" of the reactor faster than they can grow, causing the system to fail. The parameter θ_2 controls the shape of the hyperbolic Monod model, and θ_3 has this function in the exponential Tiessier model.

The Monod and Tiessier models were fitted to data from Schulze and Lipe (1964), given in Table 37.1, who pushed the experimental conditions toward the asymptotic

Table 37.1 Data of Schulze and Lipe (1964)

$\mu=$	0.059	0.091	0.124	0.177	0.241	0.302	0.358	0.425
$S=$	5.1	8.3	13.3	20.3	30.4	37	43.1	58
$\mu=$	0.485	0.546	0.61	0.662	0.725	0.792	0.852	
$S=$	74.5	96.5	112	161	195	266	386	

Note: $\hat{\mu}$ = growth rate, S = substrate concentration.

---

[1] This is shown from the steady-state material balance on bacterial solids: $QX_0 - QX + \mu XV = 0$. For $X_0 = 0$, constant flow, and constant volume, the material balance equation becomes $QX - \mu XV = 0$, and $\mu = Q/V$.

limits. We might, then, expect these data to "stress the models" and be advantageous for discriminating between them.

Nonlinear least squares (SYSTAT 1989) was used to estimate the parameters with the following results. The hats ($\hat{}$) indicate the predicted values. RSS is the residual sum of squares of the fitted model, v is the degrees of freedom, and RMS is the residual mean square (RMS = RSS/v). (The estimated parameter values have been rounded, so the values shown will not give exactly the RSS shown.)

Monod model $\quad \hat{\mu} = \dfrac{1.06\ S}{90.4 + S} \qquad\qquad$ RSS $= 0.00243\ (v = 13)$

$\qquad\qquad\qquad\qquad\qquad\qquad\qquad\qquad\quad$ RMS $= 0.000187$

Tiessier model $\quad \hat{\mu} = 0.83[1 - \exp(-0.012\ S)] \quad$ RSS $= 0.00499\ (v = 13)$

$\qquad\qquad\qquad\qquad\qquad\qquad\qquad\qquad\quad$ RMS $= 0.000384$

Having fitted two models, can we determine that one is best? Some useful diagnostic checks are to plot the data against the predictions, plot the residuals, plot joint confidence regions (Chapter 26), make a lack-of-fit test, and examine the physical meaning of the parameters.

The fitted models are plotted against the experimental data in Figure 37.2. Both appear to fit the data. Plots of the residuals, Figure 37.3, show a slightly better pattern for the Monod model, but neither model is clearly inadequate. The joint confidence regions, plotted in Figure 37.4, are small and well conditioned.

These statistical criteria do not clearly indicate one model over the other, so consider the physical significance of the parameters. (In practice, an engineer would probably consider this before anything else.) If θ_1 represents the maximum operable dilution rate, its estimated value should agree with the operating experience. Schulze and Lipe (1964) were not able to operate their chemostat at dilution rates higher than about 0.85/hr. At higher rates, all the bacteria were washed out of the reactor and the system failed. The Tiessier model predicts the maximum growth rate of 0.83/hr, which is consistent with

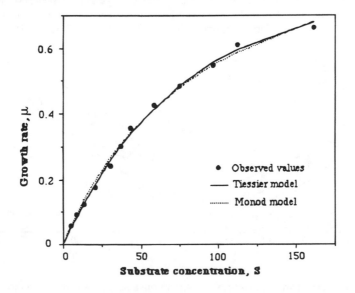

Figure 37.2 The Monod and Tiessier models fitted to the data of Schulze and Lipe (1964).

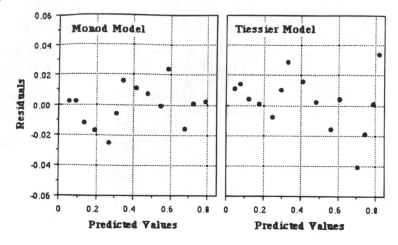

Figure 37.3 Residual plots for the Monod and Tiessier models.

experience in the reaction system studied. For the Monod model, on the other hand, estimates $\theta_1 = 1.06$, which is higher than the experimenters could attain. Based on this, one might conclude that the Monod model is inadequate because the estimated value of θ_1 is not a consistent experience in operating the biological reactor.

An estimate of the variance of the measurement error, σ_e^2, is needed in order to make a formal lack-of-fit test. Ideally, this estimate would be made using data from replicate experiments. The residual mean square (RMS = RSS/ν) of the fitted model is compared with the measurement error variance by computing the F ratio, F = (RMS/$\hat{\sigma}_e^2$), and comparing it with the tabulated value having the appropriate degrees of freedom. There were no replicates in this experiment, so we have no independent estimate of the variance of measurement error. Lacking replication, we can use the RMS as an estimate of the measurement error variance if we are willing to assume that the model is correct. To illustrate the lack-of fit-test, suppose for purposes of argument that the Monod model is correct and fits the data and that σ_e^2 can be estimated using the RSS of the Monod model. The F ratio for the Tiessier model is F = 0.000384/0.000187 = 2.05. The numerator

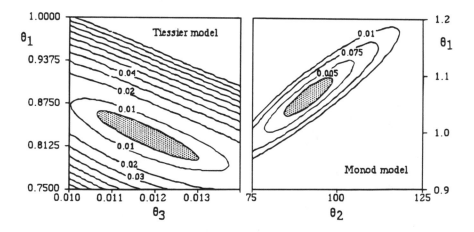

Figure 37.4 Sum of squares surface and approximate 95% joint confidence regions for the Tiessier and Monod models.

and denominator of the F ratio each have $\nu = n - p = 15 - 2 = 13$ degrees of freedom, so the critical F value is $F_{0.05,13,13} = 2.7$ for a test at the 95% confidence level. The computed F is smaller than the critical F. The interpretation of this result is the RMS of the Tiessier model is within the range that might be expected due to chance, given measurement errors with variance of $\hat{\sigma}_e^2 = 0.000187$.

CASE STUDY — FIVE RIVAL MODELS

Sometimes there is a collection of rival kinetic models. To further illustrate the problem of model discrimination, the following five models were fitted to a subset of the Schulze-Lipe data (the first 11 values). Three have no mechanistic basis in biological kinetics, but have shapes similar to the Monod model (Model 4) and the Tiessier model (Model 1). The five fitted models, ranked according to their residual sums of squares (RSS), are shown below.

Model 1 $\mu = 0.766[1 - \exp(-0.0135\ S)]$ RSS $= 0.00176$

Model 2 $\mu = 0.5619\ \tan^{-1}(0.01614\ S)$ RSS $= 0.00184$

Model 3 $\mu = 1.073\left(1 - \dfrac{1}{0.1079\ S}\right)$ RSS $= 0.00236$

Model 4 $\mu = \dfrac{1.07\ S}{92.5 + S}$ RSS $= 0.00237$

Model 5 $\mu = 0.7056\ \tanh(0.01119\ S)$ RSS $= 0.00420$

All five models fit the data very well for low values of S and diverge only at higher levels. If they had been seriously put forth as mechanistic models, it would have been difficult to select the best among the five rival models.

Biological wastewater treatment systems are supposed to operate at low concentrations, and for these systems, it may be satisfactory to use a model that fits the rising portion of the curve, but does not fit the asymptote. Many models can fit the rising part of the curve, so it is understandable why many wastewater treatment experiments have successfully used the Monod model, while many other experiments have successfully used other models. The typical biological treatment data set provides very little information for statistical model discrimination.

On the other hand, if the objective were to operate at high substrate levels, for example, in order to maximize the production rate of single cell protein, it could be important to predict performance where the curve bends and becomes asymptotic. Experiments in this region would be essential. With enough observations at the right conditions, modeling and model discrimination might be moot.

A MODEL DISCRIMINATION FUNCTION

A statistical method for model discrimination based on calculating the posterior probability in favor of each model being correct has been proposed (Box and Hill, 1967). This criterion should not be allowed to select a "correct model," but it may be used to eliminate one or more models as being unworthy of additional investigation.

"Posterior probability" is the probability computed after the models have been fitted to the data. Before the models are fitted, we may have no particular reason to favor one model over another, i.e., each model is equally likely to be correct. After the models have been fitted, evidence should emerge to favor some models and to discredit others.

In the criterion to be used, the RSS of the rival models is assumed to contain the information needed to discriminate. The posterior probabilities are calculated as a function of the RSS, adjusted for the number of parameters in the model, with better models having a lower RSS.

Here we illustrate the special case where all models have the same number of parameters (Pallesen, 1976). The method is based on an assumption that the *prior probabilities* are equal for all the candidate models. That means that before we fit any of the models, we give them equal standing on scientific grounds and are willing to accept any one of them that fits the available data.

There are k models. The posterior probability in favor of model i being correct is

$$P_i = \frac{R_i}{\Sigma R_i}$$

where the summation is over the k models and

$$R_i = \left(\frac{RSS_{min}}{RSS_i}\right)^{0.5(n-p)}$$

where
  RSS_i = the residual sum of squares of model i
  RSS_{min} = the smallest residual sum of squares in the set of k models
  n = the number of observations (all models fitted to the same data)
  p = the number of estimated parameters in each model (p equal for all k models)

The residual sums of squares (RSS_i), RSS ratios, and posterior probabilities for each model are presented in Table 37.2. An example computation of the odds in favor of Model 2 gives $R_2 = (0.176/0.184)^{0.5(11-2)} = 0.82$. The computed posterior probabilities indicate a 42% probability for the Tiessier model being correct and an 11% probability for the Monod model.

A warning is in order about using this discrimination criterion. A key assumption was that all models are equally likely to be correct *before* they are fitted (i.e., equal prior odds for each models). This is not the case in our example since the Tiessier and Monod models were put forth as having some mechanistic justification; the others were concocted for sake of example. Therefore, the prior odds are not equal, and the posterior odds given in Table 37.2 are just numbers that suggest how a model discrimination criterion might be used. It is a method with potential for abuse, like all methods that

Table 37.2 Posterior Odds for Five Rival Models (n = 11, p = 2)

Model	RSS_i (×1000)	RSS Ratio R_i	Posterior Odds $P_i = R_i / \Sigma R_i$
1 (Tiessier)	176	1.00	0.42
2	184	0.82	0.35
3	236	0.27	0.11
4 (Monod)	237	0.26	0.11
5	420	0.02	0.01
		$\Sigma R_i = 2.37$	$\Sigma P_i = 1.00$

reduce a difficult problem to a few numbers. It should not replace careful thought and common sense and using plots of various kinds, especially plots of residuals. Furthermore, it is not a substitute for doing more experiments at wisely selected critical conditions.

COMMENTS

Discriminating among rival models is difficult. Goodness of fit is a necessary, but not sufficient, condition for accepting a model as correct. One or more other models may also be adequate when judged by several common statistical criteria. If discrimination is to be attempted on statistical grounds, it is necessary to design experiments that will put the model in jeopardy by exposing it to conditions where its inadequacies will be revealed.

REFERENCES

Box, G. E. P. and W. J. Hill (1967). "Discrimination Among Mechanistic Models," *Technometrics*, 9(1) 57–71.

Boyle, W. C. and P. M. Berthouex (1974). "Biological Wastewater Treatment Model Building — Fits and Misfits," *Biotechnol. Bioeng.*, 16, 1139–1159.

Hill, W. J., W. G. Hunter, and D. Wichern (1968). "A Joint Design Criterion for the Dual Problem of Models Discrimination and Parameter Estimation," *Technometrics*, 10(1), 145–160.

Kittrel, J. R. and R. Mezaki (1967). "Discrimination Among Rival Hougan-Watson Models Through Intrinsic Parameters," *Am. Inst. Chem. Eng. J.*, 13(2), 389–470.

Monod, J. (1942). *Researches sur la Croissance des Cultures Bacteriennes*, Paris, Herman & Cie.

Pallesen, L. (1976). Unpublished paper.

Schulze, K. L. and R. S. Lipe (1964). "Relationship Between Substrate Concentration, Growth Rate, and Respiration Rate of Eschericia coli in Continuous Culture," *Arch. Microbiol.*, 48, 1.

Tiessier, C. (1936). "Les Lois Quantitatives de la Croissance," *Ann. Physiol. Physicochim. Biol.*, 12, 527.

Chapter 38

Adjustment of Survey Data

Key words: mass balance, measurement errors, data adjustment, least squares, weighted least squares

---

Natural variability, sampling, and measurement errors all cause sampled properties to be estimated imperfectly. The result is that a survey may produce a set of inconsistent data. For example, three measured angles of a triangle might not add up to 180 degrees. Such inconsistencies can be discovered only if some redundant measurements are made. If we measure two angles of the triangle, the third can be computed by difference, but there is no way to check the accuracy of the measurements. If the measurements contain some redundant information, it is possible to test the measured world against the real world and to make rational adjustments of the measured values to bring them into agreement with the known physical constraints.

This adjustment is done in the same spirit that a collection of observations are averaged to estimate a mean value. We simply want to use the measured values to extract the best possible estimates about the system being studied. The adjustment should be done so that the most precisely measured flows receive the smallest correction. The method of least squares provides a common sense adjustment procedure that is statistically justified and computationally simple.

There are three types of errors. *Gross errors* are caused by overt blunder, such as a misplaced decimal or instrument malfunction. *Bias* is a consistent displacement from the true value caused by improper instrument calibration or not following the correct analytical procedure. The purpose of quality assurance programs is to eliminate these kinds of errors. They should not exist in a well-run measurement program.

The third type is *random errors*, which result from meter and instrument noise and the inability to perfectly replicate measurement techniques. They are unavoidable. Both the sign and magnitude of the errors vary randomly, but with definite statistical properties, which can be described if repeated measurements are made. We will assume that the random errors have a mean value of zero and are normally distributed.

Data that contain only random errors are said to be in statistical control. Data that are not in statistical control cannot be usefully adjusted. The first step, then, is to purge data containing gross errors. Second, remove known bias by making properly calibrated corrections. The data should now contain only random errors. Suitable adjustments can be derived to make the data consistent with physical constraints imposed by the real world.

CASE STUDY — ERRORS IN FLOW MEASUREMENTS

Seven flows have been measured, as shown in Figure 38.1 (Schellpfeffer and Berthouex, 1972). The measured values do not satisfy the physical constraint that inflow must equal outflow. The flows measured as Q_1, Q_2, and Q_3 sum to 66, which does not agree with the value metered as $Q_4 = 62$. The total of the five measured inputs sum to 102, which does not agree with the metered quantity $Q_7 = 98$. We wish to use the available data to derive estimates of the flows that conform to the conservation of mass constraints that the real physical system must obey. The adjustment should be done so that the most precise measurements receive the smallest correction. The method of least squares provides an adjustment procedure that is statistically justified and mathematically convenient.

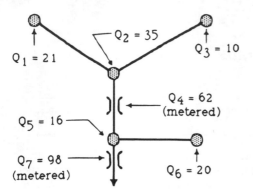

Figure 38.1 Seven measured flows are inconsistent because inflows do not equal outflows.

DATA ADJUSTMENT

Some options for using or adjusting inconsistent data are

1. Make no adjustment. Use the measured values. This may be proper if deficiencies fall within a predetermined tolerance.
2. Adjust one of the measured values. This might be proper if the accuracy of one measurement is much less than the others.
3. Distribute the discrepancy equally among the observed values.
4. Assign unequal adjustments to each observation according to some reasonable rule. One such rule is to make adjustments that are proportional to the precision of each observation (i.e., inversely proportional to the variance). This strategy is formulated as a least squares problem.

No one adjustment procedure is always the proper choice, but the method of least squares is the best method in the long run. In complex problems, it may be the only satisfying method, and it is often the simplest as well.

The objective of the method of least squares is to minimize the sum of weighted squares of the residuals.

$$S = \Sigma w_i (x_i - m_i)^2 = \Sigma \frac{1}{s_i^2} (x_i - m_i)^2$$

where x_i and m_i are the calculated and measured values of quantity i. The summation is over all observations that are subject to random error. The weight w_i indicates the relative precision of measurement m_i and is inversely proportional to its variance, $w_i = (1/s_i)^2$. The special case where all weights are equal (constant variance) gives the familiar case of ordinary least squares.

There is a natural tendency to overlook imprecise measurements in favor of those which are most precise. The weighted least squares model is in sympathy with this commonsense approach. A measurement with a large standard deviation will have a small weight and will contribute relatively little to the weighted sum of squares, thus leaving the precise measurements to dominate the overall adjustment procedure.

When the standard deviations are unknown, subjective weights can be assigned to indicate the relative reliability of the measured quantities. Weights might be assigned using knowledge of meter characteristics, chemical analysis, or other technical information. It is the ratios, and not the individual values, of weights that are important. The least reliable measurement can be assigned a weight w = 1, with proportionately larger weights being assigned to the more reliable measurements.

Formulation of the solution will be illustrated, assuming that material balance equations serve as the constraints for adjustment for the system of interest. There are M physical constraining relations involving N variables, with $N \geq M$. If $N = M$, the system has no degrees of freedom, and there is a unique solution to the set of constraint equations. If $N > M$, the system has $N - M = F$ degrees of freedom. Specifying values for F variables gives M equations in M unknowns, and a unique solution can be found. If more than F variables are measured (for example, all N variables might be measured), there is some redundant information, and inconsistencies in the measured values can be detected. This redundancy is a necessary condition for data adjustment.

Assuming that all N variables have been measured and all constraint equations are linear, the problem is to

$$\text{minimize } S = \Sigma w_i(x_i - m_i)^2 \qquad i = 1, 2, \ldots, N$$

subject to linear constraints of the form

$$c_j(x_1, x_2, \ldots, x_N) = \Sigma a_{ji}x_i \qquad j = 1, 2, \ldots, M$$

The constraints are combined with the least squares condition using the method of Lagrange multipliers to create a new objective function:

$$\text{minimize } G = \sum_{i=1}^{N} w_i(x_i - m_i)^2 + \sum_{j=1}^{M} \lambda_j c_j(x)$$

There are now $N + M$ variables, the original N unknowns (the x_i's) plus the M Lagrange multipliers (λ_j's). The minimum of this function is located at the values where the partial derivatives with respect to each unknown variable vanish simultaneously. The derivative equations have two forms:

$$\frac{\partial G}{\partial x_i} = 2 w_i(x_i - m_i) + \sum_{j=1}^{M} a_{ij}\lambda_j = 0 \qquad i = 1, 2, \ldots, N$$

$$\frac{\partial G}{\partial \lambda_j} = \sum_{i=1}^{N} \Sigma a_{ji}x_i = 0 \qquad j = 1, 2, \ldots, M$$

The quantities m_i, w_i, and a_{ji} are known, and the resulting set of linear equations are solved to compute values for the x_i's and the λs.

$$\sum_{i=1}^{N} \Sigma a_{ji}x_i = 0 \qquad j = 1, 2, \ldots, M$$

$$2 w_i x_i + \sum_{j=1}^{M} a_{ji}\lambda_j = 2 m_i w_i \qquad i = 1, 2, \ldots, N$$

CASE STUDY — SOLUTION

The system shown in Figure 38.1 will be solved for the case where all seven flows have been measured and each measured value is equally reliable, which assigns equal weights ($w_i = 1$) to all measured values. The sum of squares equation is

$$\min S = (Q_1 - 21)^2 + (Q_2 - 35)^2 + (Q_3 - 10)^2 + (Q_4 - 62)^2 + (Q_5 - 16)^2$$
$$+ (Q_6 - 20)^2 + (Q_7 - 98)^2$$

subject to satisfying the two conservation of mass constraints:

$$Q_1 + Q_2 + Q_3 - Q_4 = 0$$

and

$$Q_4 + Q_5 + Q_6 - Q_7 = 0$$

Incorporating the constraints gives a function of nine unknowns (seven Qs and two λs):

$$\min S = (Q_1 - 21)^2 + (Q_2 - 35)^2 + (Q_3 - 10)^2 + (Q_4 - 62)^2 + (Q_5 - 16)^2$$
$$+ (Q_6 - 20)^2 + (Q_7 - 98)^2 + \lambda_1(Q_1 + Q_2 + Q_3 - Q_4) + \lambda_2(Q_4 + Q_5 + Q_6 - Q_7)$$

The nine partial derivative equations are all linear:

$$2 Q_1 + \lambda_1 = 21(2) = 42 \qquad 2 Q_2 + \lambda_1 = 35(2) = 70$$
$$2 Q_3 + \lambda_1 = 10(2) = 20 \qquad 2 Q_4 - \lambda_1 + \lambda_2 = 62(2) = 124$$
$$2 Q_5 + \lambda_2 = 16(2) = 32 \qquad 2 Q_6 + \lambda_2 = 20(2) = 40$$
$$2 Q_7 + \lambda_2 = 98(2) = 196 \qquad Q_1 + Q_2 + Q_3 - Q_4 = 0$$
$$Q_4 + Q_5 + Q_6 - Q_7 = 0$$

The adjusted values are given in Table 38.1. The calculated values of λ_1 and λ_2 have no physical meaning and are not reported. The adjustments range from about 10% for Q_3 to a fraction of a percent for Q_7.

COMMENTS

The method of least squares gives the smallest possible adjustments that force measured survey data to conform with the conservation of mass constraints. Also, it agrees with common-sense in that unreliable measurements are adjusted more than precise measurements. The solution for linear constraining relations is particularly simple, even for a problem with many variables, because the set of linear equations to be solved can be

Table 38.1 **Measured Flow Rates and the Adjusted Values that Conform to the Conservation of Mass Constraints**

Quantity	Measured Value	Adjustment	Adjusted Value
Q_1	21	−1.07	19.93
Q_2	35	−1.07	33.93
Q_3	10	−1.07	8.93
Q_4	62	+0.80	62.80
Q_5	16	−0.27	15.73
Q_6	20	−0.27	19.73
Q_7	98	+0.27	98.27

written on inspection and the equations are easily solved using standard matrix algebra. For nonlinear constraining equations, the least squares computations are manageable, but more difficult, because the partial derivatives are nonlinear. A nonlinear programming algorithm must be used to solve the problem.

Measurements must be taken at enough locations to test the measured world against the real world; that is, to check whether the conservation of mass has been satisfied. Redundancies in the set of data and the constraining relations are used to adjust the inconsistencies.

The example assumed that each measured value was equally reliable. This is not usually the case. Instead, the measurement errors might be proportional to the magnitude of the flow; some measurements might have been repeated, whereas others were done only once; or instruments (or personnel) with different precision might have been used at different locations. Such problem-specific information is included by using appropriate weights that could be estimated subjectively if replicate measurements are not available from which to estimate the variances.

REFERENCES

Deming, W. E. (1943). *Statistical Adjustment of Data*, New York, John Wiley & Sons (Dover Publications edition 1964).

Ripps, D. L. "Adjustment of Experimental Data," *Chem. Eng. Prog. Symp. Ser.*, 61, 55, 8–13.

Schellpfeffer, J. W. and P. M. Berthouex (1972). "Rational Adjustment of Imbalances in Plant Survey Data," *Proc. 29th Purdue Industrial Waste Conference*, Purdue University.

Chapter 39

How Measurement Errors Are Transmitted into Calculated Values

---

Key words: error transmission, linear approximation, propagation of errors, reactor kinetics, sensitivity coefficients, titration errors, variance, uncertainty analysis

---

When observations that contain measurements are used to calculate other values, the errors are transmitted to a greater or lesser extent into the calculated values. Often it is important in monitoring and in experimentation to explore how errors will be transmitted *before* making any measurements. It is useful to know whether the errors in a particular measured variable will be magnified or suppressed. This could help in selecting measurement methods and planning the number of replications that will be required to minimize error in the variables that most strongly transmit their errors. Here are four such situations.

1. The precision with which a burette can be read is known. How is this precision related to the precision of a concentration that is computed using a titrant volume measured with the burette?

2. A study of mass transfer in a reactor involves small spheres. From time to time, a sample of n spheres is taken; their average diameter $D = \Sigma d_i/n$ is estimated; and D is used to calculate the surface area, $A = \pi D^2$, and volume of the spheres, $V = \dfrac{\pi}{6} D^3$. The precision of the estimated average diameter, D, will be expressed by giving its variance, standard deviation, or some confidence interval. It is desirable to attach a similar statement of precision to the computed values A and V.

3. An experiment to estimate the reaction rate coefficient k involves measuring reactor solids content X and effluent organics concentration S over a series of runs in which detention time $t = V/Q$ is controlled at various levels. The influent organics concentration, S_o, can be prepared precisely at the desired value. The calculation of k from the known values is $k = (S_o - S)Q/SXV$. Both X and S are difficult to measure precisely. Replicate samples must be analyzed to improve precision. Questions asked during the experimental planning stage may include, "How many replicate samples of X and S are needed if k is to be estimated within plus or minus 3%? Must both X and S be measured precisely or is one more critical than the other? Measuring S is expensive. Can money be saved if errors in S are not strongly transmitted into the calculated value of k?"

4. An air pollution discharge standard might be based on the weight of NO_x emitted per dry standard cubic foot of gas corrected to a specified CO_2 content. To calculate this quantity, one needs measurements of NO_x concentration in the actual gas, velocity, temperature, pressure, moisture content, and CO_2 concentration. The precision and bias of each measurement will affect the accuracy of the calculated value. Which variables need to be measured most precisely?

Such questions can be answered. Many experiments would be improved by finding the answers to these kinds of questions during the planning rather than the analysis phase of the experiments.

THEORY

The simplest case is for a *linear function* relating the calculated value y to the values of x_1, x_2, and x_3 obtained from measured data. The model has the form

$$y = \theta_0 + \theta_1 x_1 + \theta_2 x_2 + \theta_3 x_3$$

where the θs are coefficients (constants) and the errors in x_1, x_2, and x_3 are random. The expected value of y is

$$E(y) = \theta_0 + \theta_1 E(x_1) + \theta_2 E(x_2) + \theta_3 E(x_3)$$

and the variance of y is

$$Var(y) = \theta_1^2 \, Var(x_1) + \theta_2^2 \, Var(x_2) + \theta_3^2 \, Var(x_3) + \theta_1 \theta_2 \, Cov(x_1 x_2) + \theta_1 \theta_3 \, Cov(x_1 x_3)$$
$$+ \; \theta_2 \theta_3 \, Cov(x_2 x_3)$$

in which the covariances, $Cov(x_i x_j)$, can be positive, negative, or zero. If x_1, x_2, and x_3 are uncorrelated, the covariance terms are all zero and

$$Var(y) = \theta_1^2 \, Var(x_1) + \theta_2^2 \, Var(x_2) + \theta_3^2 \, Var(x_3)$$

The terms on the right-hand side represent the separate contributions of each x variable to the overall variance of the calculated variable y. The derived Var(y) estimates the true variance of y, σ_y^2, and can be used to compute a confidence interval for y.

This same approach can be applied to a linear approximation of *nonlinear* function. Figure 39.1 shows a curved surface $y = f(x_1, x_2)$ replaced by the linear approximation

$$Y = \theta_0 + \theta_1(x_1 - \bar{x}_1) + \theta_2(x_2 - \bar{x}_2)$$

Y is a Taylor series approximation to y where higher-order terms have been ignored. This expansion can be used to linearize any continuous function $y = f(x_1, x_2, \ldots, x_n)$.

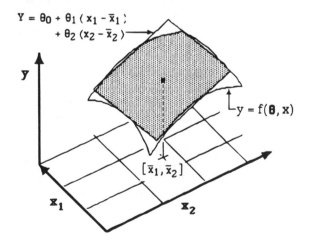

Figure 39.1 A nonlinear function $y = f(\theta, x)$ can be approximated by a linear function centered at the expected values $\bar{x}_1$ and $\bar{x}_2$.

The approximation is centered at $\bar{x}_1$, $\bar{x}_2$, and $\theta_0 = \bar{y} = f(x_1, x_2)$. The linear coefficients are the derivatives of y with respect to x_1, x_2, evaluated at $\bar{x}_1$ and $\bar{x}_2$.

$$\theta_1 = \left(\frac{\partial y}{\partial x_1}\right)_{\bar{x}_1}$$

and

$$\theta_2 = \left(\frac{\partial y}{\partial x_2}\right)_{\bar{x}_2}$$

Following the example of the linear function presented earlier, the variance of y is

$$\text{Var}(Y) = \left(\frac{\partial y}{\partial x_1}\right)_{\bar{x}_1}^2 \text{Var}(x_1) + \left(\frac{\partial y}{\partial x_2}\right)_{\bar{x}_2}^2 \text{Var}(x_2) + \left(\frac{\partial^2 y}{\partial x_1 \partial x_2}\right)_{\bar{x}_1 \bar{x}_2} \text{Cov}(x_1 x_2)$$

This can be generalized to n measured variables (x_i, i = 1, n):

$$\text{Var}(Y) = \sum_{i=1}^{n} \left(\frac{\partial y}{\partial x_i}\right)_{\bar{x}_i}^2 \text{Var}(x_i) + \sum_{i=1}^{n} \sum_{j=i-1}^{n} \left(\frac{\partial^2 y}{\partial x_i \partial x_j}\right)_{\bar{x}_i \bar{x}_j} \text{Cov}(x_i x_j)$$

There will be $n(n-1)/2$ covariance terms; each may be positive, negative, or zero. Since $\text{Cov}(x_i x_j)$ can be positive or negative, correlated variables might increase or decrease the variance of Y. If x_1, x_2, ..., x_n are independent, all covariance terms will be zero. It is tempting to assume independence automatically, but making this simplification needs some justification.

If $y = f(x)$ is formulated as a simulation model or a set of finite difference equations or if it is too unwieldy to take derivatives analytically, the linear approximation can be derived from numerical estimates of the derivatives at $\bar{x}_1$, $\bar{x}_2$. This approach may desirable for even fairly simple functions.

This region over which a linearization is desired is specified by $\bar{x}_1 + \Delta x_1$ and $\bar{x}_2 + \Delta x_2$. The function values $y_0 = f(\bar{x}_1, \bar{x}_2)$, $y_1 = f(\bar{x}_1 + \Delta x_1, \bar{x}_2)$, $y_2 = f(\bar{x}_1, \bar{x}_2 + \Delta\bar{x}_2)$, etc. are used to compute the coefficients of the approximating linear model

$$Y = \theta_0 + \theta_1(x_1 - \bar{x}_1) + \theta_2(x_2 - \bar{x}_2)$$

The coefficients are:

$$\theta_0 = y_0 = \bar{y} = f(\bar{x}_1, \bar{x}_2), \qquad \theta_1 = \frac{y_1 - y_0}{\Delta x_1}, \quad \text{and} \quad \theta_2 = \frac{y_2 - y_0}{\Delta x_2}$$

The linearization will be valid over a small region about the centering point, where "small" can be reasonably defined as $\pm 2\sigma_x$ or $\pm 3\sigma_x$.

EXAMPLE — TITRATION ERRORS

The concentration of a water specimen is measured by titration as $C = 50(y_2 - y_1)$, where y_1 and y_2 are initial and final burette readings. The coefficient 50 simply converts milliliters of titrant used $(y_2 - y_1)$ into a concentration (milligrams per liter). Assuming the variance of a burette reading σ_y^2 is constant for all y, the variance of the computed concentration is

$$\text{Var(C)} = \sigma_C^2 = (50)^2(\sigma_{y_2}^2 + \sigma_{y_1}^2) = 50^2 2\sigma_y^2$$

Suppose that the standard deviation of a burette reading is known to be $\sigma_y = 0.02$ mL, giving $\sigma_y^2 = 0.0004$. For $y_1 = 38.21$ and $y_2 = 25.73$, the concentration is

$$c = 50(38.21 - 25.73) = 624 \text{ mg/L}$$

and the variance of concentration is

$$\sigma_C^2 = (50)^2(\sigma_{y_2}^2 + \sigma_{y_1}^2) = 2500(0.0004 + 0.0004) = 2.00$$

and

$$\sigma_C = 1.4 \text{ mg/L}$$

Notice that the variance and standard deviation are not a function of the actual burette readings. Therefore, this value of the standard deviation holds for any difference $(y_2 - y_1)$. A 95% confidence interval would be approximately

$$624 \pm 2(1.4)$$

or

$$621 \text{ to } 627 \text{ mg/L}$$

Now, suppose that a water specimen is diluted by a factor D before titration (D = 2 means that the specimen was diluted to double its original volume or half its original concentration). This might be done, for example, so that no more than 15 mL of titrant is needed to reach the endpoint $y_2 - y_1 \leq 15$. The concentration is estimated as

$$C = 50\,D(y_2 - y_1)$$

and the variance of concentration is

$$\sigma_C^2 = (50\,D)^2(\sigma_{y_2}^2 + \sigma_{y_2}^2) = 50^2 2\sigma_y^2 D^2$$

D = 1 (no dilution) gives the results just shown above. For D > 1, any variation in error in reading the burette is magnified by the dilution factor. In this case, the variance may be considered uniform within a narrow range of concentration (constant D), but it becomes roughly proportional to concentration over a wider range (D varying with concentration). It is not unusual for environmental data to have a variance that is proportional to concentration. Dilution or concentration during the laboratory processing can produce this characteristic.

EXAMPLE — SPHERICAL PARTICLES

Particle diameters are to be measured and used to calculate particle volumes. Assuming that the particles are spheres, $V = \pi D^3/6$, and the variance of the volume is

$$\text{Var(V)} = \left(\frac{3\pi}{6} D^2\right)^2 \sigma_D^2 = (1.5708\ D^2)^2\sigma_D^2$$

and

$$\sigma_V = 1.5708\ D^2\sigma_D$$

The precision of the estimated volumes will depend upon the measured diameter of the particles. Suppose that $\sigma_D = 0.02$ for all diameters of interest in a particular application. Table 39.1 shows the relation between the diameter and variance of the computed volumes.

At $D = 0.798$, the variance and standard deviation of volume equal those of the diameter. For larger diameters, errors in V are magnified. For small D (<0.798), errors are suppressed. The distribution of errors in V will be stretched or compressed according to the slope of the curve that covers the range of values of D. This is shown in Figure 39.2.

Preliminary investigations of error transmission can be a valuable part of experimental planning. If, as was assumed here, the magnitude of the measurement error is the same for all diameters, a greater number of particles should be measured and used to estimate D and V if the particles are large. In contrast, the measurement error may not be constant over the range of diameters to be measured. Perhaps a different measurement method is used for small and large particles. This situation can be computed using the method of this example or it can be sketched on a curve such as Figure 39.2 to get a graphical impression of the error transmission.

Table 39.1 Propagation of Error in Measured Particle Diameter into Error in the Computed Particle Diameter

D	0.5	0.75	0.798	1	1.25	1.5
V	0.065	0.221	0.266	0.524	1.023	1.767
σ_V^2	0.00006	0.00031	0.00040	0.00099	0.00241	0.00500
σ_V	0.008	0.018	0.020	0.031	0.049	0.071
σ_V/σ_D	0.393	0.884	1.000	1.571	2.454	3.534
σ_V^2/σ_D^2	0.154	0.781	1.000	2.467	6.024	12.491

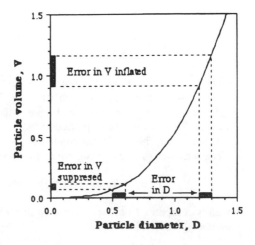

Figure 39.2 Errors in the computed volume are suppressed for small D and inflated for large D.

EXAMPLE — REACTOR KINETICS

A lab scale reactor is described by the model

$$S = \frac{S_o}{1 + kX(V/Q)}$$

where S_o is influent BOD, S is effluent BOD, X is mixed liquor suspended solids, V is reactor volume, Q is volumetric flow rate, and k is a reaction rate coefficient. Assume that the volume is fixed, $V = 1.00$. An experimental run setting $S_o = 200$, $X = 2000$, and $Q = 1$ is expected to give S approximately equal to 20. The independent variables Q, X, S_o, and S will be measured for each experiment. Their expected measurement errors, given as standard deviations, are $\sigma_Q = 0.02$, $\sigma_X = 100$, $\sigma_S = 2$, $\sigma_{S_o} = 20$. How precisely will k be estimated?

Rearranging the model gives

$$k = \frac{(S_o - S)Q}{S\,X\,V}$$

The sensitivity of k to changes in the other variables can be calculated analytically. The linearized model is

$$k = \theta_0 + \theta_{S_o}(S_o - \overline{S}_o) + \theta_S(S - \overline{S}) + \theta_X(X - \overline{X}) + \theta_Q(Q - \overline{Q})$$

The value of k at the given values of the variables is

$$k = \theta_0 = \frac{(S_o - S)Q}{S\,X\,V} = \frac{(200 - 20)(1)}{20(2000)(1)} = 0.0045$$

The other coefficients of the linearized model are

$$\theta_{S_o} = \frac{\partial k}{\partial S_o} = \frac{Q}{S\,X\,V} = \frac{(1)}{20(2000)(1)} = 2.5 \times 10^{-5}$$

$$\theta_S = \frac{\partial k}{\partial S} = \frac{Q}{S\,X\,V} = \frac{1}{(20)(2000)(1)} = 2.5 \times 10^{-5}$$

$$\theta_X = \frac{\partial k}{\partial X} = -\frac{S_o - S}{S\,X^2\,V} = -\frac{200 - 20}{(20)(2000)^2(1)} = -2.5 \times 10^{-6}$$

$$\theta_Q = \frac{\partial k}{\partial Q} = \frac{(S_o - S)}{S\,X\,V} = \frac{(200 - 20)}{(20)(2000)(1)} = 0.0045$$

The θs can be interpreted as *sensitivity coefficients* since they quantify how sensitive the response is to changes in each variable. The coefficient θ_i indicates how much a single unit change in variable i will change the value of k from the value $\theta_0 = 0.0045$. Since the θs are evaluated at particular values of the variables, they have meaning only at, or very near, this particular condition.

The variance of k, assuming covariances are all zero, is

$$\sigma_k^2 = \theta_{S_o}^2 \sigma_{S_o}^2 + \theta_S^2 \sigma_S^2 + \theta_X^2 \sigma_X^2 + \theta_Q^2 \sigma_Q^2 = 5.59 \times 10^{-7}$$

which is computed from the separate contributions of the variables using the values and variances given earlier:

$$S_o = 200, \sigma_{S_o} = 20 \qquad \theta_{S_o}^2 \sigma_{S_o}^2 = 2.5 \times 10^{-7}$$

$$S = 20, \sigma_S = 2 \qquad \theta_S^2 \sigma_S^2 = 2.5 \times 10^{-7}$$

$$X = 2000, \sigma_X = 100 \qquad \theta_X^2 \sigma_X^2 = 0.51 \times 10^{-7}$$

$$Q = 1, \sigma_Q = 20 \qquad \theta_Q^2 \sigma_Q^2 = 0.08 \times 10^{-7}$$

The estimated total variance is $\sigma_k^2 = 5.59 \times 10^{-7}$, and the standard deviation is $\sigma_k = 0.00075$. These numbers are small, but the expected value of $k = 0.0045$ is also small. A better perspective is obtained by finding an approximate 95% confidence interval.

$$k = 0.0045 \pm 2(0.00075) = 0.0045 \pm 0.0015$$

or $k = 0.003$ to 0.006.

We may be disappointed at this point that the precision is not better, say something like $k = 0.0045 \pm 0.0004$. It is better to be disappointed *before* doing the experiment than after. It allows plans to be modified at no loss of time or money.

The contributions of each variable to σ_k^2 suggest where substantial improvements might be made. Notice that the values of the sensitivity coefficients depend on the settings of the independent variables (S_o, S, X, and Q). If these are changed by moving the experiment to different settings, the values of θ_i will change. It takes some effort to see by how much and in which direction they change, but nevertheless, changing the experimental design is one way to reduce the variance of the estimated value of k.

Another possible approach to reducing σ_k^2 is to reduce the variances of the independent variables. Perhaps the measurement method can be improved, but often a more practical approach is to replicate the measurements. Making duplicate measurements ($n = 2$) would reduce σ_i to $\sigma_i/\sqrt{2}$; using $n = 4$ would give $\sigma_i\sqrt{4} = \sigma_i/2$. The corresponding values of $\theta^2\sigma^2$ would be reduced to one half (for $n = 2$) and one quarter (for $n = 4$) of the original values.

EXAMPLE — A FINITE DIFFERENCE MODEL

In the example above, the kinetic model was a simple algebraic equation. Frequently, the model of interest consists of one or more differential equations that must be solved numerically (Brown, 1987). Without worrying about the details of the system being modeled or the specific form of the equations, assume that time-consuming and expensive experiments are needed to estimate a reaction rate coefficient, k, and that preliminary tests indicates that two of the independent variables cannot be measured with a high degree of precision. We seek to understand how errors in the measurement of these variables will be transmitted into the estimate of k. The first step taken was to compute some numerical solutions of the model in the region where the experimenter thinks the first run should be done.

For our two-variable example, the estimate of variance based on the Taylor series expansion shown earlier is

$$\text{Var}(k) = \left(\frac{\Delta k}{\Delta X_1}\right)^2 \text{Var}(X_1) + \left(\frac{\Delta k}{\Delta X_2}\right)^2 \text{Var}(X_2)$$

We will estimate the sensitivity coefficients $\Delta\theta_1 = \Delta k/\Delta X_1$ and $\Delta\theta_2 = \Delta k/\Delta X_2$ by evaluating k at points located at distances ΔX_1 and ΔX_2 from the center point.

Assume that the center region of interest is located at $X_{10} = 200$, $X_{20} = 20$ and that $k_0 = 0.45$ at this point. A reasonable choice of ΔX_1 and ΔX_2 is from one to three standard deviations of the error in X_1 and X_2. We will use $2\sigma_{X_1} = \Delta X_1 = 20$ and $2\sigma_{X_2} = \Delta X_2 = 2$. Suppose that at $[X_{10} + \Delta X_1 = 200 + 20, X_{20} = 20]$ we can compute from the model that $k = 0.50$. And, that at point $[X_{10} = 200, X_{20} + \Delta X_2 = 20 + 2]$ we get $k = 0.35$. The sensitivity coefficients are

$$\frac{\Delta k}{\Delta X_1} = \frac{0.50 - 045}{20} = 0.0025$$

$$\frac{\Delta k}{\Delta X_2} = \frac{0.35 - 0.45}{2} = -0.05$$

These sensitivity coefficients can be used to estimate the expected variance of k. Suppose the estimated variances are $\mathrm{Var}(X_1) = 100$, $\mathrm{Var}(X_2) = 1$. Then,

$$\sigma_k^2 = (+0.0025)^2 100 + (-0.05)^2 1 = 0.000625 + 0.0025 = 0.003125$$

and

$$\sigma_k = 0.056$$

An approximate 95% confidence interval would be $k = 0.45 \pm 2(0.056) = 0.45 \pm 0.11$ or $k = 0.34$ to 0.526.

Eighty percent of the variance in k is contributed by X_2. This may be surprising since X_2 has the smallest variance, but it is such failures of our intuition that merit this kind of analysis. Unfortunately, *at these specified experimental settings*, the precision of the estimate of k depends almost entirely upon X_2. If the precision of k must be improved, the options are (1) try to center the experiment in another region where variation in X_2 will be suppressed, (2) improve the precision with which X_2 is measured, or (3) make a replicate measure of X_2 to average out the random variation.

COMMENTS

It is a serious disappointment to learn after an experiment has been performed that the variance of computed values is too large. Avoid disappointment by investigating this *before* running the experiment. Make an analysis of how measurement errors are transmitted into calculated values. This can be done when the model is a simple equation or when the model is complicated and must be solved by numerical approximation.

REFERENCES

Box, G. E. P., W. G. Hunter, and J. S. Hunter (1978). *Statistics for Experimenters*, New York, Wiley Interscience.

Brown, L. C. (1987). "Uncertainty Analysis in Water Quality Modeling Using QUAL2E," in *Systems Analysis in Water Quality Measurement*, (Advances in Water Pollution Control Series), pp. 309–319. Ed. M. B. Beck, Elmsford, NY, Pergamon Press.

Mandel, J. (1964). *The Statistical Analysis of Experimental Data*, New York, Dover Publications.

Chapter 40

Using Simulations to Study Statistical Problems

Key words: simulation, Monte Carlo, synthetic sampling, percentiles, t-test

---

From time to time, we are in doubt about a statistical procedure. The doubt may arise because an unfamiliar statistic has been used by a regulatory agency. Or, perhaps the data have some property that we fear may invalidate an analytical method. Also common, though often overlooked, is the situation where several measured values, each having particular statistical properties, are used to compute a measure on which decisions rest. We need to know a statistical property of the computed value that is difficult to determine analytically.

A professional statistician might be able to demonstrate from theory that our concerns are groundless. Even if this is possible, one might elect to demonstrate the statistical properties or the sensitivity of a statistical method by simulation, that is by creating a number of synthetic data sets, each having properties similar to the real data that are to be analyzed, and carrying through the proposed procedure on these synthetic data sets. This general idea is known as Monte Carlo simulation or simply simulation.

SIMULATION

Monte Carlo simulation is a method for obtaining information about the behavior of complex situations. It is a way of experimenting with a computer. The method consists of "sampling" to create many data sets that are analyzed with a statistical method to learn how it performs.

Suppose that the model of a system is $y = f(x)$, where a single input x determines the output y. It is easy to discover how variability in x translates into variability in y by putting different values of x into the model and calculating the corresponding values of y. In this way, we can see what happens for many different values of x, which we may choose to make random according to any distribution of interest. The range of values for x and the distributions of values within the specified range can be defined as a probability density function. A value for x is drawn at random from its probability distribution used to evaluate y. This process is repeated through many trials (100–5000) until the distribution of y values becomes clear. This is *simulation*, sometimes also called *synthetic sampling*.

It is economical and easy to compute uniform and normal random variates directly. The values generated actually are *pseudorandom* since they are obtained from a simple mathematical formula, but if a good program is used their properties cannot be distinguished from true random numbers. We will assume such a random number generating program is available if needed.

To obtain a value $Y_U(\alpha,\beta)$ from a uniform distribution over the interval (α,β) from a given uniform variate R_U over the interval $(0,1)$, this transformation is applied

$$Y_U(\alpha,\beta) = (\beta - \alpha)R_U + \alpha$$

In a similar fashion, a normally distributed random value $Y_N(\mu,\sigma^2)$, which has mean μ and standard deviation σ, is derived from a standard normal distribution, $R_N(0,1)$, as follows:

$$Y_N(\mu,\sigma) = \mu + \sigma R_N$$

Lognormally distributed random variates can be simulated using

$$Y_{LN}(\alpha,\beta) = e^{\mu + \sigma R_N}$$

Here, the logarithm of Y_{LN} is normally distributed with mean μ and standard deviation σ. The mean and variance of the lognormal variable Y_{LN} are

$$\alpha = \exp(\mu + 0.5\sigma^2)$$

and

$$\beta = \exp(\mu + 0.5\sigma^2)\sqrt{\exp(\sigma^2 - 1)}$$

Most statistics software programs (e.g., Minitab, Statview) will generate standard uniform and normal variates. SYSTAT has built-in commands to generate random numbers with a uniform, normal, t, F, chi-square, beta, or gamma distribution. Spreadsheet programs can also be useful; for example, EXCEL will generate uniformly distributed variables with range [0,1]. Equations for generating random values for the exponential, gamma, chi-square, lognormal, Johnson, beta, Weibull, Poisson, and binomial distributions from the standard uniform and normal variates are given in Hahn and Shapiro (1967). Another useful source is Press et al. (1992).

EXAMPLE — PROPERTIES OF A COMPUTED STATISTIC

A new regulation requires enforcement decisions to be made on the basis of 4-day averages. Suppose that preliminary sampling indicates that the daily observations are lognormally distributed with a geometric mean of 7.4 mg/L, mean = 12.2, and variance = 12.2. This corresponds to a normal distribution in log space with $\mu = 2$ and $\sigma^2 = 1$. The averages of four observations from this system should be more nearly normal than the parent lognormal population, but we want to check on how closely normality is approached. We will do this empirically by constructing a distribution of simulated averages.

It is easy to generate random values from a normal distribution, so this is the basis for our simulation. The mean and standard deviation of the log-transformed observations $y = \ln(x)$ are $\mu y = 2$ and $\sigma_y^2 = 1$. From this, the steps are

1. Generate four random, independent, normally distributed numbers having $\mu = 2$ and $\sigma = 1$.
2. Transform the normal variates into lognormal variates $x = \exp(y)$.
3. Average the four values to estimate the 4-day average, $\bar{x}$.
4. Repeat Steps 1 and 2 until sufficiently clear information is developed on the distribution of $\bar{x}$.
5. Plot a histogram of the average values.

Figure 40.1a shows the frequency distribution of the 1000 observations actually drawn in order to compute the 250 simulated 4-day averages represented by the frequency distribution of Figure 40.1b. Even though 1000 observations sounds like a large number, the frequency distributions are still not smooth, but the essential information has emerged. The distribution of 4-day averages is skewed, though not as strongly as the parent lognormal distribution. The average and standard deviation of the 1000 lognormal values

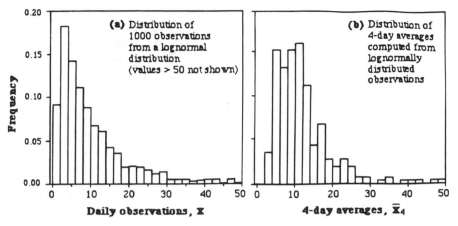

Figure 40.1 (a) Frequency distribution of 1000 daily observations that are random, independent, and have a lognormal distribution x = exp(y), where y is normally distributed with μ = 2 and σ = 1. (b) Frequency distribution of 250 4-day averages, each computed from four random values sampled from the lognormal distribution of part (a).

are 12 and 16. The average and standard deviation of the 250 4-day averages are 12 and 7.6. The range of the 4-day averages is 2.3 and 45.9. Seventy percent of the 4-day averages fall within the range of 4 to 16, while almost 20% are above 16.

EXAMPLE — PERCENTILE ESTIMATION

A state regulation requires estimation of the 99th percentile from the available historical data. The regulation recommends making this estimate by assuming that the data are generated by a lognormal process, and it allows the quantile estimate to be made parametrically or nonparametrically. (The methods for making these estimates are given in Chapter 8.)

We would like to compare these two estimation methods and also examine the spread of the estimated percentile values. This can be done by simulation, as follows.

1. Generate a set of n observations, y_i (i = 1, n), from a lognormal distribution having known statistical properties.
2. Use these n observations to estimate the pth percentile (a) parametrically and (b) nonparametrically.
3. Repeat Steps 1 and 2 many times to generate a distribution of estimated values.

The 99th percentile was estimated 100 times, each time using a sample of n = 100 observations drawn at random from the lognormal distribution shown in Figure 40.2a, which has a 99th percentile of 13.2. Figure 40.2b shows estimates made using a nonparametric method. Figure 40.2c shows 100 estimates made with the parametric method. Each dot is one estimated value; the curve is the empirical probability distribution for the 99th percentile.

The results show that the parametric estimates are less variable than the nonparametric estimates. And, they are more symmetrically distributed about the true 99th percentile value of 13.2. This pattern continues to hold when the sample size is increased (Berthouex and Hau, 1991). The reason the parametric method is better is that it uses the information that the data are from a lognormal distribution, whereas the nonparametric method assumes no prior knowledge of the distribution. Chapter 8 gives some discussion of the consequences of being willing (or unwilling) to specify a distribution.

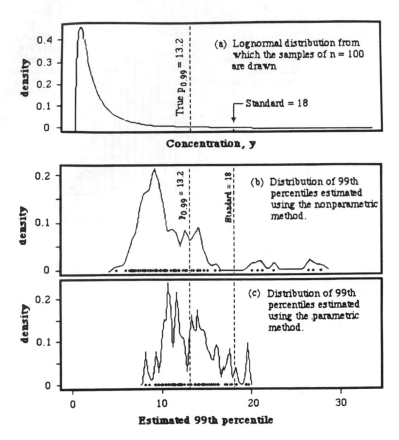

Figure 40.2 Distribution of 100 estimates of the 99th percentile, each computed using a sample of n = 100 from the lognormal distribution shown in (a). Nonparametric estimates are shown in (b) and parametric estimates in (c).

We use this example to illustrate a point regarding percentiles in regulations. Suppose that a regulation states that the 99th percentile must be less than 18, identified in Figures 40.2 by a vertical line labeled "standard = 18." Even though the true 99th percentile is well below this limit, both methods show several violations in the set of 100 estimates.

COMMENTS

Another use of simulation is to test the consequences of violation of the assumptions on which a statistical procedure rests. A good example is in Box et al. (1978) who used simulation to study how non-normality and serial correlation affect the performance of the t-test. The affect of non-normality was not very serious; in a situation where 5% of tests should have been "significant," 4.3% were significant for normally distributed data, 6.0% for a rectangular parent distribution, and 5.9% for a skewed parent distribution. The affect of modest serial correlation in the data was much greater than the differences due to non-normality. A positive autocorrelation of $\rho = 0.4$ inflated the percentage of tests found significant from the correct level of 5 to 10.5% for the normal distribution, 12.5% for a rectangular distribution, and 11.4% for a skewed distribution. They also showed that randomization would negate the autocorrelation and give percentages of significant results at the level of about 5% as it should be. Normality, which often causes people concern, turns out to be relatively unimportant, while serial correlation, which we too seldom consider, can be ruinous.

Simulation is familiar to most engineers as a design tool. Use it to explore and discover unknown properties of unfamiliar statistics and to check the performance of statistical methods that might be applied to data with nonideal properties. Sometimes, we find our worries are misplaced or unfounded.

REFERENCES

Berthouex, P. M. and I. Hau (1991). "Difficulties in Using Water Quality Standards Based on Extreme Percentiles," *Res. J. Water Pollut. Control Fed.*, 63, 5, 873–879.

Box, G. E. P., W. G. Hunter, and J. S. Hunter (1978). *Statistics for Experimenters: An Introduction to Design, Data Analysis, and Model Building*, New York, Wiley Interscience.

Hahn, G. J. and S. S. Shapiro (1967). *Statistical Methods in Engineering*, New York, Wiley.

Press, W. H., B. P. Flannery, S. A. Tenkolsky, and W. T. Vetterling (1992). *Numerical Recipes in FORTRAN: The Art of Scientific Computing*, 2nd ed., Cambridge, England, Cambridge University Press.

Chapter 41

Intervention Analysis

Key words: time series, intervention analysis, detergent ban, phosphorus, random walk, white noise

---

Environmental regulations are intended to cause changes in environmental quality. Often it is not easy to detect the resulting change, let alone estimate its magnitude, because a collection of environmental factors are acting so the system does not vary randomly about a fixed level before or after the intervention. If the variation was random about fixed levels, the difference between these two levels would be an appropriate estimate of the magnitude of the intervention effect. A confidence interval for this estimated effect could be calculated using the t statistic.

When the variation in the data is not random, serial correlation must be accounted for in the intervention analysis. When the system exhibits drift (nonstationarity) or seasonality, that must also be taken into account. *Intervention analysis*, a method originally proposed by Box and Tiao (1965), estimates the effect of a known change in conditions affecting a time series of serially correlated data. We will present a simple model that should fit a number of environmental time series that have a slow drift, but no regular seasonality.

CASE STUDY — BAN ON PHOSPHATE DETERGENTS

Wisconsin passed a law in 1978 that resulted in two interventions that were intended to change the amount of phosphorus entering the environment. The law required that after July 1, 1979, household laundry detergents could contain no more than 0.5% phosphorus (P) by weight. Before this virtual ban on phosphate detergents went into effect, detergents contained approximately 5% phosphorus. In 1982, the legislature allowed the phosphate ban to lapse, and in July 1982 detergents containing phosphates started to reappear in Wisconsin, although continuing reformulation and marketing changes may have reduced the average detergent phosphorus content from preban levels. A few years later, a new ban was imposed.

From the mid-1970s until the mid-1980s, there was controversy about phosphates in detergents. One was over how much the ban actually reduced the phosphate loading to wastewater treatment plants. Of course the detergent manufacturer's knew how much less phosphate had been sold in detergents, but some confirmation of their values from treatment plant data was desirable. Making this estimate was difficult because the potential effect was relatively small compared to the natural random fluctuations of the relevant environmental series.

The largest treatment plants in Wisconsin are the Jones Island and South Shore plants in Milwaukee. Both plants have expert management and reliable measurement processes. The data showing mass load of influent phosphorus to these two plants combined are plotted in Figure 41.1. The Wisconsin phosphate detergent ban stretched from July 1979 to June 1982, and indeed the lowest phosphorus concentrations are recorded during this period. The record shows a downward trend starting long before the imposition of the ban, and it is not obvious from the plot how to distinguish the effect of the ban from this general trend. If an average value for the ban period were to be compared to a preban average, the difference would depend heavily on how far the pre-ban average were extended into the past.

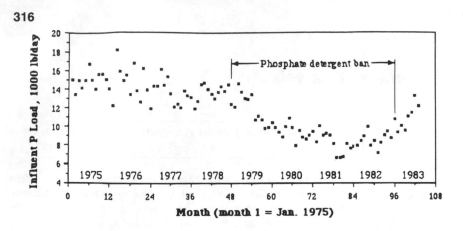

Figure 41.1 Monthly average mass flow of total phosphorus for the Milwaukee South Shore and Jones Island plants combined. The ban on phosphate detergents extended from July 1979 to June 1982.

Somehow, allowance must be made for the seemingly random drift in influent P load that is unrelated to the ban. Intuition suggests that in order to minimize the influence of random drift or trends on the estimation, the averages of observations made to estimate the effect of the ban should not stretch too far away from time of the intervention. On the other hand, relying on too few observations around the intervention ban might impair the estimation because there was too little averaging of random fluctuations. It would be appealing if the observations entering into the averages were not necessarily given the same weight; those closest to the expected shift should get more weight than observations more distant in time.

The goal is to estimate, as precisely as possible, the level of the system before and after the intervention. The effect of the intervention is estimated by the difference in these two levels. Exactly how to proceed depends on what mathematical model is considered to be an adequate representation of the data.

Before presenting a model that appears useful as the basis for analysis of many environmental series, such as the present one, a brief discussion is given of two other models which have often been relied upon (explicitly or implicitly) to estimate the effects of changes.

TWO SIMPLE MODELS FOR INTERVENTION ANALYSIS

Two simple models that represent the extremes are presented to contrast the approaches that would be used if (1) the data series consists of random independent variation about a fixed mean level and (2) the data series consists of pure random walk in which there is no fixed level. Assume that an intervention only shifts the data series upward or downward and does not change the pattern of variation. If this is true, then in the first case, the intervention would be estimated by taking the difference of the mean levels before and after the shift. In the second case, only the observations at the intervention would be used to estimate the shift. In the first case, all the data contribute to the estimation; in the second case, almost all the data are disregarded in making the estimation.

The White Noise Model

When the data vary about a fixed level and contain non-negligible random measurement error, sampling error, etc., the statistical analysis often relies on the model

$$y_i = \mu + e_i \qquad (1)$$

where

y_i = the ith measurement i = 1, 2, . . . , n

μ = the true (and unobservable) mean value of y_i

e_i = random error, assumed to be independently distributed according to normal distribution with mean 0 and variance σ_e^2, i.e., $e_i \approx N(0, \sigma_e^2)$

Figure 41.2a illustrates n_1 observations made under one condition, followed by n_2 observations made under a new condition and represents a case where Equation 1 would be appropriate. A deliberate invention has caused the change in conditions at time T. The intervention is

$$y_t = \mu_1 + \delta I + e_t \qquad (2)$$

where

y_t = value observed at time t

T = time the intervention takes place

δ = the effect of the intervention, $\delta = \mu_2 - \mu_1$

I = an indicator function equal to 1 after the intervention and 0 before it

μ_1 = the mean value of y_t for t $\leq$ T

μ_2 = the mean value of y_t for t $>$ T

e_t = the independently distributed error, $N(0, \sigma_e^2)$

Based on this *white noise model*, the effect of the intervention is estimated as the difference of the averages before and after the intervention:

$$\hat{\delta} = \bar{y}_2 - \bar{y}_1 = \frac{1}{n_2} \sum_{t=T+1}^{T+n_2} y_t - \frac{1}{n_1} \sum_{t=1}^{T} y_t \qquad (3)$$

and the estimated variance of δ is

$$\mathrm{Var}(\hat{\delta}) = \mathrm{Var}(\bar{y}_2) - \mathrm{Var}(\bar{y}_1) = \sigma_e^2 \left(\frac{1}{n_2} + \frac{1}{n_1} \right) \qquad (4)$$

where σ_e^2 is the pooled variance of the deviations from the two average levels.

The reader has by now probably recognized that this computation is the same as that done to compare two averages using a t-test. The ratio $\hat{\delta}/\sqrt{\mathrm{Var}(\hat{\delta})}$ could be compared to

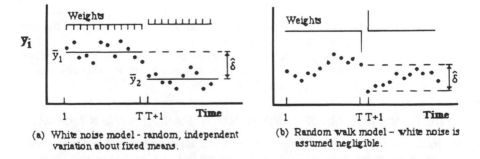

(a) White noise model - random, independent variation about fixed means.

(b) Random walk model - white noise is assumed negligible.

Figure 41.2 Illustration of the weights used to estimate a shift caused by an intervention according to two models that represent extreme assumptions regarding the pattern of random noise in the data.

a t distribution (with $n_1 + n_2 - 2$ degrees of freedom) to test whether δ is significantly different from zero.

Estimating δ from the *white noise model*, Equation 3, means giving equal weight to all observations before the intervention, regardless of whether they are close to time T or far from it. Also, all observations after the intervention are given equal weight. This uniform weighting, shown in Figure 41.2a, is a natural consequence of Equations 1 and 2, where the assumption of statistical independence among observations is crucial.

The Random Walk Model

Figure 41.2b illustrates observations made in time sequence, but where the true value of the response variable Y_t is observable at any time. This true value tends to drift randomly over time. Each value is independent of the values near it. Note the subtle, but important, difference from the usual statistical assumptions about observed values Y_t. In this case, Y_t is assumed to have no error (which is why we represent it by capital Y in contrast to lower case y, which is a value observed with error). It is the difference $Y_t - Y_{t-1}$ (i.e., the drift) that is subject to random variation. Data series of this nature can often be described by a *random walk model*:

$$Y_t = Y_{t-1} + \varepsilon_t \tag{5}$$

where Y_t = the true level of the response at time t, and ε_t = the random drift between time t −1 and t; the ε_t's are independent $N(0, \sigma_e^2)$

The estimated effect of the intervention at time T is

$$\hat{\delta} = Y_{T+1} - Y_T \tag{6}$$

Although the observations Y_{T+1} and Y_T are known without error, $\hat{\delta}$ is subject to uncertainty, because it represents not only the intervention, but also some random drift ε_{T+1}. Hence,

$$\text{Var}(\hat{\delta}) = \text{Var } \varepsilon_{T+1} = \sigma_\varepsilon^2 \tag{7}$$

and σ_ε^2 is estimated as

$$\hat{\sigma}_\varepsilon^2 = \frac{\sum_{t=1}^{T} (Y_t - Y_{t-1})^2 + \sum_{t=T+1}^{T+n_2} (Y_t - Y_{t-1})^2}{(n + n - 2)} \tag{8}$$

The ratio $\delta/\sigma_\varepsilon$ can be referred to a t distribution (with $n_1 + n_2 - 2$ degrees of freedom) to test whether δ is significantly different than zero.

In estimating δ according to the *random walk model*, Equation 6, the observations just before and just after the intervention are given full weight and all other observations are given zero weight, as illustrated in Figure 41.2b.

THE COMBINED MODEL (WHITE NOISE-RANDOM WALK MODEL)

For most environmental time series, both the white noise and the random walk models are too simple to be realistic. Real systems are subject to ever-changing conditions. It is usually unreasonable to assume that the mean level of the variable in question remains constant for all times, except for a shift caused by the deliberate intervention of the study. Hence, observations in the remote past carry little information about the level of the response variable prior to a particular intervention. Similarly, developments immediately after the intervention gradually became irrelevant to future observations. On the other hand, any single observation is usually affected by temporary random fluctuations, which implies that some sort of averaging would increase the precision of estimation.

A model that suitably reflects these conditions is a combination of the *white noise* and *random walk* models:

$$y_t = Y_t + e_t \tag{9}$$

$$Y_t = Y_{t-1} + \varepsilon_t$$

where

y_t = the observation at time t
Y_t = the underlying unobservable "true" value for y_t
e_t = the white noise component of time series and $e_t \approx N(0, \sigma_e^2)$
ε_t = the random walk component of the time series and $\varepsilon_t \approx N(0, \sigma_\varepsilon^2)$

The two error series e_t and ε_t are independent of each other. If the true values of y_T and y_{T+1} were known, the effect of the intervention would be estimated as $\hat{\delta} = y_T - y_{T+1}$ with $Var(\hat{\delta}) = \sigma_\varepsilon^2$. Since the true values are unknown, their estimates are used and

$$\hat{\delta} = \bar{y}_T - \bar{y}_{T+1} + \varepsilon_T \tag{10}$$

Sometimes the full effect of the intervention is not fully realized within the interval T and T + 1. A practical consideration then is how to represent the transition period over which the intervention is realized. A dynamic model (an exponential decay, for example) can be written for the transition period (Box and Tiao, 1965). Here, we will simply say that the transition takes place during an interval of length G, being fully realized in the gap between T and T + G. In this case, the estimate of the shift becomes

$$\hat{\delta} = \bar{y}_T - \bar{y}_{T+G} + \sum_{j=1}^{G} \varepsilon_{T+j} \tag{11}$$

The last term represents the expected magnitude of the random walk during the transition period over which the intervention is fully realized.

The averages $\bar{y}_T$ and $\bar{y}_{T+G}$ that represent the levels before and after the intervention are exponentially weighted averages (EWA). In the equations below, θ is a weighting factor that has a value between 0 and 1. If θ approaches 1, observations near the intervention are emphasized and observations farther away are forgotten (See Chapter 4).

$$\bar{y}_T = \theta y_T + \theta(1 - \theta)y_{T-1} + \theta(1 - \theta)^2 y_{T-2} + \ldots \tag{12}$$

$$\bar{y}_{T+G} = \theta y_{T+G} + \theta(1 - \theta)y_{T+G+1} + \theta(1 - \theta)^2 y_{T+G+2} + \ldots$$

The variance of the estimated shift is

$$Var(\hat{\delta}) = Var(\bar{y}_T) + Var(\bar{y}_{T+G}) + G\sigma_\varepsilon^2 \tag{13}$$

If G = 1, the intervention is modeled as being fully realized in the interval between T and T + 1. If a longer transition period is needed, one or more observations will be omitted. The value of G will be one plus the number of omitted observations. The number of random walk steps acting over the gap will be G + 1, and this is accounted for by the last terms in Equations 11 and 13.

The exponential weighting factor θ in the EWA (Equation 12) is related to the variances in the model according to

$$\frac{\sigma_e^2}{\sigma_\varepsilon^2} = \frac{\theta}{(1 - \theta)^2} \tag{14}$$

In the situation where $\sigma_e^2 \ll \sigma_\varepsilon^2$, θ approaches 0 and the weights die out slowly. In this case, the weights die out slowly and the model approaches the white noise model of Equation 1 and Figure 41.2a. Note that for θ = 0, each observation would be given equal weight, and the time series would be considered to have no drift. On the other hand, $\sigma_e^2 \gg \sigma_\varepsilon^2$ implies θ approaching 1. Thus, observations away from the intervention rapidly become irrelevant to estimating the intervention; the model approaches the random walk model described by Equation 6 and Figure 41.2b.

If we accept this model, the intervention estimation problem essentially becomes how to determine the weighting factor θ. Fortunately, an alternate formulation of the model makes this reasonably simple. The white noise-random walk model, Equation 9, is equivalent to an ARIMA (0, 1, 1) model (Box and Jenkins, 1976; Box et al., 1994)

$$y_t = y_{t-1} + a_t - \theta a_{t-1} \tag{15}$$

where
  y_t = the observation at time t, as before
  θ = a parameter, $0 \le \theta \le 1$
  a_t = an independent random noise model distributed as N(0, σ_a^2).

The white noise and the random walk error components are related to σ_a^2 as follows:

$$\sigma_e^2 = \theta \sigma_a^2 \tag{16}$$

$$\sigma_\varepsilon^2 = (1 - \theta)^2 \sigma_a^2 \tag{17}$$

The equivalence of the two models and the derivation of Equations 16 and 17 are given in Pallesen et al. (1985).

There are advantages to each way of expressing the model. For the purpose of estimating the model parameters, Equation 15 is convenient. Those not familiar with the ARIMA form of time series models often find Equation 9 intuitively clear, whereas Equation 15 has the somewhat confusing aspect of including a fraction of the random noise from the previous observation, as reflected by the last term θa_{t-1}.

A recursive iteration is done separately for each section of data before and after the intervention to estimate θ. The method is to

1. Choose a starting value for θ and use Equation 15 to recursively calculate the residuals a_t at t = 2, 3, . . . , T, using $a_t = y_t - y_{t-1} + \theta a_{t-1}$, where a_1 is set equal to zero to start the calculations.
2. Calculate the residual sum of squares, RSS(θ) = Σa_t for each section. Add these to get the total RSS for the entire series. If a gap has been used to account for a transition period, data in the gap are omitted from these calculations.
3. Search over a range of θ to minimize RSS(θ).
4. Use the minimum RSS to estimate

$$\hat{\sigma}_a^2 = \frac{RSS(\theta)}{n}$$

where n is the total number of residuals used to compute $RSS(\theta)$.

5. Use the estimated variance $\hat{\sigma}_a^2$ in Equations 16 and 17 to estimate σ_e^2 and σ_ε^2.

The variance of the estimated shift is

$$Var(\hat{\delta}) = Var(\bar{y}_T) + Var(\bar{y}_{T+G}) + G\sigma_\varepsilon^2$$

The last term is the variance contributed by the random drift over the transition gap G. Since the minimum value of G is 1, this term cannot be omitted. Unless the series are short on either side of the intervention, we can assume that

$$Var(\bar{y}_T) = Var(\bar{y}_{T+G}) = \sigma_e^2(1 - \theta)$$

where the estimated values for σ_e^2 and σ_ε^2 are found from Equations 11 and 12. This gives

$$Var(\hat{\delta}) = \sigma_e^2(1 - \theta) + \sigma_e^2(1 - \theta) + G\sigma_\varepsilon^2 = 2(1 - \theta)\sigma_e^2 + G\sigma_\varepsilon^2$$

CASE STUDY — SOLUTION

The case study is a problem presented in Pallesen et al. (1985). The data record is 103 observations long, running from January 1975 to July 1983. The phosphorus ban imposed on July 1, 1979, may already have had some effect in June, since shipment of phosphorus-free detergent to retailers in Wisconsin had already started in May. Hence, the analysis disregards the months of June, July, August, and September, which are taken as being transition months during which existing stocks of phosphate detergents were used up. The transition was handled by leaving a gap of four months; the data for June–September 1979 were disregarded in the calculations. The gap is G = 5. The ban was lifted in July 1982, and shipment of detergent containing phosphorus began during the month of July. This makes the June 1982 last observation unaffected by the lifting of the ban. Again, relying on information from the detergent manufacturers, a transition period of four months (July to October) was disregarded (again G = 5).

The calculations were done on the natural logarithms of the monthly average loads. The results are shown in Figure 41.3 and Table 41.1. Note that the variance of the

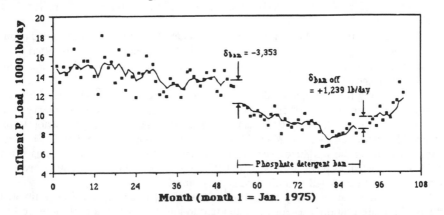

Figure 41.3 Estimated effects of the interventions and the model (solid line) fitted to the data.

Table 41.1 **Results of Intervention Analysis on Phosphorus Loads to Milwaukee Wastewater Treatment Plants**

Event	n	On Natural Logarithmic Scale		Est. Levels (lb/day)
		Levels	Variances	
1979 preban, $\bar{y}_T$	53	9.4835	0.002324	13,141
1979 postban, $\bar{y}_{T+G}$	33	9.1889	0.002324	9,788
1982 prelift, $\bar{y}_T$	33	9.0427	0.002324	8,457
1982 postlift, $\bar{y}_{T+G}$	9	9.1795	0.002326	9,696
Estimated effects	1979 ban	$\hat{\delta} = -3,353 \pm 1200$ lb/day		
	1982 ban	$\hat{\delta} = +1,239 \pm 960$ lb/day		
Variances (log scale)	$\sigma_e^2 = 0.006440$	$\sigma_{\hat{\varepsilon}}^2 = 0.001312$		
ARIMA model parameter		$\theta = 0.64$		

November 1982 level is virtually identical to the others, even though it is estimated from a very short series after the 1982 lifting of the ban. The estimated levels were transformed back to the original metric of pounds per day. For example, the estimated level for May 1979 is 9.4835 on the logarithmic scale and the variance of this estimate is Var(May 1979) = 0.002324. The preban level (May 1975) is exp(9.4835) = 13,141 lb/day.

The variances of the effects are calculated using the values in Table 41.1.

$$\text{Var}(\hat{\delta}_{1979 \text{ ban}}) = 0.002324 + 0.002324 + 5(0.006440) = (0.106)^2$$

$$\text{Var}(\hat{\delta}_{1982 \text{ ban off}}) = 0.002324 + 0.002326 + 5(0.006440) = (0.106)^2$$

In both interventions, the standard deviation is $\sigma = 0.106$ in terms of the natural logarithm, which translates to a relative standard deviation of 10.6% of the geometric average of the two levels (pre and post) involved on the original scale. For the 1979 ban, this is 10.6% of $(13,141 \times 9788)^{1/2} = 1200$ lb/day. For the 1982 ban lifting, 10.6% of $(8457 \times 9696)^{1/2} = 960$.

Based on the estimated shifts, phosphorus-based laundry detergents contributed about 25% of the influent P load prior to the ban and 13% after the ban was lifted. The similarity of phosphate-based products which reentered the market to those that were banned is not known.

The weighting factor of the model was estimated to be $\theta = 0.64$. The weights used to estimate the preban and postban averages $\bar{y}_T$ and $\bar{y}_{T+G}$ decay away from the intervention as follows: 0.640, 0.230, 0.083, 0.03, and so on, reducing each day by the factor $1 - \theta = 0.36$. The first three days on either side of the gap account for more than 95% of the estimated value of the weighted average. This is why the level after the ban was lifted could be estimated so precisely (var = 0.002326), even though the estimated level in November 1982 was based on only nine observations.

Having noted that relatively few observations actually entered into the estimated effects of an intervention, it is equally important to make clear that a fairly long record is necessary for computing reliable values of σ_e^2 and $\sigma_{\hat{\varepsilon}}^2$.

COMMENTS

Assuming random variation about fixed levels often provides a very poor model for assessing the magnitude of environmental interventions. This method is inappropriate when it is necessary to account for drift, seasonality, and perhaps other nonrandom patterns in the data. This chapter has suggested one such time series model that may fit a number of types of environmental data. Box and Tiao (1965, 1975) provide a suite of models that can be used when the simple model discussed in this chapter proves inadequate.

The technique presented here may be useful for objectives other than analysis of already collected data. Given a sufficiently long data record, it would be possible to find σ_e^2 and σ_ε^2 before a planned or proposed intervention is made. The values could be used to determine the variance of the effect of the expected intervention, thereby establishing beforehand how small an effect could be found statistically significant with reasonable probability. One could also decide ahead of time how many observations would be needed after the intervention in order to estimate the magnitude of its effect with essentially full precision.

REFERENCES

Box, G. E. P. and G. M. Jenkins (1976). *Time Series Analysis, Forecasting, and Control*, San Francisco, Holden Day.

Box, G. E. P. and G. C. Tiao (1965). "A Change in Level of a Non-Stationary Time Series," *Biometrika*, 52, 181–192.

Box, G. E. P. and G. C. Tiao (1975). "Intervention Analysis With Applications to Economic and Environmental Problems," *J. Am. Stat. Assoc.*, 70, 70–79.

Box, G. E. P., G. M. Jenkins, and G. C. Reinsel (1994). *Time Series Analysis: Forecasting and Control*, Englewood Cliffs, NJ, Prentic-Hall.

Pallesen, L., P. M. Berthouex, and K. Booman (1985). "Environmental Intervention Analysis: Wisconsin's Ban on Phosphate Detergents," *Water Res.*, 19, 353–362.

Statistical Tables

Table A   **Table of Normal Distribution Function** $F(z) = \dfrac{1}{\sqrt{2\pi}} \displaystyle\int_{-\infty}^{z} e^{-t^2/2}\, dt$

z	0.00	0.01	0.02	0.03	0.04	0.05	0.06	0.07	0.08	0.09
0.0	0.5000	0.5040	0.5080	0.5120	0.5160	0.5199	0.5239	0.5279	0.5319	0.5359
0.1	0.5398	0.5438	0.5478	0.5517	0.5557	0.5596	0.5636	0.5675	0.5714	0.5753
0.2	0.5793	0.5832	0.5871	0.5910	0.5948	0.5987	0.6026	0.6064	0.6103	0.6141
0.3	0.6179	0.6217	0.6255	0.6293	0.6331	0.6368	0.6406	0.6443	0.6480	0.6517
0.4	0.6554	0.6591	0.6628	0.6664	0.6700	0.6736	0.6772	0.6808	0.6844	0.6879
0.5	0.6915	0.6950	0.6985	0.7019	0.7054	0.7088	0.7123	0.7157	0.7190	0.7224
0.6	0.7257	0.7291	0.7324	0.7357	0.7389	0.7422	0.7454	0.7486	0.7517	0.7549
0.7	0.7580	0.7611	0.7642	0.7673	0.7704	0.7734	0.7764	0.7794	0.7823	0.7852
0.8	0.7881	0.7910	0.7939	0.7967	0.7995	0.8023	0.8051	0.8078	0.8106	0.8133
0.9	0.8159	0.8186	0.8212	0.8238	0.8264	0.8289	0.8315	0.8340	0.8365	0.8389
1.0	0.8413	0.8438	0.8461	0.8485	0.8508	0.8531	0.8554	0.8577	0.8599	0.8621
1.1	0.8643	0.8665	0.8686	0.8708	0.8729	0.8749	0.8770	0.8790	0.8810	0.8830
1.2	0.8849	0.8869	0.8888	0.8907	0.8925	0.8944	0.8962	0.8980	0.8997	0.9015
1.3	0.9032	0.9049	0.9066	0.9082	0.9099	0.9115	0.9131	0.9147	0.9162	0.9177
1.4	0.9192	0.9207	0.9222	0.9236	0.9251	0.9265	0.9279	0.9292	0.9306	0.9319
1.5	0.9332	0.9345	0.9357	0.9370	0.9382	0.9394	0.9406	0.9418	0.9429	0.9441
1.6	0.9452	0.9463	0.9474	0.9484	0.9495	0.9505	0.9515	0.9525	0.9535	0.9545
1.7	0.9554	0.9564	0.9573	0.9582	0.9591	0.9599	0.9608	0.9616	0.9625	0.9633
1.8	0.9641	0.9649	0.9656	0.9664	0.9671	0.9678	0.9686	0.9693	0.9699	0.9706
1.9	0.9713	0.9719	0.9726	0.9732	0.9738	0.9744	0.9750	0.9756	0.9761	0.9767
2.0	0.9772	0.9778	0.9783	0.9788	0.9793	0.9798	0.9803	0.9808	0.9812	0.9817
2.1	0.9821	0.9826	0.9830	0.9834	0.9838	0.9842	0.9846	0.9850	0.9854	0.9857
2.2	0.9861	0.9864	0.9868	0.9871	0.9875	0.9878	0.9881	0.9884	0.9887	0.9890
2.3	0.9893	0.9896	0.9898	0.9901	0.9904	0.9906	0.9909	0.9911	0.9913	0.9916
2.4	0.9918	0.9920	0.9922	0.9925	0.9927	0.9929	0.9931	0.9932	0.9934	0.9936
2.5	0.9938	0.9940	0.9941	0.9943	0.9945	0.9946	0.9948	0.9949	0.9951	0.9952
2.6	0.9953	0.9955	0.9956	0.9957	0.9959	0.9960	0.9961	0.9962	0.9963	0.9964
2.7	0.9965	0.9966	0.9967	0.9968	0.9969	0.9970	0.9971	0.9972	0.9973	0.9974
2.8	0.9974	0.9975	0.9976	0.9977	0.9977	0.9978	0.9979	0.9979	0.9980	0.9981
2.9	0.9981	0.9982	0.9982	0.9983	0.9984	0.9984	0.9985	0.9985	0.9986	0.9986
3.0	0.9987	0.9987	0.9987	0.9988	0.9988	0.9989	0.9989	0.9989	0.9990	0.9990
3.1	0.9990	0.9991	0.9991	0.9991	0.9992	0.9992	0.9992	0.9992	0.9993	0.9993
3.2	0.9993	0.9993	0.9994	0.9994	0.9994	0.9994	0.9994	0.9995	0.9995	0.9995
3.3	0.9995	0.9995	0.9995	0.9996	0.9996	0.9996	0.9996	0.9996	0.9996	0.9997
3.4	0.9997	0.9997	0.9997	0.9997	0.9997	0.9997	0.9997	0.9997	0.9997	0.9998
3.5	0.9998									
4.0	0.99997									
5.0	0.9999997									
6.0	0.999999999									

Table B **Probability Points of the *t* Distribution with *v* Degrees of Freedom**

ν	Tail Area Probability						
	0.4	**0.25**	**0.1**	**0.05**	**0.025**	**0.01**	**0.005**
1	0.325	1.000	3.078	6.314	12.706	31.821	63.657
2	0.289	0.816	1.886	2.920	4.303	6.965	9.925
3	0.277	0.765	1.638	2.353	3.182	4.541	5.841
4	0.271	0.741	1.533	2.132	2.776	3.747	4.604
5	0.267	0.727	1.476	2.015	2.571	3.365	4.032
6	0.265	0.718	1.440	1.943	2.447	3.143	3.707
7	0.263	0.711	1.415	1.895	2.365	2.998	3.499
8	0.262	0.706	1.397	1.860	2.306	2.896	3.355
9	0.261	0.703	1.383	1.833	2.262	2.821	3.250
10	0.260	0.700	1.372	1.812	2.228	2.764	3.169
11	0.260	0.697	1.363	1.796	2.201	3.718	3.106
12	0.259	0.695	1.356	1.782	2.179	2.681	3.055
13	0.259	0.694	1.350	1.771	2.160	2.650	3.012
14	0.258	0.692	1.345	1.761	2.145	2.624	2.977
15	0.258	0.691	1.341	1.753	2.131	2.602	2.947
16	0.258	0.690	1.337	1.746	2.120	2.583	2.921
17	0.257	0.689	1.333	1.740	2.110	2.567	2.898
18	0.257	0.688	1.330	1.734	2.101	2.552	2.878
19	0.257	0.688	1.328	1.729	2.093	2.539	2.861
20	0.257	0.687	1.325	1.725	2.086	2.528	2.845
21	0.257	0.686	1.323	1.721	2.080	2.518	2.831
22	0.256	0.686	1.321	1.717	2.074	2.508	2.819
23	0.256	0.685	1.319	1.714	2.069	2.500	2.807
24	0.256	0.685	1.318	1.711	2.064	2.492	2.797
25	0.256	0.684	1.316	1.708	2.060	2.485	2.787
26	0.256	0.684	1.315	1.706	2.056	2.479	2.779
27	0.256	0.684	1.314	1.703	2.052	2.473	2.771
28	0.256	0.683	1.313	1.701	2.048	2.467	2.763
29	0.256	0.683	1.311	1.699	2.045	2.462	2.756
30	0.256	0.683	1.310	1.697	2.042	2.457	2.750
40	0.255	0.681	1.303	1.684	2.021	2.423	2.704
60	0.254	0.679	1.296	1.671	2.000	2.390	2.660
120	0.254	0.677	1.289	1.658	1.980	2.358	2.617
∞	0.253	0.674	1.282	1.645	1.960	2.326	2.576

Table C   Percentage Points of the F Distribution: Upper 5% Points

v_2 \ v_1	1	2	3	4	5	6	7	8	9	10	12	15	20	24	30	40	60	120	∞
1	161.4	199.5	215.7	224.6	230.2	234.0	236.8	238.9	240.5	241.9	243.9	245.9	248.0	249.1	250.1	251.1	252.2	253.3	254.3
2	18.51	19.00	19.16	19.25	19.30	19.33	19.35	19.37	19.38	19.40	19.41	19.43	19.45	19.45	19.46	19.47	19.48	19.49	19.50
3	10.13	9.55	9.28	9.12	9.01	8.94	8.89	8.85	8.81	8.79	8.74	8.70	8.66	8.64	8.62	8.59	8.57	8.55	8.53
4	7.71	6.94	6.59	6.39	6.26	6.16	6.09	6.04	6.00	5.96	5.91	5.86	5.80	5.77	5.75	5.72	5.69	5.66	5.63
5	6.61	5.79	5.41	5.19	5.05	4.95	4.88	4.82	4.77	4.74	4.68	4.62	4.56	4.53	4.50	4.46	4.43	4.40	4.36
6	5.99	5.14	4.76	4.53	4.39	4.28	4.21	4.15	4.10	4.06	4.00	3.94	3.87	3.84	3.81	3.77	3.74	3.70	3.67
7	5.59	4.74	4.35	4.12	3.97	3.87	3.79	3.73	3.68	3.64	3.57	3.51	3.44	3.41	3.38	3.34	3.30	3.27	3.23
8	5.32	4.46	4.07	3.84	3.69	3.58	3.50	3.44	3.39	3.35	3.28	3.22	3.15	3.12	3.08	3.04	3.01	2.97	2.93
9	5.12	4.26	3.86	3.63	3.48	3.37	3.29	3.23	3.18	3.14	3.07	3.01	2.94	2.90	2.86	2.83	2.79	2.75	2.71
10	4.96	4.10	3.71	3.48	3.33	3.22	3.14	3.07	3.02	2.98	2.91	2.85	2.77	2.74	2.70	2.66	2.62	2.58	2.54
11	4.84	3.98	3.59	3.36	3.20	3.09	3.01	2.95	2.90	2.85	2.79	2.72	2.65	2.61	2.57	2.53	2.49	2.45	2.40
12	4.75	3.89	3.49	3.26	3.11	3.00	2.91	2.85	2.80	2.75	2.69	2.62	2.54	2.51	2.47	2.43	2.38	2.34	2.30
13	4.67	3.81	3.41	3.18	3.03	2.92	2.83	2.77	2.71	2.67	2.60	2.53	2.46	2.42	2.38	2.34	2.30	2.25	2.21
14	4.60	3.74	3.34	3.11	2.96	2.85	2.76	2.70	2.65	2.60	2.53	2.46	2.39	2.35	2.31	2.27	2.22	2.18	2.13
15	4.54	3.68	3.29	3.06	2.90	2.79	2.71	2.64	2.59	2.54	2.48	2.40	2.33	2.29	2.25	2.20	2.16	2.11	2.07
16	4.49	3.63	3.24	3.01	2.85	2.74	2.66	2.59	2.54	2.49	2.42	2.35	2.28	2.24	2.19	2.15	2.11	2.06	2.01
17	4.45	3.59	3.20	2.96	2.81	2.70	2.61	2.55	2.49	2.45	2.38	2.31	2.23	2.19	2.15	2.10	2.06	2.01	1.96
18	4.41	3.55	3.16	2.93	2.77	2.66	2.58	2.51	2.46	2.41	2.34	2.27	2.19	2.15	2.11	2.06	2.02	1.97	1.92
19	4.38	3.52	3.13	2.90	2.74	2.63	2.54	2.48	2.42	2.38	2.31	2.23	2.16	2.11	2.07	2.03	1.98	1.93	1.88
20	4.35	3.49	3.10	2.87	2.71	2.60	2.51	2.45	2.39	2.35	2.28	2.20	2.12	2.08	2.04	1.99	1.95	1.90	1.84
21	4.32	3.47	3.07	2.84	2.68	2.57	2.49	2.42	2.37	2.32	2.25	2.18	2.10	2.05	2.01	1.96	1.92	1.87	1.81
22	4.30	3.44	3.05	2.82	2.66	2.55	2.46	2.40	2.34	2.30	2.23	2.15	2.07	2.03	1.98	1.94	1.89	1.84	1.78
23	4.28	3.42	3.03	2.80	2.64	2.53	2.44	2.37	2.32	2.27	2.20	2.13	2.05	2.01	1.96	1.91	1.86	1.81	1.76
24	4.26	3.40	3.01	2.78	2.62	2.51	2.42	2.36	2.30	2.25	2.18	2.11	2.03	1.98	1.94	1.89	1.84	1.79	1.73
25	4.24	3.39	2.99	2.76	2.60	2.49	2.40	2.34	2.28	2.24	2.16	2.09	2.01	1.96	1.92	1.87	1.82	1.77	1.71
26	4.23	3.37	2.98	2.74	2.59	2.47	2.39	2.32	2.27	2.22	2.15	2.07	1.99	1.95	1.90	1.85	1.80	1.75	1.69
27	4.21	3.35	2.96	2.73	2.57	2.46	2.37	2.31	2.25	2.20	2.13	2.06	1.97	1.93	1.88	1.84	1.79	1.73	1.67
28	4.20	3.34	2.95	2.71	2.56	2.45	2.36	2.29	2.24	2.19	2.12	2.04	1.96	1.91	1.87	1.82	1.77	1.71	1.65
29	4.18	3.33	2.93	2.70	2.55	2.43	2.35	2.28	2.22	2.18	2.10	2.03	1.94	1.90	1.85	1.81	1.75	1.70	1.64
30	4.17	3.32	2.92	2.69	2.53	2.42	2.33	2.27	2.21	2.16	2.09	2.01	1.93	1.89	1.84	1.79	1.74	1.68	1.62
40	4.08	3.23	2.84	2.61	2.45	2.34	2.25	2.18	2.12	2.08	2.00	1.92	1.84	1.79	1.74	1.69	1.64	1.58	1.51
60	4.00	3.15	2.76	2.53	2.37	2.25	2.17	2.10	2.04	1.99	1.92	1.84	1.75	1.70	1.65	1.59	1.53	1.47	1.39
120	3.92	3.07	2.68	2.45	2.29	2.17	2.09	2.02	1.96	1.91	1.83	1.75	1.66	1.61	1.55	1.50	1.43	1.35	1.25
∞	3.84	3.00	2.60	2.37	2.21	2.10	2.01	1.94	1.88	1.83	1.75	1.67	1.57	1.52	1.46	1.39	1.32	1.22	1.00

Source: Merrington, M. and C. M. Thompson (1943) Tables of Percentage Points of the Inverted Data (F) Distribution, *Biometrika*, 33, 73. With permission.

INDEX

A

Aberrant values, 4
Accuracy, 11
Acidification, 235
Additivity, 154
Air pollution, 145
Aliases, 173
Analysis of variance (ANOVA), 129, 141, 142, 145, 146, 148
ANOVA, see Analysis of variance
Approximate confidence region, 202
Approximate joint confidence region, 207, 211
a priori parameter estimates, 268
ARIMA model, 320
Augmented design, 261
Augmented model, 236
Autocorrelation, 185, 243
  coefficient, 185, 187, 245
  effect of on regression, 243–249
    autocorrelation and trend analysis, 248
    case study, 243–244
    consequences of , 244–245
    example of autocorrelated errors, 246–247
    statistic to indicate possible autocorrelation, 247–248
    why autocorrelation distorts variance estimates, 245
  function, 187
Average of differences, assessing, 103–110
  benefits of paired design, 106–108
  paired t-test analysis, 105
  study of dissolved oxygen, 104

B

Background noise, 76
Bacteria, 57
Between-run precision, 12
Bias, 11, 12, 243, 273, 275, 295
Binomial distribution, normal approximation, 118
Binomial probability distribution, 115
Bioassay, 115
Biochemical oxygen demand (BOD), 28, 41, 52, 210, 265, 273
Biokinetics, 275, 279, 288, 306

Biological counts, 59
BOD, see Biochemical oxygen demand
Bonferroni error rate, 123
Box and Whisker plot, 31
Box-Draper determinant criterion, 280
Box graph, 30
Box-Jenkins models, 320

C

Calculated values, how measurement errors are transmitted into, 301–308
  finite difference model, 307–308
  reaction kinetics, 306–307
  spherical particles, 304–305
  theory, 302–303
  titration errors, 303–304
Calibration, 204, 213–220
  case study, 214
  case study solution, 216–218
    fitting calibration line, 216–217
    using calibration curve to predict concentrations, 217–218
  standards, 213
  theory, 214–216
Categorical variables, 235
Causality, 233
Causation, 181
Censored data, 4, 91
Censored samples, estimating mean of, 91–95
  Cohen's maximum likelihood estimator method, 91–93
Center points, 261
Central limit effect, 13, 18
Chi-square distribution, 18
Coefficient of determination, 219, 224, 229–234
  assurance of valid relation, 229
  definition of "explained", 233
  low, 229–230
  magnitude of, 231–232
  significant, 230–231
  statistically significant, 229
  ways to examine model, 232–233
Cohen's method, 91
Components of variance, nested design to estimate, 139

329

Confidence band, 215
Confidence interval, 22, 97, 99, 106, 113, 166,
    198, 203, 307
Confidence limits, 61, 68, 119
Conformance with a standard, assessing, 97–101
    case study, 98
    solution, 99–100
    theory, 98–99
Confounded effects, 162
Confounding pattern, 173
Constant variance, 58, 108, 147, 273
Contour map, 259
Contour plot, 261
Copepod, 106
Correlation, 181, 185
Correlation coefficients, 181–184, 219
    case study, 181
    case study solution, 182–183
    covariance and correlation, 181–182
Covariance, 181, 302
Critical level, 54
Critical sum of squares, 198
Cube plots, 152, 154, 157
Cumulative frequency distribution, 9, 44
Cumulative frequency plot, 41
Cumulative probability plot, 86
Curve fitting, 191

D

Data
    density plot, 8
    dredging, 123
    plotting, 25–33
        in search of trends, 26–28
        original data record, 25
        plot of residuals, 30
        scatterplots, 28–29
        showing precision of replicated
            measurements, 31–32
        statistical variation, 30–31
        Youden plot, 25–26
    smoothing, 35–40
        exponentially weighted moving average,
            37–39
        methods, 35–36
        moving average, 37
        reexpressing data, 36
    snooping, 123
Defining relation, 164, 173
Degrees of freedom, 11, 18

Dependent variable, 191
Derivative matrix, 267
Design matrix, 153, 163
Detergents, 315, 321
Determinant criterion, 267, 280
Difference of proportions, assessing, 115–121
    assessing difference between two proportions,
        118–120
    binomial model, 115–118
    case study, 115
    solution, 120–121
Difference of two averages, assessing, 111–114
    mercury data, 113
    mercury in domestic wastewater, 111
    t-test to compare average of two samples,
        111–112
Digidot plot, 25
Dissolved oxygen (DO), 98, 104, 161
Distribution, seeing shape of, 41–47
    case study, 41
    dot diagrams, 42–43
    probability plots, 44–45
    randomness and independence, 45–46
    use and misuse of probability plots, 45
DO, see Dissolved oxygen
Dot diagrams, 41–43
Double reciprocal plot, 275
Drift, 315
Dummy variables, 236, 238
Dunnett's method of multiple comparisons, 124
Durbin-Watson statistic, 247

E

Empirical models, 151, 191, 221
Environmental problems and statistics, 1–6
    special problems, 4–5
    statistics and environmental law, 1–2
    truth and statistics, 2–4
Error, 7
    bars, 30, 31
    propagation of, 305
    sum of squares, 147
    transmission, 305
Errors-in-variables problem, 183
Estimated parameters, precision of, 201–211
    bacterial growth model, 206–208
    concept of joint confidence region, 201–202
    contribution of observations to precision,
        208–210
    linear model, 204–206

parameter correlation, 210
size and shape of confidence region, 208
theory, 202–204
EWA, see Exponentially weighted averages
EWMA, see Exponentially weighted moving
averages
Expected value, 10
Experiment, 3
Experimental design, 151, 208, 210, 251
Experimental error, 7, 240
Experimental error variance, 197
Experimental measurements, estimating variance
components in, 137–144
foundry wastes, 137–138
sampling waste foundry sand, 140–143
variance components analysis, 139–140
Exploratory data analysis, 32, 40
Exponentially weighted averages (EWA), 319
Exponentially weighted moving averages
(EWMA), 35, 37, 39
External reference distributions, 49–55
to compare mean values, 50–52
constructing, 49–50
for monitoring, 52–53
setting critical levels, 53–54

F

Factor, 152
Factorial design, 145, 255, 258
Factorial experimental designs, 151–160
case study, 151–152
case study solution, 157–159
fractional, 161–169
case study, 161–162
case study solution, 164–168
method, 162–164
full 2^k factorial design, 152–157
data analysis, 153–157
experimental design, 153
Family error rate, 123
F distribution, 131, 226
Fit, lack of, 282
Flow measurement, 295
Fly ash, 171
Foundry waste, 137, 142
Fractional factorial design, 153, 162, 171, 172,
255
Frequency distribution, 311
F test, 226, 241
Full factorial design, 152

G

Gaussian distribution, 14
Generating function, 164
Geometric average, 322
Geometric mean, 61, 310
Grand average, 131
Graphical analysis, 86
Graphing data, 25
Graphing residuals, 30
GREG software program, 282
Gross errors, 295
Groundwater monitoring, 115

H

Haldane model, 288
Half fraction, 162, 172
Happenstance data, 4, 148, 233
Histograms, 8
Hypothesis test, 20

I

IDL, see Instrument limit of detection
Incineration, 145
Independence, 14, 45, 47, 49, 147, 189, 243
Independent estimates, 160
Independent t-test, 103, 111, 114
Independent variables, 191
Indicator variables, 236
Individual error rate, 123
Initial data analysis, 32, 40
Instrument limit of detection (IDL), 72–73
Interaction, 149, 152, 156, 165, 168, 259
Interlaboratory study, 104
Interval estimates, 203
Intervention analysis, 315–323
case study, 315–316
case study solution, 321–322
combined model, 318–321
simple models, 316–318
random walk model, 318
white noise model, 316–318
Iterative design, 255, 258, 268
Iterative experimental strategy, 169
Iterative process, 3

J

Joint confidence region, 199, 201, 206, 254,
270, 282, 285, 290

K

k averages, analysis of variance to compare,
129–135
case study, 129
case study solution, 133–134
one-way analysis of variance, 129–133
k averages, multiple paired comparisons of,
123–128
case study, 124
Dunnett's method, 126–127
Tukey's method solution, 126
Tukey's paired comparison method, 125
Kinetics, 306

L

Lack-of-fit test, 290
Lagrange multipliers, 297
Landfill, 137
LCL, see Lower confidence limits
Lead, measurements of, 124
Least squares, 193, 223, 280, 296
estimate, 194
parameter estimates, 223, 267
Limit of detection, 4, 71–79, 81
alternate model for MDL, 75–77
case study, 71
case study solution, 77–78
method detection limit, 71–73
methods for analyzing data below, 81–90
graphical methods, 84–86
median, 82–83
regression on order statistics, 87
replacement or deletion methods, 81–82
trimmed mean, 83
Winsorized mean, 83–84
U.S. EPA approach to estimating MDL, 73–75
Linear approximation, 302
Linearization, 273, 275
Linear model, 155
Linear regression, 183, 191, 205, 215, 273
Linear regression, empirical model building by,
221–227
case study, 221–223
case study solution, 223–227
starting from complicated model, 224–227
starting from simple model, 224
method, 223
Lineweaver-Burke plot, 275
Logarithmic scale, 36
Logarithmic transformation, 109

Lognormal distribution, 52, 63, 65, 66, 310, 311
Log scale, 47
Log transformation, 28, 58, 59, 93, 108, 273
Lower confidence limits (LCL), 68
Lurking variables, 4

M

MA, see
Main effect, 149, 152, 154, 156, 175, 259
calculation of, 157
variance of, 157
Mass transfer, 265
Material balance, 297
Maximum likelihood estimate (MLE), 91
MDL, see Method detection limit
Mean, 82
residual sum of squares, 226, 240
squares, 132
Mechanistic model, 191
Median, 82
Median of 3 smooth, 40
Mercury, 111
Method detection limit (MDL), 71–73, 81
Method of least squares, estimating parameters
using, 191–199
examples, 194–197
method of least squares, 193–194
precision of estimates of linear model,
197–198
precision of estimates of nonlinear model, 198
regression model, 191–193
Michaelis-Menten model, 275
Minimum variance design, 267, 268
Minitab, 5
Missing values, 40
MLE, see Maximum likelihood estimate
Model(ing), 151, 155
building, 191, 251
discrimination, 287–293
five rival models, 291
function, 291–293
two rival biological models, 287–291
equivalence, 240
iterative approach to, 251–256
matrix, 155, 163, 176
simplification, 240
Monod kinetics, 251
Monod model, 206, 251, 279, 288
Monte Carlo simulation, 309
Moving averages (MA), 35, 37, 53
Multiple paired comparison, 123

Multiresponse data, fitting models to, 279–285
   case study, 279–280
   case study solution, 282–284
   method, 280–282
Multiresponse parameter estimation, 282

N

Nested design, 139, 141
NLLS, see Nonlinear least squares estimates
Noise, 3, 7
Nonadditivity, 154
Nonconstant variance, 5, 28, 57
Nonlinear least squares, 265, 273, 275, 288
Nonlinear least squares (NLLS) estimates, 277
Nonlinear models, designing experiments to
         estimate parameters in, 265–272
   case study, 265
   case study solution, 269–272
   method, 265–269
Nonlinear regression, 191, 207
Nonnormal distributions, 5
Non-normality, 49, 312
Nonparametric estimation, 67
Nonstationarity, 315
Normal distribution, 9, 14, 16, 44, 66, 309
Normal equation, 194
Normality, 13
Normal order scores, 87, 159
Normal plot, 159, 175, 177
Normal probability plot, 45, 159
Normal scores, 45
Null hypothesis, 20, 97, 99, 109, 111, 119

O

One-sided hypothesis test, 21
Optimal design, 268, 270
Optimization, 258, 261
Optimum conditions, 260
Orthogonal design, 151
Orthogonality, 151
Outliers, 4
Overfitting, 221
Oxygen transfer, 265

P

Paired experimental design, 106
Paired t-test, 103, 109, 114, 123, 183
Pairwise comparisons, 123
Parameter, 9

confidence intervals of, 205
   correlation, 201, 270, 277
   estimates, 191, 244, 254, 265, 273
Parameter estimates, why linearization can bias,
         273–278
   bacterial growth, 273
   first-order kinetic model, 273–275
   Michaelis-Menten model, 275–278
Parsimonious model, 221
Parsimony, 287
Percentiles, estimating, 65–69
   definition of quantile and percentile, 65
   nonparametric estimates of quantiles, 67–68
   parametric estimates of quantiles, 65–67
Permeability, 151, 171
pH, 235
   concentration of, 49
   observations, 26
Phenol, 258
Phosphorus, 28, 315, 321
Plankton, 59
Polynomial function, 222
Pooled variance, 74, 112, 149
Population, 7
   distribution, 65
   mean, 10
Posterior probability, 291
Power plant, 106, 151
Power transformation, 63
Precision, 11, 12, 296
Precision of parameter estimates, 197
Prediction, 217, 232
Prediction interval, 233
Primary standards, 97
Prior probability, 292
Probability frequency distribution, 9
Probability plot, 9, 41, 44, 85
Proportions, 115
Pseudorandom numbers, 309
Pure experimental error, 226

Q

Quadratic effects, 261
Quadratic model, 258, 261
Quality improvement, 134
Quantile, 65

R

Random, definition of, 13
Random error, 76, 192, 295

Random error variance, 148
Randomization, 13, 148, 153, 189, 312
Randomness, 45
Random numbers, 309
Random variable, 7–9
Random walk, 316
Random walk model, 317, 318
Rank, 44
Rankits, 45, 87
Ratios, 115
Redundant measurements, 295
Reference distribution, 19, 49, 127, 167
Regression, 57, 87, 191, 244
    analysis, 191, 229
    joint confidence region, 266
    models, 235
    sum of squares, 225, 240
Regression analysis, with categorical variables,
        235–242
    case study, 235–236
    case study solution, 238–242
    method, 236–238
Regressors, 191
Relative frequency, 9
Repeatability, 12
Replicate experiments, 13
Replicate measurements, 138
Replication, 58, 208
Reproducibility, 12
Residual error, 192
Residual mean square, 224
Residual plots, 224, 282, 290
Residuals, 11, 30
Residual sum of squares, 196, 202, 225
Response surface methodology, seeking
        optimum conditions by, 257–264
    case study, 258–262
        first iteration, 258–259
        iteration 4, 262
        second iteration, 259–260
        third iteration, 261–262
    optimum location, 262
    response surface methodology, 257–258
Robust, definition of, 13
Ruggedness testing, 161

S

Sample(s), 7, 145
    average, 10
    correlation coefficient, 185
    covariance, 181

    distribution, 17
    standard deviation, 11, 18
    variance, 10
Saturated factorial design, 160
Scaleless covariance, 181
Scatterplots, 29
Seasonality, 315
Sedimentation, 221
Sensitivity analysis, 306
Sensitivity coefficients, 306
Serial correlation, 49, 243, 245, 248, 312, 315
Serial correlation, assessing, 185–189
    case study, 185
    case study solution, 187–188
    correlation and autocorrelation coefficients,
        185–187
Serial dependence, 46, 185
Significance, 229, 230
    effect, 159
    level, 20
    test, 106, 110
Simulations, using to study statistical problems,
        309–313
    percentile estimation, 311–312
    properties of computed statistic, 310–311
    simulation, 309–310
Skewness, 41
Smoothing, 35
Spurious correlations, 234
Square root transformation, 59, 108
SS, see Suspended solids
Standard deviation, 10, 15
Standard error, 18
    of effects, 166
    of the estimate, 232
    of the mean, 17, 18
Standard normal deviate, 15, 66
Star points, 261
Statistic, 9, 74
Statistical control, 295
Statistical inference, 20
Statistical significance, 22, 134
Statistical tables, 325–327
Statistics, review of, 7–23
    accuracy, bias, and precision, 11–13
    average, variance, and standard deviation,
        9–11
    experimental errors, 7–8
    normal distribution, 14–15
    normality, randomness, and independence,
        13–14
    plotting data, 8–9

population and sample, 7
probability distributions, 9
random variable, 7
sampling distribution of average and
   variance, 17–20
significance tests and confidence intervals,
   20–22
t distribution, 15–17
Staview, 5
Steepest ascent, 259
Stem-and-leaf plot, 25
Studentized range, 125
Student's *t* distribution, 15
Sum of squares, 132, 202
   function, 193
   surface, 196, 206, 208, 290
Suspended solids (SS), 28
Synthetic data, 309
Synthetic sampling, 309
SYSTAT, 5
System memory, 188

T

Taylor series, 302, 307
t distribution, 15, 17–19, 22, 166
Thomas slope method (TSM), 273, 274
Three-level factorial design, 153
Tiessier model, 288
Time sequence, 25
Time series, 40, 315
Time series models, 320
Total phosphorus (TP), 28
Total sum of squares, 147, 225
TP, see Total phosphorus
Transformations, 28, 57
Transformations, using, 57–64
   Box-Cox power transformations, 63
   confidence intervals on original scale, 63
   confidence intervals and transformations,
      61–62
   transformations for linearization, 57–59
   variance stabilizing transformations, 59–61
Transition period, 319
Trend analysis, 28
Trickling filter, 31
Trimmed mean, 83
TSM, see Thomas slope method
t statistics, 240
t-test, 123, 189, 246, 312, 317
Tukey's multiple t-test, 124

Two-factor interactions, 154
Two level factorial design, 151
Two-sided hypothesis test, 21

U

UCL, see Upper confidence limits
Unbiased estimates, 273
Uniform distribution, 43, 44, 309
Uniform variate, 309
Upper confidence limits (UCL), 68

V

Variables, screening of important, 171–179
   case study, 171
   case study solution, 175–177
   method, 171–174
Variance, 129, 273, 296, 302
   components, 139, 141
   -covariance matrix, 267, 280
   decomposition, 148
   of effects, 165
   of intercept, 203
   multiple factor analysis of, 145–150
      dioxin case study results, 148–149
      method, 146–148
      sampling dioxin and furan emissions,
         145–146
   of the average, 17
   of parameter estimates, 267
   of predicted value, 203
   of slope, 203
   serial correlation, 245

W

Warning levels, 54
Wastewater, 258
   survey, 41
   treatment, 28, 29, 315
Weighted least squares, 296
Weighting factor, 319
White noise, 316
White noise model, 317
Winsorization, 83
Winsorized mean, 83
Within-run precision, 12
Within-treatment variance, 130
Working-Hotelling confidence region, 216